# Register for Free Membership to

## solutions@syngress.com

Over the last few years, Syngress has published many best-selling and critically acclaimed books, including Tom Shinder's *Configuring ISA Server 2000*, Brian Caswell and Jay Beale's *Snort 2.0 Intrusion Detection*, and Angela Orebaugh and Gilbert Ramirez's *Ethereal Packet Sniffing*. One of the reasons for the success of these books has been our unique **solutions@syngress.com** program. Through this site, we've been able to provide readers a real time extension to the printed book.

As a registered owner of this book, you will qualify for free access to our members-only solutions@syngress.com program. Once you have registered, you will enjoy several benefits, including:

- Four downloadable e-booklets on topics related to the book. Each booklet is approximately 20-30 pages in Adobe PDF format. They have been selected by our editors from other best-selling Syngress books as providing topic coverage that is directly related to the coverage in this book.
- A comprehensive FAQ page that consolidates all of the key points of this book into an easy to search web page, providing you with the concise, easy to access data you need to perform your job.
- A "From the Author" Forum that allows the authors of this book to post timely updates links to related sites, or additional topic coverage that may have been requested by readers.

Just visit us at **www.syngress.com/solutions** and follow the simple registration process. You will need to have this book with you when you register.

Thank you for giving us the opportunity to serve your needs. And be sure to let us know if there is anything else we can do to make your job easier.

SYNGRESS®

SYNGRESS®

Managing and Securing a

# Cisco® Structured Wireless-Aware Network


**David Wall** CCSI, Technical Editor

**Jan Kanclirz Jr.** CCIE #12136

**Youhao Jing** CCIE#5253

**Jeremy Faircloth**

**Joel Barrett**

| **KEY** | **SERIAL NUMBER** |
|---|---|
| 001 | HJIRTCV764 |
| 002 | PO9873D5FG |
| 003 | 829KM8NJH2 |
| 004 | TLP678MA21 |
| 005 | CVPLQ6WQ23 |
| 006 | VBP965T5T5 |
| 007 | HJJJ863WD3E |
| 008 | 2987GVTWMK |
| 009 | 629MP5SDJT |
| 010 | IMWQ295T6T |

PUBLISHED BY
Syngress Publishing, Inc.
800 Hingham Street
Rockland, MA 02370

**Managing and Securing a Cisco® Structured Wireless-Aware Network**



Printed in the United States of America
1 2 3 4 5 6 7 8 9 0
1-932266-91-7

Acquisitions Editor: Christine Kloiber
Technical Editor: David Wall
Page Layout and Art: Patricia Lupien
Cover Designer: Michael Kavish
Copy Editor: Judy Eby
Indexer: J. Edmund Rush

Distributed by O'Reilly & Associates in the United States and Jaguar Book Group in Canada.

# Acknowledgments

We would like to acknowledge the following people for their kindness and support in making this book possible.

Syngress books are now distributed in the United States by O'Reilly & Associates, Inc. The enthusiasm and work ethic at ORA is incredible and we would like to thank everyone there for their time and efforts to bring Syngress books to market: Tim O'Reilly, Laura Baldwin, Mark Brokering, Mike Leonard, Donna Selenko, Bonnie Sheehan, Cindy Davis, Grant Kikkert, Opol Matsutaro, Lynn Schwartz, Steve Hazelwood, Mark Wilson, Rick Brown, Leslie Becker, Jill Lothrop, Tim Hinton, Kyle Hart, Sara Winge, C. J. Rayhill, Peter Pardo, Leslie Crandell, Valerie Dow, Regina Aggio, Pascal Honscher, Preston Paull, Susan Thompson, Bruce Stewart, Laura Schmier, Sue Willing, Mark Jacobsen, Betsy Waliszewski, Dawn Mann, Cindy Wetterlund, Kathryn Barrett, and to all the others who work with us. A thumbs up to Rob Bullington for all his help of late.

The incredibly hard working team at Elsevier Science, including Jonathan Bunkell, Ian Seager, Duncan Enright, David Burton, Rosanna Ramacciotti, Robert Fairbrother, Miguel Sanchez, Klaus Beran, Emma Wyatt, Rosie Moss, Chris Hossack, and Krista Leppiko, for making certain that our vision remains worldwide in scope.

David Buckland, Daniel Loh, Marie Chieng, Lucy Chong, Leslie Lim, Audrey Gan, Pang Ai Hua, and Joseph Chan of STP Distributors for the enthusiasm with which they receive our books.

Kwon Sung June at Acorn Publishing for his support.

Jackie Gross, Gayle Voycey, Alexia Penny, Anik Robitaille, Craig Siddall, Iolanda Miller, Jane Mackay, and Marie Skelly at Jackie Gross & Associates for all their help and enthusiasm representing our product in Canada.

Lois Fraser, Connie McMenemy, Shannon Russell, and the rest of the great folks at Jaguar Book Group for their help with distribution of Syngress books in Canada.

David Scott, Tricia Wilden, Marilla Burgess, Annette Scott, Geoff Ebbs, Hedley Partis, Bec Lowe, Andrew Swaffer, Stephen O'Donoghue and Mark Langley of Woodslane for distributing our books throughout Australia, New Zealand, Papua New Guinea, Fiji Tonga, Solomon Islands, and the Cook Islands.

Winston Lim of Global Publishing for his help and support with distribution of Syngress books in the Philippines.

# Contributors

**Jan Kanclirz Jr.** (CCIE #12136-Security, CCSP, CCNP, CCIP, CCNA, CCDA, INFOSEC Professional) is a Senior Network Information Security Engineer working for IBM Global Services. Currently, he is responsible for strategic and technical evolution of a large, multi-customer/multi-data center networks and their security environment. Jan specializes in multi-vendor, hands-on implementations and architectures of network technologies such as routers, switches, firewalls, intrusion sensors, content networking, and wireless networks. Beyond network design and engineering, Jan's background includes extensive experience with Linux and BSD administration and security implementations.

In addition to Jan's full time position at IBM G.S., he is involved in many different projects such as MakeSecure.com, where he dedicates his time to security awareness. Jan also runs a small Internet Service Provider (ISP), where he provides several services such as network consulting and Linux server hosting solutions.

Jan would like to acknowledge the understanding and support of his family and friends during the writing of the book, "Thank You".

**Youhao Jing** (CCIE#5253) is currently Director of Product Management and Consulting at Alcatel IP Division, responsible for defining the company's carrier class IP product strategy with a focus on the Asia Pacific market. He has held various senior level consulting positions at AT&T, Procket, Juniper Networks, and ICG Netcom, where he was responsible for new service and solution development, network and product architecture, design consulting for large-scale converged multi-service IP/MPLS networks.

Youhao Jing received his M.S degree from UC Berkeley and pursued further study on high performance networking systems at Stanford University. He lives with his wife Jane and two sons, Albert and Geoffrey, in Sunnyvale, CA

**Jeremy Faircloth** (Security+, CCNA, MCSE, MCP+I, A+, etc.) is a Staff Systems Administrator for EchoStar Satellite L.L.C., where he architects and maintains enterprise-wide client/server and Web-based technologies. He also acts as a technical resource for other IT professionals, using his expertise to help others expand their knowledge. As a systems engineer with over 12 years of real world IT experience, he has become an expert in many areas including Web development, database administration, enterprise security, network design, and project management. Jeremy has contributed to several Syngress books including *C# for Java Programmers* (ISBN: 1-931836-54-X), *Snort 2.0 Intrusion Detection* (ISBN: 1-931836-74-4), and *Security+ Study Guide & DVD Training System* (ISBN: 1-931836-72-8).

Jeremy currently resides in Colorado Springs, CO and wishes to thank his wife and son, Christina and Austin, for their support in his various technical endeavors.

**Joel Barrett** (CCNP, CCDP, CWNA, MCSE, and Novell's Master CNE) is a wireless specialist with Cisco Systems, Inc. He supports Cisco's wireless partners and developers throughout the southeast United States, assisting partner executives to develop technical go-to market strategies. Joel also educates partner engineering teams with a full understanding of wireless LAN technologies and solutions

With over fifteen years of IT experience, Joel has earned Cisco's and Planet3's certifications. Joel serves as the team leader for the Channels Technology Advisory Team for Mobility, an advisor for the Enterprise Mobility Virtual Team, and a member of Cisco's Enterprise Mobility Technology Leadership Program. He is a board member for the Wireless Technology Forum in Atlanta, and a speaker for the Georgia Wireless Users Group. He is also the facilitator for the Atlanta Cisco Study Group, helping over 200 network engineers attain Cisco certifications.

Joel was co-author and principle technical editor for several wireless LAN and IT books, including *Certified Wireless Security Professional (CWSP) Official Study Guide*, *Wireless Networks First-Step*, and the *Cisco Advanced Wireless* training course. Joel and his wife,

Barbara Kurth, live near Atlanta, Georgia with Barbara's son and daughter, Shane and Paige, and Joel's daughter, Ashley.

**Donald Lloyd** (CISSP), author of Syngress Publishing's *Designing a Wireless Network* (1-928994-45-8) is a senior consultant for International Network Services, Inc. (INS) and a regional leader for their Fixed Wireless Practice. His specialties include network security architecture and wireless network design. In addition to "unwiring" corporate offices, Donald spends considerable time designing and deploying secure wireless networks in remote oil and gas fields, airports, municipalities, and warehouses.

This is the third book that Donald has co-authored with Syngress, and Donald wishes to thank INS for their patience while finishing this book. He also sends a BIG hug to the pride and joy of his life, his son.

**Lev Shklover** (CCNP, CCDP, Cisco WLAN Design and Support Specialist, Certified Solaris Administrator, Nortel Networks Router and Network Management Specialist) is a Senior Consultant with International Network Services, Inc (INS); a leading global computer networking and security consultancy. He has over 13 years of experience in designing and implementing large computer networks for major U.S. and International corporations.

Lev's other specialization is lab testing of network designs, network devices and network protocols to maintain network reliability. He started working with Cisco WLAN hardware in early 2000, right after Cisco's acquisition of Aironet Communications. As a member of INS's Wireless Networking Practice, Lev has designed and deployed numerous Cisco 802.11a/b/g solutions for various clients, including a WLAN for a 44-story building.

Lev graduated from the Technical University of Radio Electronics and Automation in Moscow, Russia with a MS Degree in Optical Engineering. He currently resides in NJ with his wife and two children.

# Technical Editor

**David Wall** (CCSI #22530), author of *Multi-Tier Application Programming with PHP: A Practical Guide for Architects and Programmers*, contributes regularly to technical and general-interest publications and reviews books for online bookseller Amazon.com. David also works as a consultant, specializing in voice over IP applications and network design. A Cisco Certified Systems Instructor, David teaches engineers and salespeople about technologies from Cisco Systems.

David's other professional interests include hosting applications for small businesses, and the integration of disparate systems using open-source technologies.

A pilot, David enjoys flying around eastern Australia. David maintains a Web presence at http://www.davidwall.com.

# Contents

**Foreword** xxiii

**Chapter 1 Wired versus Wireless and Wireless-aware LANs** 1

Introduction 2
What is a WLAN? 2
How does a Wireless LAN Work? 3
WLAN Benefits 9
WLAN Design Considerations 12
Attenuation 12
Attenuation Due to Antenna Cabling 13
Attenuation Due to Exterior Considerations 13
Accounting for the Fresnel Zone and Earth Bulge 18
Radio Frequency Interference 19
Interference from Radio Transmitters 20
Harmonics 21
Application Considerations 22
Structural Considerations 22
Security Considerations 25
Network Management Considerations 26
WLAN Modes of Operation 27
What is a Wireless-aware LAN? 30
Wireless-aware LAN Benefits 31
Integrated Wired and WLAN Services using the Cisco Infrastructure and Cisco IOS Software 32
CiscoWorks WLAN Solution Engine 32
Wireless Domain Services for IEEE 802.1X Local Authentication Service and Fast Secure Roaming Support 33

Rogue AP Detection and Location 34
Interference Detection to Isolate and Locate Network Interference 35
Simplified WLAN Deployment Processes with Assisted Site Surveys 35
Streamlined WLAN Management and Operations Support 36
Seamless Delivery of Enhanced Network Security Solutions 38
Wireless-aware Design Considerations 39
Summary 40
Solutions Fast Track 41
Frequently Asked Questions 44
**Chapter 2 Designing Wireless-Aware LANs 47**
Introduction 48
Radio Frequency (RF) Basics 48
Transmitting Radio Signals over EM Waves 48
Anatomy of a Waveform 49
Propagating a Strong Radio Signal 57
Understanding Signal Power and S/N Ratio 57
Attenuation 58
Bouncing 61
Refracting 63
Line of Sight 64
Penetration 64
Understanding the Wireless Elements 66
Generic Radio Components 66
Laws, Regulations, and Environmental Considerations 70
Regulatory Agencies 70
The Need to Know 71
Regulations for Low Power, Unlicensed Transmitters 71
Environmental Considerations 72
IEEE 802.11 Standards 73
Does the 802.11 Standard Guarantee Compatibility across Different Vendors? 77
DSSS 78

IEEE 802.11b Direct Sequence Channels 78
IEEE 802.11a OFDM Physical Layer 80
IEEE 802.11a Channels 80
Planning for RF Deployment 81
WLAN Coverage 81
WLAN Data Rates 85
Client Density and Throughput 85
Antenna Options 87
Omnidirectional Antennas 87
Directional Antennas 88
Interference Detection 93
Conducting Site Surveys 93
Preparation 94
Other Preparations 97
Infrastructure Awareness 100
Preparing a Site Survey Kit 105
Performing an Interior Wireless Site Survey 115
Performing an Exterior Wireless Site Survey 124
Summary 129
Solutions Fast Track 129
Frequently Asked Questions 131

**Chapter 3 WLAN Roaming 133**

Introduction 134
Cisco L2 Roaming Solutions 135
Beacon Frames 137
Probe Frames 143
Roaming Decisions and Criteria 149
Roaming Target Selection Process 151
Roaming Behavior of Cisco 7920 WVoIP Phones 153
Cisco Solutions to Speed the L2 Roaming Process 160
Improved Client Channel Scanning 160
Fast Reauthentication Using CCKM 162
Cisco L3 Roaming Solutions 165
Mobile IP 166
Proxy Mobile IP 170
WLAN Design Considerations 171

Summary 174
Solutions Fast Track 174
Frequently Asked Questions 176

**Chapter 4 IP Multicast in a Wireless LAN 179**
Introduction 180
The OSI Model Overview 180
Data Communication Methods 182
The Unicast Method 182
Multicast WLAN Deployment Recommendations 186
Configuring Multicast and Broadcast Minimum Data Rate Settings in IOS 188
IP Multicast WLAN Configuration 190
Controlling IP Multicast in a WLAN with APs 191
Protocol Filters 192
Controlling IP Multicast in a Peer-to-peer WLAN using Bridges 193
Point-to-point Bridging 193
Point-to-multipoint Bridging 194
Configuring Reliable Multicast for Workgroup Bridges 195
Summary 197
Solutions Fast Track 197
Frequently Asked Questions 199

**Chapter 5 WLAN Guest Network Access 201**
Introduction 202
Guest WLANs 202
Designing a Guest VLAN 202
Design 203
Topology 203
Deployment 204
Guest WLAN Recommendations 204
Configuring Guest WLANs 205
Access Point and Switch Configuration 207
WLAN Guest VLAN Filtering 208
Summary 209
Solutions Fast Track 209
Frequently Asked Questions 210

**Chapter 6 Implementing Cisco Wireless LANs 211**
Introduction 212
The Cisco Wireless and Wireless-aware Vision 212
The Cisco Structured Wireless-aware Network
Product Line 213
APs 214
Aironet Bridges 214
Client Adapters 215
Cisco IOS 215
Wireless LAN Solution Engine 215
Wireless Security Suite 216
Access Control Server 216
Cisco Wireless LAN Switches and Routers 216
Cisco Wireless Antennas and Accessories 217
Ceiling Mount Omnidirectional
Antenna 2.4 GHz (AIR-ANT1728) 219
Mast Mount Omnidirectional
Antenna 2.4 GHz (AIR-ANT2506) 219
High-Gain Mast Mount Omnidirectional
Antenna 2.4 GHz (AIR-ANT24120) 219
Pilar Mount Diversity Omnidirectional
Antenna 2.4 GHz (AIR-ANT3213) 220
POS Diversity Dipole Omnidirectional
Antenna 2.4 GHz (AIR-ANT3351) 220
Diversity Ceiling Mount Omnidirectional
Patch Antenna 2.4 GHz (AIR-ANT5959) 221
Directional Wall Mount Patch Antenna
2.4 GHz (AIR-ANT3549, AIR-ANT1729) 221
Diversity Directional Wall Mount
Patch Antenna 2.4 GHz (AIR-ANT2012) 222
Yagi Antenna 2.4 GHz (AIR-ANT1949) 222
Dish Antenna 2.4 GHz (AIR-ANT3338) 222
Cisco's 2.4 GHz Antennas Summary 223
5 GHz Antennas 224
Cisco Wireless IP Phone 225
Cisco IOS and WLANs 226

Upgrading from VxWorks to IOS 227
Using the Aironet Conversion Tool for Cisco IOS Software v2.0 227
Using Browser and VxWorks 228
Using CiscoWorks WLSE for IOS Conversion 229
Cisco Aironet Access Points (APS) 229
Aironet 1200 AP 229
First-Time Basic Configuration 233
Aironet 1100 AP 235
Aironet 350 AP 237
Cisco Aironet WLAN Client Adapters 240
Cisco Aironet 350 Series Client Adapters 240
Cisco Aironet 5GHz Client Adapter 242
Cisco Aironet 802.11a/b/g Client Adapters 242
CiscoWorks WLSE 2.x 243
Fault Monitoring 245
Device Management 247
Device Configuration and Firmware Upgrades 247
Configuration Tab 247
Firmware Tab 249
Reports 249
Radio Manager 251
Cisco Wireless Security Suite 252
Mitigating Vulnerabilities with the Cisco Security Suite 253
Cisco Secure Access Control Server (ACS) 3.2 254
Enhanced Client Network Management Features with Extended Client Support 257
Workgroup Bridges 259
Aironet 350 Workgroup Bridge 259
Aironet 340 Workgroup Bridge 262
Wireless Bridges 263
Cisco Aironet 1400 Wireless Bridge 263
Cisco Aironet 350 Wireless Bridge 266

Summary 270
Solutions Fast Track 271
Frequently Asked Questions 275

**Chapter 7 WLAN Security Considerations 277**

Introduction 278
Network Security in General 279
Why We Need Security 280
Security Threats 281
Understanding Reconnaissance Attacks 281
Unauthorized Access 284
Understanding DoS and DDoS Attacks 286
Data Sniffing 287
Data Manipulation 288
Continuing the Security Cycle 288
Securing 289
Monitoring 290
Testing 290
Improving Security 291
How Wireless Technology Changes Network Security 292
Overview of 802.11 Standards 292
Shared Network Model 293
Protecting the Data Link and Physical Layers 294
Tracking and Attacking Anonymity 294
Attacks on Wireless Networks 294
Authentication 295
Physical Security 295
Preventing War Driving and Unauthorized Use of Legitimate Access Points 296
Devices Required in War Driving 297
Wi-Fi Client Adapter 297
Antenna 297
GPS 298
War Driving Software 299
Protecting Against War Drivers 300
Disabling SSID Broadcasts 300
Protecting Against Unauthorized Wireless Access 301

Open Authentication 301
Shared Authentication 301
EAP Authentication 302
MAC Address Authentication 302
VPN Authentication 303
Protecting Against Unauthorized Access Point Access 304
Changing the Default Settings 304
Disabling Unwanted Services 304
Exploring Rogue Access Points 304
Detecting and Protecting against Rogue Access Points 305
Corporate Policy and User Awareness 305
Mutual Authentication 305
Sniffers 306
Cisco Rogue Detection by Client Reports 306
Physical Detection 307
Wired Detection 307
Designing for Security 308
Creating a Security Policy 308
Risk Assessment 308
The Big Three 308
Logging and Accounting 309
Hot Standby 309
Configuring Hot Standby 310
Implementing Firewalls for Additional Security 311
Public Secure Packet Forwarding 312
Filters 313
WLAN LAN Extension 802.1x/EAP 313
EAP 314
EAP Packet Format 314
EAP Request and Response 315
EAP Success and Failure 315
802.1x 316
EAP Types 317
EAP Message Digest 5 317
EAP Generic Token Cards 317
EAP TLS 317

Cisco EAP 318
Implementing LEAP 320
Configuring ACS 321
Configuring Access Points 323
Configuring the Client 326
WLAN LAN Extension IPSec 328
Standards Used in IPSec 329
IKE 329
AH 331
ESP 332
Implementing IPSec over WLAN 332
VPN Device List in WLAN 334
Configuring VPN Gateway 334
Configuring an Access Point 336
Configuring a VPN Client 339
WLAN Static WEP Keys 341
WEP 341
IV WEP Vulnerable 342
IV and RC4 Vulnerabilities 342
Mitigating WEP Vulnerability 343
TKIP 343
Configure Static 128-bit WEP with TKIP 344
Using a Web Browser for Access Point Configuration 344
Configuring the Client 345
WLAN Security with VLANs 346
VLAN in Aironet 347
VLAN by SSID 347
VLAN by RADIUS 347
Summary 348
Solutions Fast Track 349
Frequently Asked Questions 353

**Chapter 8 WLAN Rogue Access Point Detection and Mitigation 355**
Introduction 356
The Problem with Rogue Access Points 356
A Rogue Access Point is Your Weakest Security Link 358

An Intruder's Rogue Access Point 359
Preventing and Detecting Rogue Access Points 360
Preventing Rogue Access Points with a Security Policy 360
Provide a Secure, Available Wireless Network 360
Sniffing Radio Frequency to Detect and Locate Rogue Access Points 361
Cisco's Rogue Access Point Detection 363
Central Management withWLSE to Detect Rogue Access Points 364
IEEE 802.1x Port-based Security to Prevent Rogue Access Points 367
Prevent Users from Using Rogue Access Points with 802.1x 367
Preventing Rogue Access Point from Connecting to Wired Network with 802.1x 369
Understanding Devices and their Roles in Wired 802.1x Implementation 370
Configuring 802.1x Authentication on a Supported Switch 371
Detecting a Rogue Access Point from the Wired Network 374
Detecting a Rogue Access Point with a Port Scanner 374
Using Catalyst Switch Filters to Limit MAC Addresses per Port 376
MAC Addresses in Port Security 376
Static MAC 377
Dynamic MAC 377
Sticky MAC 377
Security Violation 377
Protect Mode 378
Restrict Mode 378
Shutdown Mode 378
Configuring Port Security in an IOS Catalyst Switch 378
Summary 382
Solutions Fast Track 383
Frequently Asked Questions 385

**Chapter 9 Wireless LAN VLANs** **387**
Introduction 388
Understanding VLANs 389
VTP in a Wired Network 393
VTP Modes 393
Dealing with Trunk Ports 394
VLANs in a Wireless Environment 395
Per-VLAN Settings 396
VTP in a Wireless Network 398
Trunk Ports 398
Trunk Ports between Bridges 398
Wireless VLAN Deployment 399
Native VLAN 399
Routing between VLANs 399
Per-VLAN Filters 400
Per-VLAN QOS 401
Per-VLAN Authentication and Encryption 401
Configuring Wireless VLANs Using the IOS: A Case Study 401
Broadcast Domain Segmentation 408
Traffic Types 408
Unicast 408
Broadcast 408
Multicast 409
Broadcast Domain in Wireless 409
Primary (Guest) and Secondary SSIDs 410
Guest SSID 411
Using RADIUS for VLAN Access Control 411
Configuring RADIUS Control 413
Summary 415
Solutions Fast Track 416
Frequently Asked Questions 418

**Chapter 10 WLAN Quality of Service (QoS)** **421**
Introduction 422
The Requirement for Service Quality 423
Bandwidth 423
Latency 423

Jitter 424
Packet Loss 424
Network Availability 424
QoS 425
How Wireless Changes QoS 427
Extending QoS Support to WLAN Wireless Network 428
Integrating QoS in Wireless and Wired Networks 430
WLAN QoS Design Guidelines 432
Dimensioning WLAN Network for Sufficient Capacity 432
Handling Roaming Introduced Delay 434
Configuring for Wireless QoS in IOS 435
MetroWiFi integrated QoS Design 435
Configuring EDCF Frame Prioritization Scheme 438
Configuring Traffic Classification for EDCF Prioritization Scheme 439
Using Existing Network QoS Configuration 442
Summary 443
Solutions Fast Track 443
Frequently Asked Questions 445

**Index 449**

# Foreword

Wireless networking is all about freedom. Freedom is a worthy goal in its own right, and freedom in an organizational context pays dividends.

It's tremendously cool to be able to sit down in an office or conference room, open your portable computer, and have instant connectivity to the resources you need to do your job. With wireless networking in place, there's no need to fiddle with Ethernet cables and the associated software settings — you just get on with whatever you want to do. Simple, mobile connectivity like this makes you more efficient and more capable.

Considered from a manager's point of view, wireless networking is an empowering technology, a "force multiplier" if you want to use the military term. A given number of people, given a method of working more efficiently than usual, can do the work of a larger number of people. The fact that wireless networking enables people to connect to the network from anywhere in their work area, without any overhead, means that they can spend more time doing their jobs and not getting stressed out about hooking up to the network.

If you consider that wireless networking makes possible new modes of work, the advantages become even more apparent. Quite a few organizations are using wireless devices to scan items (for inventory management and tracking, to cite the two main examples) and report the scanned data to a central repository in near-real time. This makes more current—and therefore more valuable—information available to the people who need it. That wouldn't be possible without wireless networking. A few shops are already experimenting with wireless telephone service, in which users can roam all over their workplace while maintaining connectivity to the organizational telephone switch via a portable handset. This means they can be reached at their usual phone extension wherever they go, with no need to configure forwarding. The next generation of such portable phones will be able to connect to the wireless LAN when a suitable one is available, and switch over to the public cellular network when the user goes out of range. Wireless networking is not a gimmick. It is a productivity-enhancing, profit-enhancing infrastructure element that makes businesses stronger.

What's more, wireless networking makes it easy for an organization to do a bit of public service. By installing an access point that covers a portion of a public park or a nearby café—where the organization's employees would probably want to work from time to time anyway—it's possible to give wireless Internet access (under proper control, of course) to anyone who cares to connect. That's the kind of corporate philanthropy that people appreciate, and it'll do more for the company image than putting the logo on a theater program.

Critics of wireless networking call attention to security worries. A network with wireless service is subject to attack over the airwaves by anyone within radio range. Stripping unauthorized modems from the network and keeping intruders out of the building is no longer sufficient. The problem is complicated by the fact that anyone with 50 bucks to spare can plug a cheap access point, meant for home use, into an Ethernet port and start providing potentially unregulated access to the company network. (The fact that people are very often willing to part with 50 bucks to provide themselves with wireless service at the office says a lot about the value of wireless connectivity, by the way.)

That is why you need to take a managed approach to wireless networking. Rather than let rogue access points sprout haphazardly, and rather than allow access to your corporate network without proper authentication and auditing, you need to design your network to give the people of your organization what they need to do their jobs happily (and maybe a bit more, just to see what they do with the extra capability). If you're going to allow guest access, and maybe even some public use, you need to build that in. You need, however, to be precisely aware of what you're allowing, and use the best tools you can to prevent the kind of network activity that would do harm to your organization.

The Structured Wireless-Aware Networking (SWAN) solutions from Cisco Systems are excellent tools, and they're the subject of this book. With Cisco's hardware and software—some of it specific to the wireless portions of your network, some of it part of the Internetwork Operating System (IOS) and the routers on which it runs—you can allow the kind of access you want to allow, and prevent unauthorized access. It's a technical challenge, and Cisco's products go a long way toward providing a solution.

The capabilities of the Cisco SWAN products are pretty astounding. Not only will they allow you to cover a work area (office, factory, outdoor campus, whatever) with wireless LAN service, they'll manage authentication seamlessly even when a user roams from the coverage area of one access point to that of another. The access points can monitor each other, determine when one of their number is no longer on the air, and "heal" the network by increasing the power outputs of the active antennas. They'll also tell the difference between legitimate access points and rogues, and use triangulation to estimate the location of rogue APs—which the Wireless LAN Solution Engine (WLSE) can then display graphically on a scanned floor plan. SWAN is powerful.

Regardless of how powerful Cisco SWAN is, you have to cover the rest of the ground yourself, with knowledge and forethought. A tool is only as good as the person who wields it, and the quality of your Cisco wireless implementation will be proportional to your level of knowledge about the relevant equipment and protocols. That's where this book comes in. It covers the whole range of knowledge you need to design and deploy a good Cisco SWAN installation, from the basics of communicating data over radio waves to specific IOS commands to be used in implementing quality of service (QoS) for packet telephony and VLANs for compartmentalized access. A lot of experts have worked hard to make this book the best on the market.

I hope you enjoy this book, and that it's useful in your work with Cisco SWAN. Please visit my Web site (www.davidwall.com) or contact me by electronic mail (david@davidwall.com) if you have any comments or ideas for the next edition.

*—David Wall*
*Sydney, Australia*

Chapter 1

# Wired versus Wireless and Wireless-aware LANs

## Solutions in this Chapter:

- What is a Wireless LAN (WLAN)?
- WLAN Benefits
- WLAN Design Considerations
- WLAN Modes of Operation
- What is a Wireless-aware LAN?
- Wireless-aware LAN Benefits
- Wireless-aware Design Considerations

☑ Summary

☑ Solutions Fast Track

☑ Frequently Asked Questions

# Introduction

This chapter provides an introduction to wireless local area networks (WLANs). It explains what a WLAN is and how it is different from both hard-wired and purely wireless local area network (LAN) solutions. There is also an introduction to the inherent security problems associated with wireless and wireless-aware networks as contrasted with their wired ancestors.

This chapter also details what a WLAN is and how it should be designed. It covers some of the pitfalls that you can run into when designing a WLAN. There are many factors that can affect a WLAN design, many of which are covered in this chapter providing an excellent understanding of the best way to create a secure, reliable, and useful WLAN.

# What is a WLAN?

A WLAN is a LAN that uses radio waves as the physical medium. In a traditional wired LAN, individual network stations are linked via some type of physical cabling. This cabling can vary from shielded copper wires to fiber-optic cables. Most office structures are wired throughout to facilitate networking using this medium.

The three main problems with physical cabling are cost, distance limitations, and mobility. The installation of any type of physical cabling requires a great deal of effort and is therefore very expensive. The cost to wire an average sized office building can be thousands of dollars. In addition, there are physical limitations as to the length of any given physical cabling scheme. These distances vary depending on the type of cable used, but there is always a defined maximum distance that the signal can travel along the cable before it deteriorates. To send a signal any farther than this maximum distance requires additional hardware to boost the power of the signal. Lastly, using physical cables becomes inconvenient when network users need to be mobile. A good example is a sales representative that must carry a laptop to different conference rooms to make presentations using data on the LAN. Assuming that all of the conference rooms are wired into the LAN, the sales representative would have to carry a cable to connect into any conference room that they visit, find the appropriate wall jack, and connect into the network.

With a WLAN, most of the physical cabling (such as Cat 5 for client desktop connections) becomes unnecessary as you are now using radio waves to carry your signal. In a typical WLAN design, the only cables used are those necessary to connect devices that do not support wireless networks. As this technology evolves, devices that support wireless networking are becoming more prevalent and easier to find.

# How does a Wireless LAN Work?

The standards used for Wireless LAN (WLAN) communications are based on the Institute of Electrical and Electronic Engineers (IEEE's) 802.11 series. The 802.11 standards are responsible for defining the Physical and Media Access Control (MAC) layers of operation in a WLAN. The primary standard used in this chapter is 802.11b, which is an extension of the original 802.11 standard. 802.11b's primary objective defines the use of the 2.4 Gigahertz (GHz) band in radio frequency (RF) for high-speed data communications. 802.11b supports the original 802.11 data rate of 2 Mbps up to 11 Mbps. There is also an 802.11g standard that supports speeds up to 54 Mbps and also uses the 2.4 GHz frequency band.

The frames generated by a WLAN device differ in many ways from the frames generated by an Ethernet device. WLANs are not physically connected by cables like an Ethernet LAN is, so new fields in the frames must be created to describe aspects of the WLAN. The following section examines a typical 802.3 Ethernet frame and compares it to an 802.11b frame.

An 802.3 Ethernet frame is comprised of seven fields each with a specific function (Figure 1.1 illustrates an 802.3 Ethernet frame):

**Figure 1.1** 802.3 Frame Format

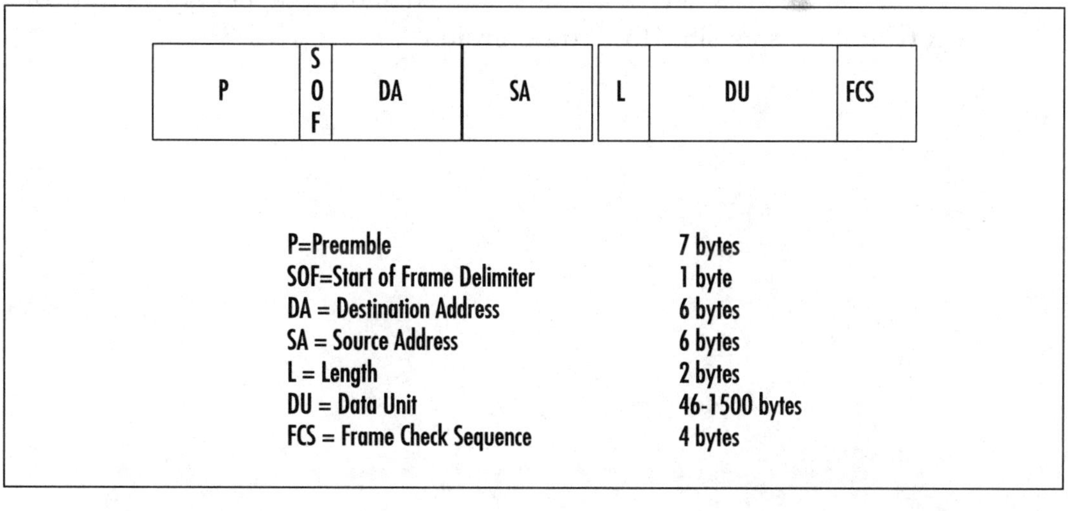

- **Preamble** The Preamble field is a 7-byte long alternating pattern of 0s and 1s that tells receiving devices that a new frame is arriving.
- **Start of Frame Delimiter** The byte before the destination address in both an Ethernet and an IEEE 802.3 frame is a Start of Frame (SOF) Delimiter. This byte ends with two consecutive 1 bits, which serve to synchronize the frame reception portions of all stations on the LAN.
- **Destination Address and Source Address** The Destination Address (DA) and Source Address (SA) fields are 2 or 6 bytes long and contain the MAC address of the source device on the network and the DA. The DA may be a single MAC address in the case of a unicast, a broadcast to all nodes on the network, or a multicast to a group of nodes on the network.
- **Length** This field is 2 bytes long and describes the number of bytes of data following this field.
- **Data Unit** The Data Unit field contains the user data of the frame and is 46 to 1500 bytes long. This is where the data being encapsulated into the frame is located (for example, a graphic in a Web page requested by your system). This field will vary in length based on the data encapsulated.
- **Frame Check Sequence** The Frame Check Sequence (FCS) field is 4 bytes long. The FCS is a cyclic redundancy check (CRC) that allows the receiver of a frame to perform basic error control on the frame. If a frame fails the CRC check, it is discarded and the upper layer protocol is typically responsible for retransmission.

An 802.11b frame (illustrated in Figure 1.2) is comprised of nine fields:

**Figure 1.2** 802.11b Frame Format

| FC | D/ID | A1 | A2 | A3 | SC | A4 | FB | FCS |
|---|---|---|---|---|---|---|---|---|

| | |
|---|---|
| FC=Frame Control | 2 bytes |
| D/ID = Duration/ID | 2 bytes |
| A1 = Address 1 | 6 bytes |
| A2 = Address 2 | 6 bytes |
| A3 = Address 3 | 6 bytes |
| SC = Sequence Control | 2 bytes |
| A4 = Address 4 | 6 bytes |
| FB = Frame Body | 0-2312 bytes |
| FCS = Frame Check Sequence | 4 bytes |

- The first field in an 802.11b frame is the Frame Control (FC) field, which is 2 bytes long. The FC field contains the following 11 subfields, which are some of the prime differentiators in an 802.11b frame:
    - **Protocol Version** The Protocol Version field is the first field within the FC field and is 2 bits long. The default value for this field is 0 with all other values being reserved.
    - **Type** The Type field is 2 bits long and works in conjunction with the 4-bit Subtype field to identify the function of the frame. The possible combinations and their descriptions are illustrated in Table 1.1.
    - **Subtype** The Subtype field is 4 bits long and works in conjunction with the 2-bit Type field to identify the function of the frame. The possible combinations and their descriptions are illustrated in Table 1.1.
    - **To DS** The To DS field is 1 bit long and is set to 1 in all frames sent by an associated station with an access point (AP) to signify that the frame is destined for the network behind the AP, such as a server connected to the same Ethernet network as the AP. All other frames have the To DS bit set to 0.
    - **From DS** The From DS field is 1 bit long and is set to 1 on all frames exiting the DS. All other frames have the From DS bit set to 0.

- **More Fragments** The More Fragments (MF) field is 1 bit long and is set to 1 in all frames that contain another fragment of the current MAC Service Data Unit (MSDU) or MAC Management Protocol Data Unit (MMPDU). All other frames have the MF bit set to 0.
- **Retry** The Retry field is 1 bit long and is set to 1 in all frames, data or management, which are retransmissions of earlier frames. Frames that are not retransmissions of a previous frame are set to 0.
- **Power Management** The Power Management (PM) field is 1 bit long and is used to indicate the power management mode of a station. The value is used to indicate the state that the station will be in after the successful completion of the frame exchange sequence. A value of 1 is used to indicate that the station will be in power-save mode, whereas 0 indicates that the station is in active mode.

**NOTE**

The PM field in frames transmitted by a wireless AP will always be set to 0, indicating active mode. It would not be desirable for an AP on your network to go into power-save mode.

- **More Data** The More Data (MD) field is 1 bit long and used to tell an associated station in power-save mode that one or more frames are buffered for the station on the AP. The MD field is set to 0 for all other directed frames.
- **WEP** The WEP field is 1 bit long and is set to 1 if the frame body contains data that has been processed by the WEP algorithm. Frames that have not been processed by WEP have a WEP field value of 0.
- **Order** The Order field is 1 bit long and is set to 1 in any data frame that contains data using the *StrictlyOrdered* service class. All other frames have a value of 0 in the Order field.

**NOTE**

The StrictlyOrdered service class is a mechanism built into the 802.11 standard that provides additional protection against out-of-order frames. This is accomplished by holding any multicast or broadcast traffic that matches addresses for frames that are already queued. Without this mechanism, it would be possible for broadcast or multicast traffic to reach a recipient out of order and create communications problems.

**Table 1.1** 802.11 Type and Subtype Combinations in the FC Field

| Type Value | Type Description | Subtype Value | Subtype Description |
|---|---|---|---|
| 00 | Management | 0000 | Association Request |
| 00 | Management | 0001 | Association Response |
| 00 | Management | 0010 | Reassociation Request |
| 00 | Management | 0011 | Reassociation Response |
| 00 | Management | 0100 | Probe Request |
| 00 | Management | 0101 | Probe Response |
| 00 | Management | 0110-0111 | Reserved |
| 00 | Management | 1000 | Beacon |
| 00 | Management | 1001 | Announcement traffic indication message (ATIM) |
| 00 | Management | 1010 | Disassociation |
| 00 | Management | 1011 | Authentication |
| 00 | Management | 1100 | Deauthentication |
| 00 | Management | 1101-1111 | Reserved |
| 01 | Control | 0000-1001 | Reserved |
| 01 | Control | 1010 | Power Save (PS) Poll |
| 01 | Control | 1011 | Request To Send (RTS) |
| 01 | Control | 1100 | Clear To Send (CTS) |
| 01 | Control | 1101 | Acknowledgement (ACK) |
| 01 | Control | 1110 | Contention-Free (CF) End |
| 01 | Control | 1111 | CF-End + CF-ACK |

**Continued**

**Table 1.1** 802.11 Type and Subtype Combinations in the FC Field

| Type Value | Type Description | Subtype Value | Subtype Description |
|---|---|---|---|
| 10 | Data | 0000 | Data |
| 10 | Data | 0001 | Data + CF-ACK |
| 10 | Data | 0010 | Data + CF-Poll |
| 10 | Data | 0011 | Data + CF-ACK + CF-Poll |
| 10 | Data | 0100 | Null function (no data) |
| 10 | Data | 0101 | CF-ACK (no data) |
| 10 | Data | 0110 | CF-Poll (no data) |
| 10 | Data | 0111 | CF-ACK + CF-Poll (no data) |
| 10 | Data | 1000-1111 | Reserved |
| 11 | Reserved | 0000-1111 | Reserved |

- The next field in an 802.11b frame is the Duration/ID field, which is 16 bits long and is used to carry the association ID of a station with an AP.
- The next fields in the 802.11b frames are address fields. If you review an Ethernet frame, you see that there are only two fields for addresses: destination and source. In 802.11b frames, there may be up to four, which include the following:
  - The basic service set identifier (BSSID) is the MAC address of the AP.
  - The DA is the MAC address of the final recipient.
  - The SA is the MAC address of the sending station on the WLAN.
  - The receiver address (RA) is the MAC address of the intended immediate recipient station on the WLAN.
  - The transmitter address (TA) is the MAC address of the sending station on the WLAN.
- The next field in an 802.11b frame is the Frame Body field, which is 0 to 2312 bytes long. The frame body is the payload, or data contained within the frame. This is where the data being encapsulated into the frame is located (for example, the graphic in a Web page requested by your system). This field will vary in length based on the data encapsulated.

- The final field in the 802.11b frame format is the FCS.

As you can see, there are a number of differences between Ethernet and 802.11b frames. These differences are required to enable high-speed communications on a physical medium of radio waves rather than standard copper or fiber media.

# WLAN Benefits

There are many obvious benefits to using a WLAN design, most of which hinge around the problems with typical wired LANs. Convenience is certainly a benefit to using wireless communications. With wireless, as long as you are in range of an AP, you have a connection to the network. This is a tremendous advantage to mobile sales forces, personnel performing physical inventories of a warehouse, or IT professionals who may need to get access to data from anywhere in a building. Using wireless technology makes it easy and effective to let people physically go wherever they need to go and still be able to access any data that they need from the network.

Another benefit to using a WLAN is that cable distance limitations become less of an issue. There are many situations where the distance between the network link and the end user is such that the signal strength is degraded by the time the cable has been routed up walls, through floors, and around permanent objects. Wireless communications negate this by doing direct "line-of-sight" connections to a system. The signal strength from a wireless AP or network card is typically between 150 to 300 feet indoors (depending on the design and structure of the building) and up to 1000 feet outdoors. Obviously, the 1000-foot outdoor range outdistances the maximum unshielded twisted pair (UTP) cable length of 328 feet. In addition, a wireless signal can be boosted by using more than one AP or by using a wireless relay to extend the range even farther. Figure 1.3 shows an example of a wireless network design with an additional AP.

**Figure 1.3** Wireless Network with Additional AP

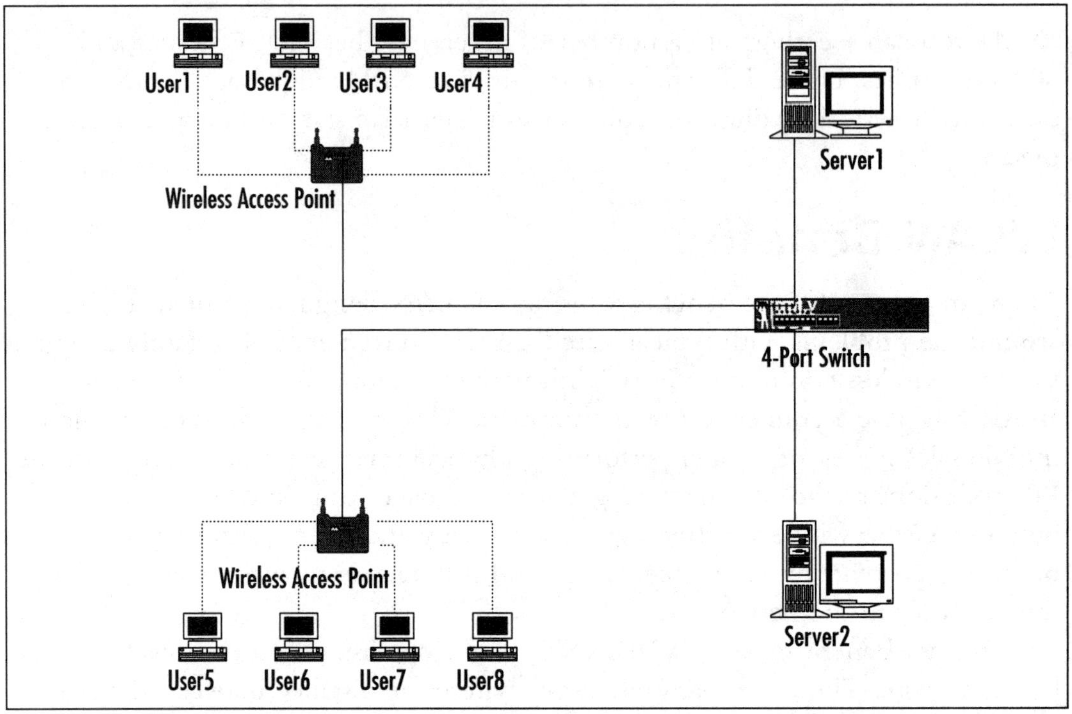


When the total cost of the elements of a traditional wired LAN such as switches, cables, cable racks, routers, and implementation time are added up, the price can be very high. A small office network for ten workstations and a couple of servers can cost thousands of dollars just in wiring and networking equipment. When setting up a WLAN, most of the costs associated with a traditional network are negated by the simple fact that not as many cables are used and not as much equipment is needed to support the LAN. For a good example, take a look at the designs shown in Figure 1.4.

**Figure 1.4** Wired vs. WLAN Designs

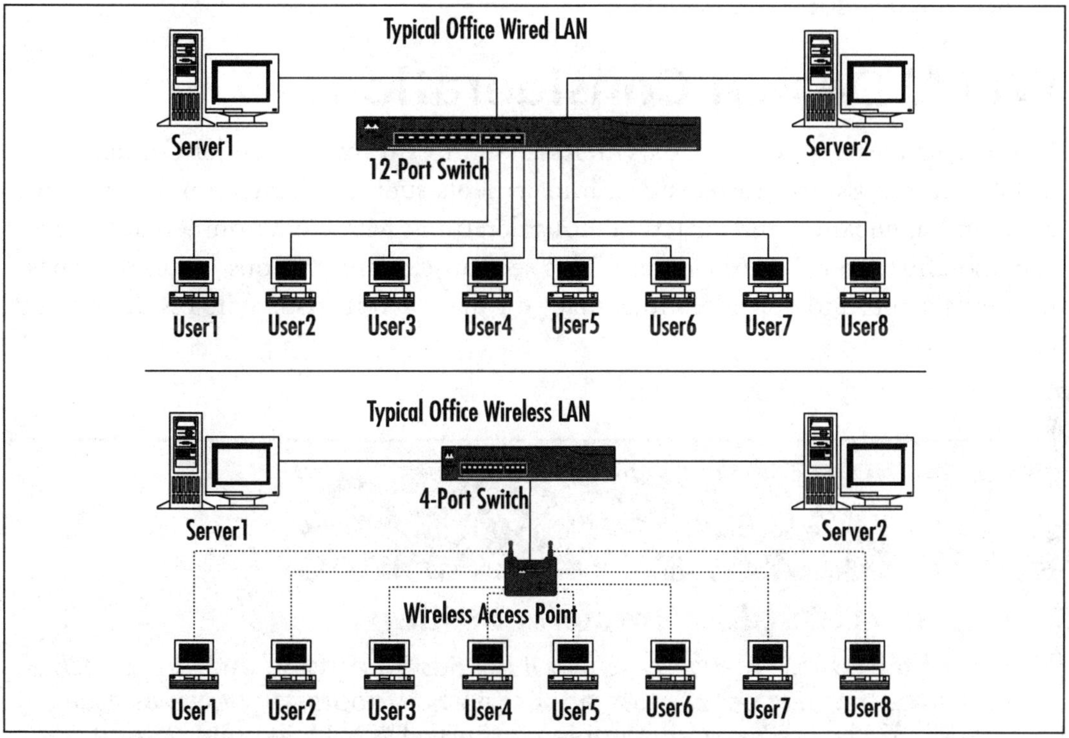


In Figure 1.4, there are two LAN designs—one for a wired LAN and one for a WLAN. In the wired LAN design, a 12-port switch is required due to the number of connections to workstations and servers. In addition, Cat-5 cabling must be laid in the building to support these systems. All systems must also have network interface cards (NICs). This design is pretty typical for an average small office.

A more cost-effective solution is shown in the second design. This design is considered a hybrid WLAN as it contains elements of both wireless and wired networking. The workstations use wireless communications to connect to the network; however, the servers are using traditional wiring into a switch. A 4-port switch could be used rather than a 12-port to save on cost. The only wiring necessary is used to connect the servers to the switch. The servers would require standard NICs and the workstations would require wireless NICs. The only additional requirement is the wireless AP. The savings are in the cost of networking hardware, wiring, and maintenance. For example, if the office was to be rearranged and the systems moved, there would be no additional wiring changes necessary. The systems would simply connect to the wireless network regardless of their physical location.

# WLAN Design Considerations

In order to create a design for a wireless system, you must consider common WLAN transmission and reception impairments such as attenuation, RF interference, and application and structural considerations. Many environmental factors can also affect your WLAN design. This section explains various common types of impairments and considerations that you may face in your wireless design and testing efforts.

**Designing & Planning...**

### Indoor vs. Outdoor Implementation

A lot of the issues covered here as WLAN design considerations are based on factors that can occur both indoors and outdoors. However, as a general rule, if you are looking at implementing a WLAN in a building that has been built within the last ten years and does not have any special structural considerations (i.e. concrete shielding for radiation labs), then implementing a WLAN should be pretty straight forward. There are a lot of additional factors when extending a WLAN over long distances outdoors, but in a standard office implementation these are usually not an issue.

## Attenuation

Attenuation is the decrease in strength of a radio wave between a transmitter and a receiver; the strength decreases as the distance from the antenna increases. It can be caused by natural conductivity or by resistance of all sorts of physical matter, but the greatest resistor to radio waves is the Earth. Radiated energy from the Earth and interference from trees and buildings will cause attenuation of a signal's ground waves, just as radiated energy and interference from water and dust particles in the atmosphere will affect a signal's sky waves. You must plan your design and equipment use based on affecting ground and sky wave propagation such as transmitter height, distance between transmitters, and solar radiation factors.

## Attenuation Due to Antenna Cabling

Loss due to antenna cable length must always be considered when designing a wireless system. Cisco cabling produces 6.7 decibels (dBi, also referred to as dB) of loss per 100 feet of cabling. The reason for this is that the radio wave actually starts at the radio device. The radiated energy traveling through the cabling from the radio device to the antenna induces a voltage in the cabling, decreasing the strength of the wave as the distance from the radio device to the antenna becomes greater.

## Attenuation Due to Exterior Considerations

If you plan on coverage outdoors that is point-to-point or point-to-multipoint, you will need to pay particular attention to considerations that are distance-related. For example, Earth bulge will come into play only if you are implementing a point-to-point or point-to-multipoint WLAN, whereas weather is a consideration for any outdoor implementation.

All matter produces attenuation (loss) to some degree. Because weather can produce rain, snow, or fog, all of which are matter, weather must be considered in a WLAN design.

Researching any unusual weather conditions that are common to the site location is important. These conditions can include excessive amounts of rain or fog, wind velocity, or extreme temperature ranges. If extreme conditions exist that may affect the integrity of the radio link, you should take these conditions into consideration early in the planning process.

### *Rain, Snow, and Fog*

Except in extreme conditions, attenuation (weakening of the signal) due to rain does not require serious consideration for frequencies up to the range of 6 or 8 GHz. When microwave frequencies are at 11 or 12 GHz or above, attenuation due to rain becomes much more of a concern, especially in areas where rainfall is of high density and long duration.

The attenuation rate for snow is generally higher, due in large part to the size of the particles of snow or for that matter rain and fog as well, in compared to the wavelength of the signal. For example, a 2.4 GHz signal will have a wavelength of approximately 125 millimeters, or 4.9 inches. A 23 GHz signal will have a wavelength of approximately 0.5 inches. A raindrop approaches 0.25 of an inch. At 2.4 GHz, heavy rain or snow should not have much of an impact on the wireless system; however, in a 23 GHz system, the wavelength is reduced to half by this rain. At this size, the rain or snow becomes a reflective surface and disperses the 23 GHz signal.

In most cases, the effects of fog are considered to be much the same as rain. However, fog can adversely affect the radio link when it is accompanied by atmospheric conditions such as temperature inversion, or very still air accompanied by stratification (layers of significantly differing air temperatures). Temperature inversion can negate clearances, and still air along with stratification can cause severe refractive or reflective conditions, with unpredictable results. Temperature inversions and stratification can also cause ducting, which may increase the potential for interference between systems that do not normally interfere with each other. Where these conditions exist, use shorter paths and adequate clearances.

## *Atmospheric Absorption*

A relatively small effect on the wireless link is from gases and moisture in the atmosphere. It is usually significant only on longer paths and particular frequencies. Attenuation (loss) in the 2 to 14 GHz frequency range is approximately 0.01 dB/mile. You may have to include atmospheric absorption in your design consideration if you are planning on implementing a wireless system above 10 GHz where atmospheric absorption is prevalent. There are some wireless systems on the market today licensed in the 23 GHz band, that are significantly impacted by this type of loss. Antenna height has some impact on loss related to atmospheric absorption, because the density of the air decreases as altitude increases. Thus, a 23 GHz system with an antenna significantly elevated over a similar implementation at a lower elevation will suffer less from attenuation due to atmospheric absorption. Table 1.2 depicts attenuation due to atmospheric absorption versus path distance. Attenuation is listed as negative decibels, or –dB.

**Table 1.2** Attenuation (Absorption) over Distance

| Path Distance (In Miles) | 2–6 GHz | 8 GHz | 10 GHz | 12 GHz | 14 GHz |
|---|---|---|---|---|---|
| 20 | –0.20 dB | –0.26 dB | –0.32 dB | –0.38 dB | –0.48 dB |
| 40 | –0.40 dB | –0.52 dB | –0.64 dB | –0.76 dB | –0.96 dB |
| 60 | –0.60 dB | –0.78 dB | –0.96 dB | –1.14 dB | –1.44 dB |

Continued

**Table 1.2** Attenuation (Absorption) over Distance

| Path Distance (In Miles) | 2–6 GHz | 8 GHz | 10 GHz | 12 GHz | 14 GHz |
|---|---|---|---|---|---|
| 80 | –0.80 dB | –1.04 dB | –1.28 dB | –1.52 dB | –1.92 dB |
| 100 | –1.00 dB | –1.30 dB | –1.60 dB | –1.90 dB | –2.40 dB |

## *Multipath Distortion*

Multipath distortion is caused by the transmitted signal traveling to the receiver via more than one path: A common cause of this is reflection of the signal from bodies of water, hills, or tall buildings. Figure 1.5 shows an example of multipath distortion caused by reflection. The antennas are the same height. In the worst case, the reflected signal arrives at the receiving antenna at the same time as the intended signal, but out of phase with the intended signal, both signals will cancel each other out, resulting in complete loss of data. Best case, the reflected signal arrives a moment later than the intended signal causing distortion and therefore reduced performance. Examples of reflective surfaces include water, asphalt, fields, metal roofs, or any smooth, relatively flat surface. Dispersing extraneous radio waves is better than reflecting them. Examples of dispersal surfaces include rough, rocky surfaces, shrubbery, trees, and so on. In a big city, more people receive an echoed distortion of the wireless signal than receive the actual signal, because the original signal bounces off buildings.

**Figure 1.5** Multipath Distortion Diagram

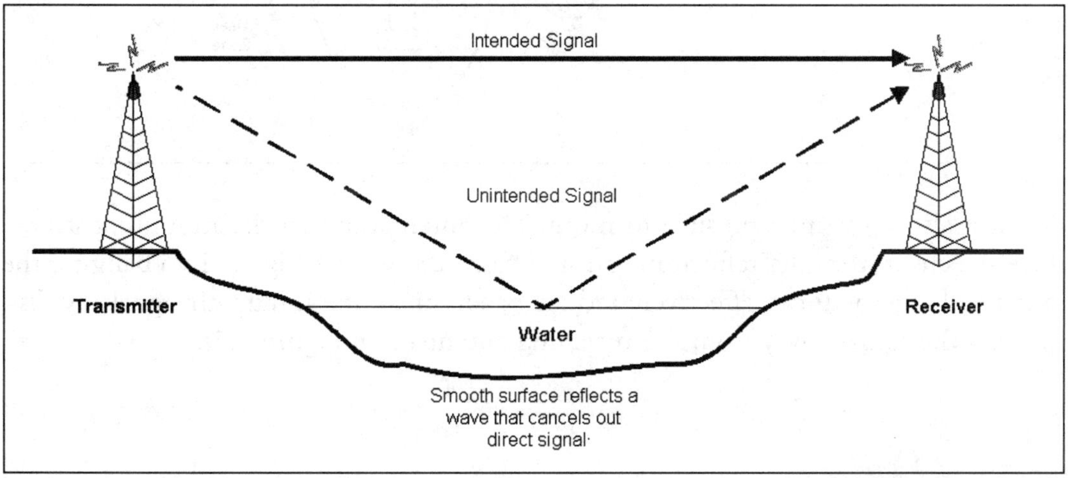

The best way to reduce multipath distortion is to use a directional rooftop antenna (for example, a directional antenna that will only pick up signals coming from the direction of the transmitter and will reject reflections that arrive at its sides or its back). A Yagi antenna is one example of a directional antenna that will help reduce or eliminate multipath distortion (see Figure 1.6).

**Figure 1.6** Directional Antenna to Reduce or Eliminate Multipath Distortion (Birds-eye View)

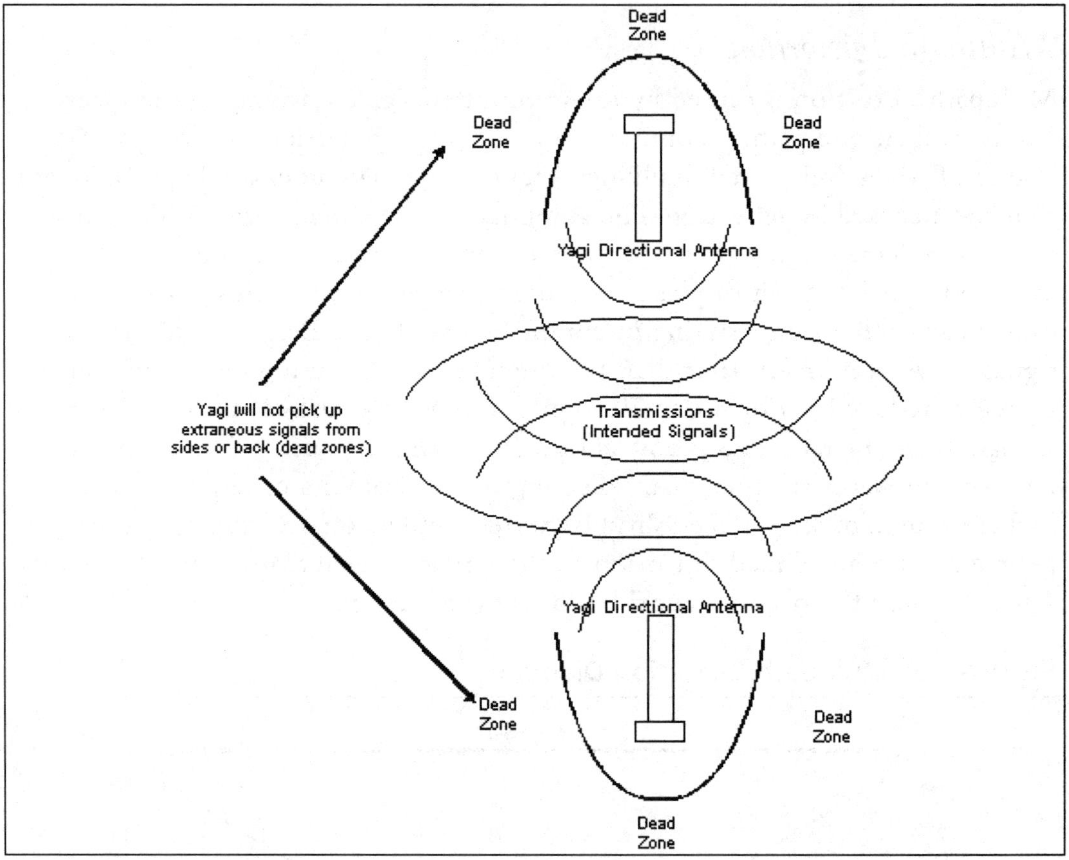


It is also sometimes possible to mount the antenna so that the mounting structure screens it from the reflections but not from the wanted signal. By changing the antenna height you can effectively reduce or eliminate the multipath signals by dispersing the signals away from the receiving antenna (see Figure 1.7).

**Figure 1.7** Dispersing Multipath Reflections

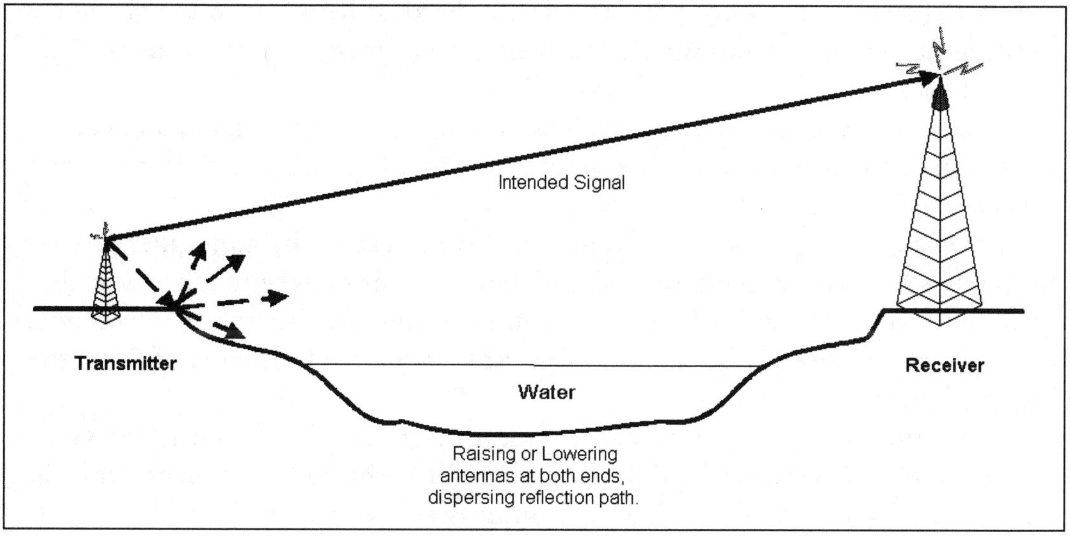


## *Refraction*

When a radio wave travels between two substances of different densities, the wave will bend or refract because electromagnetic signals move slower through substances of greater density. This phenomena impacts a radio wave as it travels through the atmosphere. The density of the Earth's atmosphere decreases as altitude increases. Therefore, the bottom of the radio wave travels through a denser atmosphere than the top of the wave. This means the bottom of the wave will move slower than the top of the wave, causing the signal to bend towards the Earth's surface and follow the curvature of the Earth, but at an arc radius approximately 1.33 times greater than the Earth's arc radius (see Figure 1.8).

**Figure 1.8** Refraction

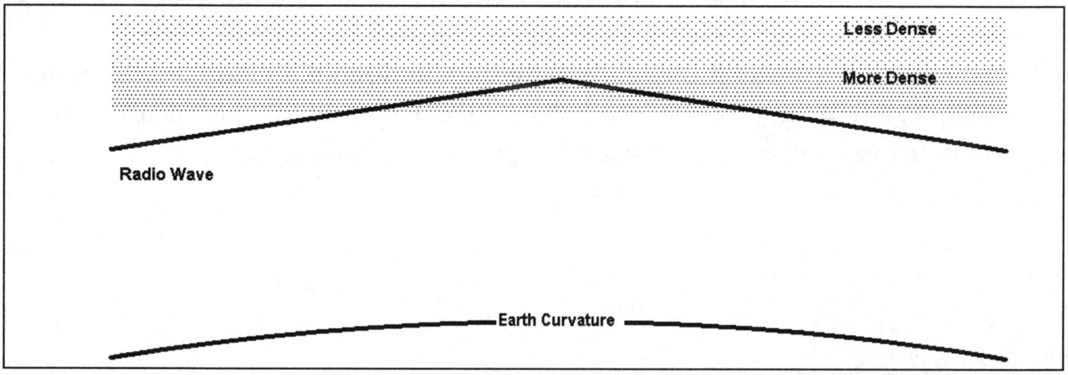

At night, the air cools and much of the moisture in the air moves closer to the Earth's surface. The cool, wet air near the Earth is denser than the air higher in the atmosphere, so radio signals can bend farther than they do in the daylight hours. This is known as *super refraction*.

Other refraction phenomena, such as ducting or bending, can also occur. *Ducting* happens when radio waves are trapped in a high-density duct between two areas of lower density.

*Bending* is similar to super refraction, but is not caused by atmospheric conditions related to day or night. Instead, differences in air density in a horizontal plane, like cooler air over a lake or field and warmer air over a shore or highway, cause the radio waves to bend in the direction of the cooler, denser air over the lake or field.

Refraction is one reason why radio line-of-sight is not necessarily the same as optical line-of-sight. Refraction is minimal for paths under 10 miles, with the exception of hot, humid areas like the southeastern section of the U.S.

## Accounting for the Fresnel Zone and Earth Bulge

A main consideration of any point-to-point design is the Fresnel zone. An electromagnetic signal traveling between two antennas does not travel in a straight line. The wave spreads out as it propagates. The individual waves that make up the signal do not travel at a constant velocity. A pair of antennas define a three-dimensional elliptical path for the radio waves that propagate between them. This elliptical path is divided into several zones based on the phase and speed of the propagating waves. These zones are referred to as Fresnel zones. Each Fresnel zone differs in phase by up to half a wavelength, or 180 degrees. This Fresnel zone is commonly thought of as line-of-sight (see Figure 1.9). Radio line-of-sight is not the same as visual line-of-sight. In visual line of sight, a direct line exists between two points; it is easy to think this way between two antennas in a point-to-point design. However, radio line of sight is not a straight line between the antennas; it is more of an ellipse. In a good point-to-point design, this ellipse should be calculated to determine its size and clear of obstacles to provide a good signal.

**Figure 1.9** Fresnel Zone (Radio Line of Sight)

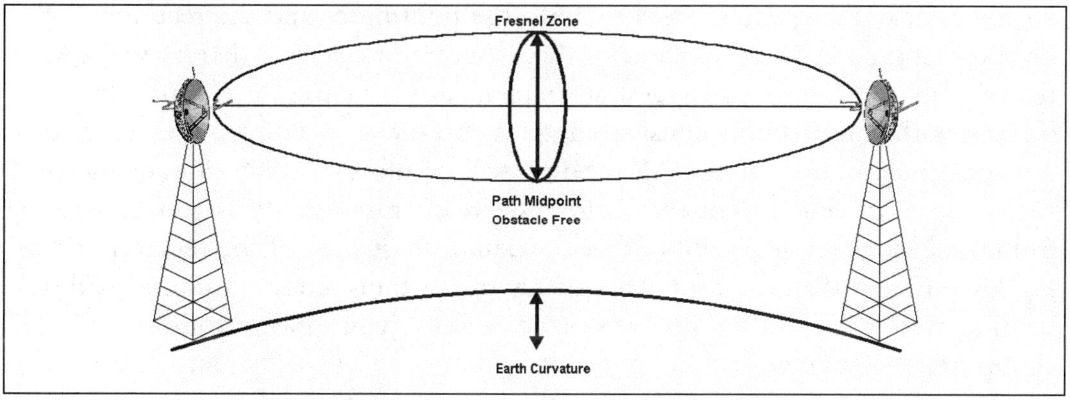


Because of the elliptical shape of the Fresnel zone, the antennas used in a point-to-point design must be high enough to provide clearance of the Fresnel zone's radius at the midpoint. As the distance increases, other factors must be considered such as the curvature of the Earth, where the line of sight becomes difficult at 6 miles (for a 6-foot tall person) and disappears altogether at 16 miles (for two structures at 10 feet) (see Figure 1.10). Paths over 20 miles are extremely difficult to align and install, so take caution when recommending these types of configurations.

**Figure 1.10** Minimum Clearance for Long Distances

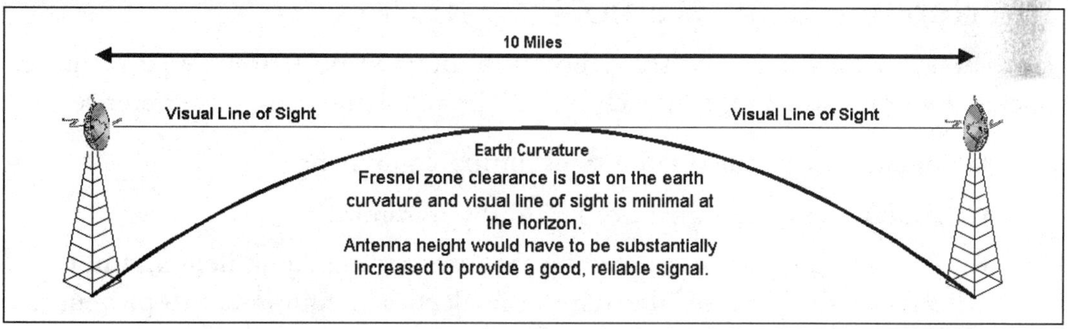


# Radio Frequency Interference

Radio frequency interference is extraneous energy that impedes the reception of signals. It can be caused by a radio emission from another transmitter at approximately the same frequency.

When we talk about the frequency that an AP radio operates on, we really mean its "center frequency," because both the transmitter and the receiver operate within a band of frequencies that is several megahertz (MHz) wide. AP transmitters will transmit strongest at frequencies very close to its center frequency, with a decrease in signal strength as you move away from the center frequency. Similarly, the wireless AP receiver will be most sensitive to frequencies very close to its center frequency, with a decrease in sensitivity as you move away from the center frequency. Note that the center frequency of the receiver can be slightly different than the center frequency of the transmitter and things will still seem to work okay, but because power decreases as you move out from the center frequency, range will be reduced.

The width of this band of frequencies around the center frequency is a major factor in determining the effects of radio interference. If your receiver encounters a second signal that is too close to its center frequency and the two bandwidths end up overlapping too much, interference will result. The closer the interfering signal is to the receiver's center frequency, the less power is needed to cause interference. In the extreme case, if somebody turns on their microwave oven and its emanations are on exactly the same frequency as yours, you may drop down in speed even if the signal is very weak. Conversely, if something is operating on a frequency that is quite far away from the center frequency of your AP's receiver, it can still interfere if its signal is strong enough.

## Interference from Radio Transmitters

Interference usually occurs when radio transmitters and electronic equipment are operated within close range of each other. The following causes interference:

- Incorrectly installed radio transmitting equipment
- An intense radio signal from a nearby transmitter
- Unwanted signals generated by the transmitting equipment and not enough shielding or filtering in the electronic equipment to prevent it from picking up those unwanted signals

Any signal other than the desired signal is called an unwanted signal, or *spurious radiation*. Spurious radiation includes harmonic radiation, usually in the form of standing or traveling waves. Use a spectrum analyzer, a calibrated field intensity meter, or a frequency-selective voltmeter to measure unwanted radiation. A spectrum analyzer is a device that measures the frequency components of a radio signal. It provides a visual image of how the amplitude of a radio signal varies in relation to its frequency. If adjusting the channel does not solve the problem completely, you should permanently install a low-pass band filter in the transmitter antenna feed line after all the other accessories.

Standing waves are a form of spurious radiation causing undesired effects that occur when two or more waves of the same frequency are present at the same time and do not travel away from their source. This may happen, for example, when the transmitter, transmission line (antenna cabling), or antenna are not properly matched to each other. Incorrectly terminated or damaged antenna cabling is a typical source of standing waves. When this happens, the transmitted signal to the antenna is reduced because the damaged cable is transmitting unwanted signals.

## Harmonics

Harmonics occur when signals are produced at two or three times the station's operating frequency in addition to the desired signals (see Figure 4.8). If the harmonics fall on another locally used frequency, such as an AP channel, they are likely to cause interference. Figure 1.11 shows how a signal from a radio device may interfere with an AP set to channel 1.

These undesired transmissions occur at multiples of the original frequency. In the example, harmonics of Device A, which is transmitting at 804 MHz, may occur at 1.608 MHz (frequency × 2) or 2.412 MHz (frequency × 3). In addition, Device A's second harmonic is reduced in power by roughly half of the originating signal's power. The third harmonic's power is roughly half of the second harmonic and so on.

**Figure 1.11** Harmonics

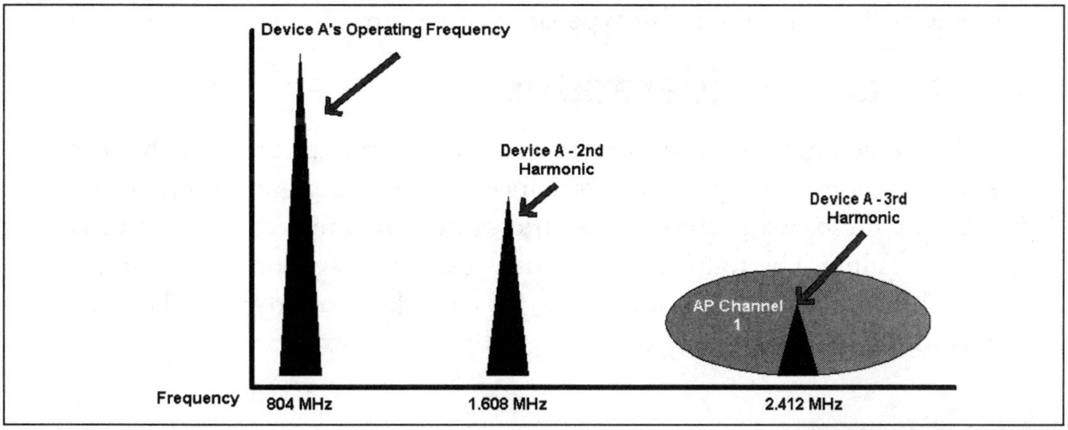

As you can see in Figure 1.11, the harmonic frequencies of Device A could present a potential problem in a wireless design. Although Device A's second harmonic does not interfere with the AP's frequency channel, the third harmonic, although weaker in strength, can affect the transmission and reception for the channel 1 AP.

# Application Considerations

Applications play a significant role in the determination of a wireless implementation. Due to the high bandwidth utilization of some applications, you may need to modify or completely remove a wireless design as an infrastructure solution. If the high bandwidth and/or high traffic application is not necessary on the wireless system you intend to deploy, you should filter that application's traffic from the wireless network by installing a router between the wireless segment and the wired segment.

Graphics-intensive applications, such as desktop publishing and computer aided design (CAD) programs, can have significant impact on a wireless design depending on how the applications are used on the network. For example, if these files are stored on a shared network device such as a file server, there will be some network impact each time a user stores or retrieves a file. Typically, these files are large (20 MB or more), and if the number of users is large and/or they store or retrieve files frequently, the impact on the network becomes greater. The same is true for file transfer applications. To determine if the impact on the network is significant enough to justify changing or removing your wireless design, you should perform a baseline of network utilization for the existing network. A device or program known as a protocol analyzer can give you this information by monitoring the number, type, and size of packets traversing the network over a period of time. Network management programs, such as Ciscoworks and HP Openview, will also give you this type of information.

# Structural Considerations

Physical considerations are classified into two major groups of issues that can affect wireless connectivity and performance—path fading and propagation losses. Below these two groups are specific causes that produce either propagation loss or path fading. The following sections explain propagation losses and path fading and then explore common causes for each, keeping in mind their affect on wireless design.

As discussed in Chapter 2, radio waves are propagated through space at the speed of light. This speed is attained assuming that there are no obstructions for the electromagnetic wave to pass through. The reason for this is because electromagnetic waves pass through different substances at different speeds. The greater the density of the substance, the slower the wave propagates through it. For example, a radio wave will travel faster through the air than it will water and faster through water than a concrete building.

Under normal circumstances, as the signal radiates out from an antenna and encounters objects within the environment, it will exhibit one or more of the following reactions: the signal may penetrate the object, reflect off the object, or be absorbed by the object.

In most cases, all of these reactions will occur to varying degrees, depending on the density and type of object encountered. This is the *propagation* of the signal. The strength of the signal decreases as it propagates. Penetration, reflection, and absorption all factor into the signal as it travels, each taking with it some amount of signal strength. These actions not only weaken the signal, but they may affect the direction in which the electromagnetic wave travels and the speed at which it travels.

As the radio wave propagates through the Earth's atmosphere and encounters objects within the environment, the strength of the signal will decrease. Any distortion of a wave's amplitude, phase, or direction can affect the strength of the received signal. This is known as *path fading*. The strength of the received signal is equal to the strength of the transmitted signal minus path fading.

As you can see, propagation loss and path fading are very similar. The difference is really a matter of perspective. Receivers can suffer from path fading and transmitters can suffer from propagation losses. Ideally, because most WLAN radios both send and receive, elimination or minimization of propagation losses and path fading are extremely desirable.

Differing environments can have substantial structural considerations to work around or overcome to successfully implement a WLAN solution. The following list takes a look at some of the common problems encountered in various environments and the solutions or alternatives available for each.

- **Hospitals** The most obvious issue that comes to mind in any medical environment is compatibility of wireless networks with existing medical equipment and, more importantly, medical diagnostic devices.

  Another consideration is the need for many healthcare providers to meet federal regulations in terms of their information systems. The

Health Insurance Portability and Accountability Act of 1996 (HIPAA) is causing a major reassessment regarding privacy and related issues in healthcare information systems. Because of the lack of security in previously implemented WLANs used in these organizations, data encryption is a must for compliance in these facilities.

Structurally, hospitals offer a variety of radio frequency obstacles. You should be aware of X-ray areas in particular, because most hospitals have lead-lined or extremely thick walls surrounding these areas to prevent X-ray bleed-through. Consider these areas "dead zones" to radio frequency. If coverage is necessary in these areas, you will have to install your antenna directly in each room requiring coverage. Full site surveys are recommended for all areas of a hospital, and all equipment normally used in the hospital should be on during the survey.

- **Warehouses** Warehouses generally contain stock and rows of shelving. This presents coverage problems due to the density of the stock items and the metal construction of the shelving. Be sure to find out what the current stock levels are when performing your site survey. If the stock levels are high, you will get a much more accurate picture of coverage in your wireless implementation. The type of product that is stored in the warehouse also makes a difference . For example, a fully stocked warehouse that contains only cases of empty plastic water jugs will have better coverage with fewer APs than the same warehouse containing cases of full plastic water jugs.
- **Metal Construction** In general, WLAN radio devices do not penetrate metal construction very well, if at all. Keep this in mind when designing your WLAN. In most cases, you will need to place antennas in each area that is contained by metal construction. Your site survey will aid you in verifying this requirement.
- **Other Construction** The materials used in the construction of walls, pillars, and supports can also cause radio frequency impairments and impair coverage in a given area. Exterior walls tend to be thicker and contain more reinforcement materials than interior walls and partitions. Rebar (metal rods used to increase the strength of concrete construction) reinforced cinder block or concrete walls and pillars can present a design issue in most facilities that use them. In general, the denser the material, the more difficult it will be for radio waves to penetrate.

- **RF-producing Devices** In addition to the construction and application considerations in your site survey and wireless design, you should constantly be on the lookout for potential interference from other electronic devices. There are many devices that can potentially cause interference and require you to change your AP channel assignments. The most common culprits are 2.4 GHz cordless phones and microwave ovens. Be sure to have someone use these devices if they are inside a coverage area in your design. This will help you to determine the best channel to use on your AP.

  Other potential interfering devices can be arc welding and telemetry equipment, 2.4 GHz lighting systems, and Spectralink phone systems. Spectralink phone systems are used to provide cellular phone coverage within a company and are based on the IEEE 802.11b standard—the same standard used for WLANs.

## Security Considerations

Some of the major drawbacks to the implementation of a wireless network are the inherent security issues. In a typical wired LAN, access to the physical network is controlled by the physical locations of switches, wall ports, and so on. With a WLAN, radio waves are the physical medium; therefore, anybody can easily gain access to the physical portion of the network.

That is not to say that a wireless network is completely insecure. With the proper implementation of WEP, protocol filtering, and Virtual Private Networks (VPNs), a wireless network can be almost as secure as a normal wired network. However, it takes a great deal of planning and effort to implement a WLAN with this level of security. This additional overhead of work plus the general fear of a possible security breach is one of the primary reasons that wireless networks are not as prominent in the corporate world as one might expect.

When designing a WLAN, security should be one of the most important factors in the design plans. Whenever an AP or any other wireless network device is implemented, it should be configured to be as secure as possible, use the highest level of encryption that it can, and meet the security policies implemented throughout the rest of the WLAN. Ensuring the implementation of high security standards throughout the WLAN is the only way to mitigate the risks involved in using radio waves as the physical medium.

To ensure the highest possible security of a WLAN, it is critical to encrypt wireless transmissions by using WEP. If WEP is not implemented in a WLAN, every wireless transmission on the WLAN can be intercepted using a simple sniffer on a laptop with a wireless card. Any network traffic not encrypted by default will be available in cleartext to an intruder.

Using protocol filtering or limiting access to a network based on wireless card MAC addresses are two additional methods of ensuring the security of a WLAN. Protocol filtering allows you to prevent the use of various unsafe or non-preferred protocols on your WLAN. Limiting WLAN access based on MAC addresses will prevent any intruders from connecting to the APs unless they have gained access to an authorized wireless network card. Both of these methods can help a great deal in securing a wireless network.

Another design that helps secure a WLAN is the use of VPNs. Using a VPN to connect a wireless network to a wired network can help prevent any successful intruders of the WLAN from getting access to the network devices or systems on the wired LAN. This added layer of separation can also make it easy to add a firewall between the WLAN and the wired LAN to prevent various attacks against the wired LAN devices that could potentially come from the WLAN.

## Network Management Considerations

When implementing a WLAN, the concepts generally associated with network management come into a whole new light. With a traditional wired LAN, most network management can be done through configuration of switches, routers, and the layout of the physical cable plant. Monitoring takes place through management tools that watch the network through the devices and statistics or network-related information can be gathered through these tools.

In a WLAN, it has been difficult to port these types of management tools and procedures over to a whole new network infrastructure. Most of the existing tools are not designed to handle WLANs, and the tools that are designed for WLANs are not necessarily designed to interface well with existing network tools. This has led to difficulty in WLANs being adopted into corporate environments where network management is a critical and necessary part of daily operations.

As the wireless technologies improve, more and more tools are available to help integrate WLANs into an existing wired LAN infrastructure. Obviously, one of these tools is the subject of this book, the Cisco Wireless-Aware LAN. This concept is covered later in this chapter and throughout the remainder of the book. The important point to remember is that wireless technologies evolve almost daily and as time goes on, the integration of WLANs and wired LANs will become easier to administer and control.

# WLAN Modes of Operation

WLANs can function in two primary modes of operation—"ad-hoc mode" (also referred to as "peer-to-peer mode") and "AP mode." Each one functions in a slightly different way and each has advantages within specific situations. Some networks actually make use of both modes and are thus called "hybrid mode" networks.

The first mode for WLANs is ad-hoc mode. This is the easiest WLAN mode to configure and requires the least hardware. An ad-hoc mode WLAN is comprised of two or more computers communicating directly with each other using wireless network cards. There is no hardware or software AP used in this type of WLAN; each computer simply joins the network on an ad-hoc basis. Figure 1.12 shows the design of an ad-hoc mode WLAN.

**Figure 1.12** Ad-hoc Mode WLAN

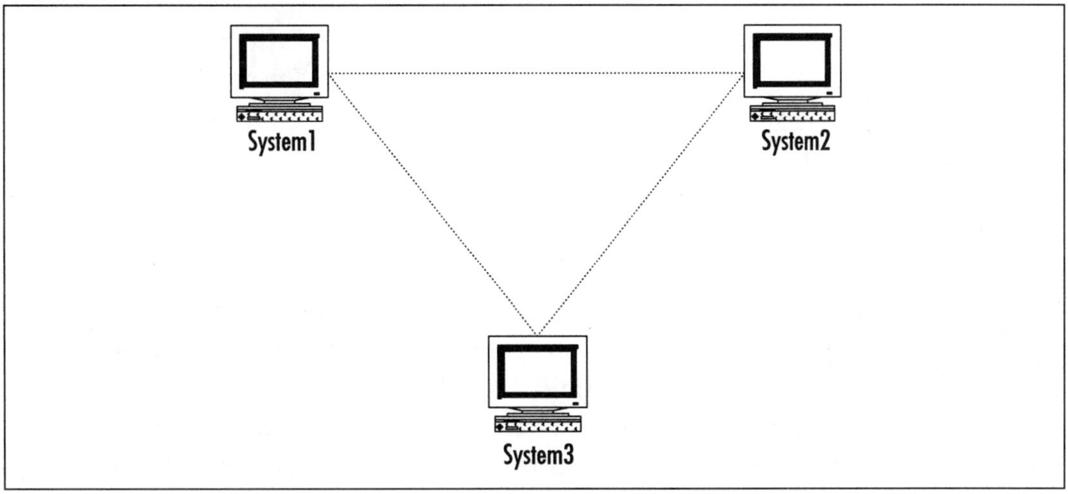


The second WLAN mode is AP mode. In this WLAN mode, a hardware or software AP is configured as part of the WLAN design. This AP then provides connectivity for all of the systems on the WLAN. The wireless network card on each computer is configured to use a specific AP to connect to a specific WLAN and all traffic to other computers on the WLAN is brokered through the AP.

Figure 1.13 shows the design of an AP WLAN with both software and a hardware APs on the network.

**Figure 1.13** AP Mode WLAN

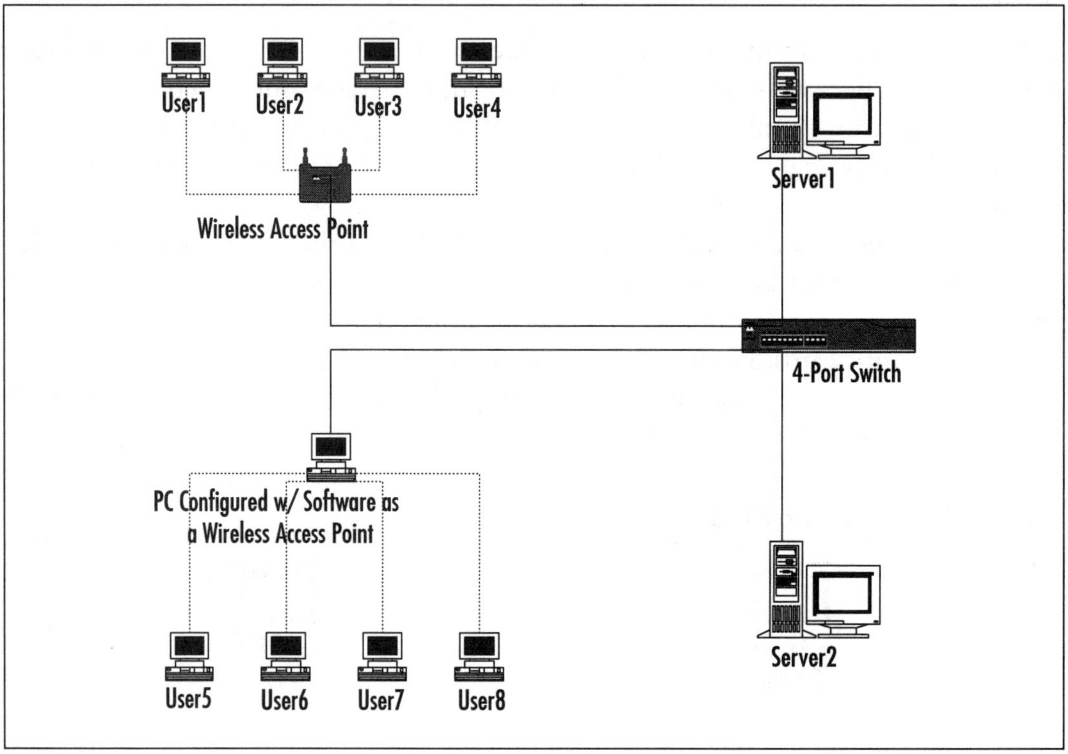


Each of these WLAN modes offers benefits and detriments depending on the WLAN environment that you are trying to configure. Ad-hoc mode WLANs are very easy to configure and do not require a great deal of effort to set up. There is no real administrative management required for the network and additional systems can be added in or removed with almost no effort. While this works well for small home networks or very small offices, this mode of WLAN does not work well in most business environments, as there is no ability to centrally manage the WLAN or supply security for the WLAN. In an ad-hoc network, each computer is responsible for its own security and the WLAN itself is unmanaged.

AP mode WLANs are slightly more difficult to set up than ad-hoc mode WLANs, but provide a single point of control and security for the WLAN. This helps a great deal in keeping the network safe and controlled. Using an AP mode WLAN does also require either a hardware or software AP in the WLAN design.

This adds cost over an ad-hoc WLAN, but is usually the best way to go for most WLAN designs.

A hybrid mode WLAN makes use of both ad-hoc mode and AP mode. This configuration is fairly rare because once an AP is added to the WLAN, most administrators reconfigure all of the systems to use the AP. The design for this type of WLAN is shown in Figure 1.14. Typically the only time you will run into this type of configuration is when a WLAN is being switched from ad-hoc mode to AP mode, and the transition is not yet complete. This certainly is not a design that an administrator should intentionally introduce to a network.

**Figure 1.14** Hybrid Mode WLAN

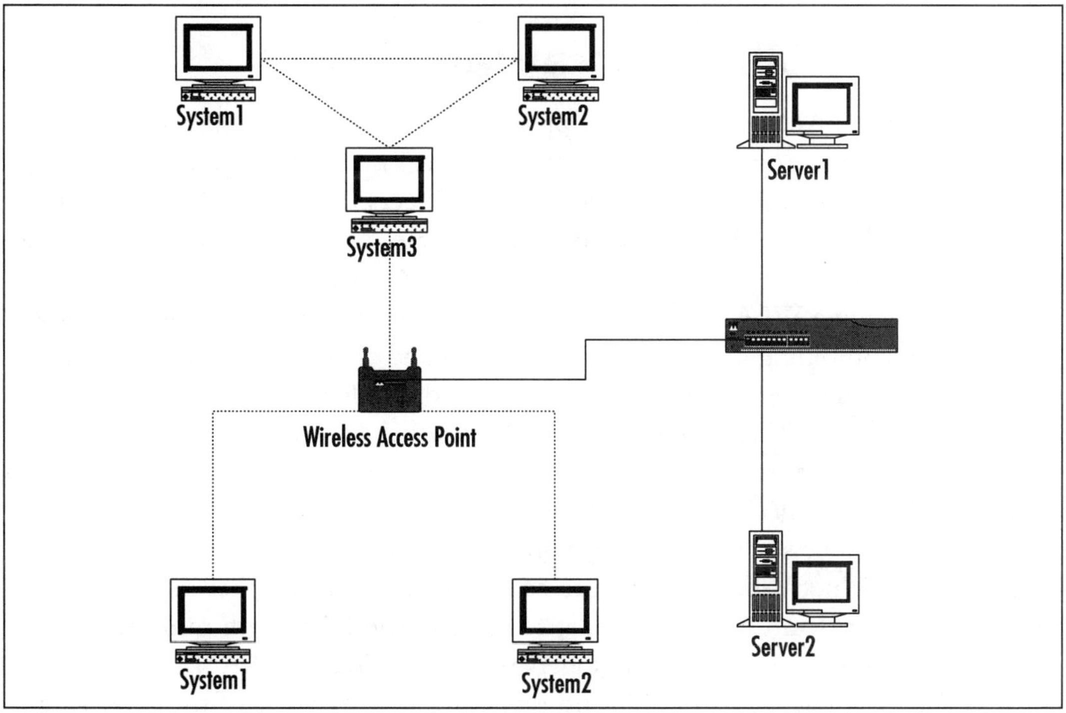

Configuring & Implementing...

**Hybrid Mode WLANs**

In reality, any wireless network with more than three systems should probably be designed as an AP mode WLAN. If you are just looking for a quick way to transfer some data between a few systems with wireless network cards, an ad-hoc mode WLAN would suit your needs perfectly.

# What is a Wireless-aware LAN?

A wireless-aware LAN is Cisco's new approach to designing and implementing WLANs into an existing Cisco LAN infrastructure. This approach is an attempt to address and eliminate the issues that network engineers and management face in the implementation of a WLAN. Some of these issues were discussed earlier in this chapter, such as the lack of centralized management, security issues, and the difficulty of deploying a WLAN within an existing infrastructure. Cisco has minimized this effort by combining an existing Cisco switch and router infrastructure with Cisco wireless infrastructure to make an integrated wireless-aware LAN.

Cisco brings the WLAN and wired LAN together by using a combination of their switches, routers, APs, compatible client adapters, and LAN management software. This allows you to build on an existing Cisco hardware-based infrastructure and easily add wireless networking support to your design. In a wireless-aware LAN, all of these network infrastructure components work together seamlessly to provide both wired and wireless services to users of the LAN with minimal additional workload for the network engineers who administer the LAN.

The Cisco wireless-aware LAN infrastructure is comprised of the following eight components:

- Cisco IOS Software
- Cisco Aironet Series WLAN APs
- Cisco Aironet Series WLAN Client Adapters
- Cisco Compatible Client Adapters
- Cisco Works WLAN Solution Engine 2.x (WLSE)

- Cisco Wireless Security Suite
- Cisco Secure Access Control Server 3.2 (ACS)
- Cisco Wireless-Aware LAN Switching and Router Products

Each of these work together to make s WLAN a simple extension of pre-existing wired LAN. The following sections discuss some of the benefits of this approach and some special design considerations to keep in mind when implementing a Cisco wireless-aware LAN.

# Wireless-aware LAN Benefits

Cisco defines their wireless-aware LAN framework as "a highly scalable, secure, and manageable solution that simplifies WLAN deployment and management and maximizes wireless network uptime." To this end, the wireless-aware LAN framework provides the following features:

- Integrated wired and WLAN services using the Cisco infrastructure and Cisco Internet Operating System (IOS) Software
- Simplified management of hundreds to thousands of central or remotely located APs
- Wireless Domain Services for IEEE 802.1X local authentication service and fast secure roaming support
- Rogue AP detection and location
- Air/RF scanning and monitoring
- Interference detection to isolate and locate network interference
- Simplified WLAN deployment processes with assisted site surveys
- Streamlined WLAN management and operations support
- Enhanced troubleshooting and diagnostic tools for proactive performance and fault monitoring
- High availability with self-healing WLANs
- Security policy monitoring
- Seamless delivery of enhanced network security solutions

This section goes through some of these features and describes the benefits that they offer to both the end user and the network administrator. Obviously, as a package, they offer a plethora of new features not previously available to network engineers attempting to implement WLANs. By using the wireless-aware framework, network engineers can vastly improve the implementation time and the overall security and reliability of the WLAN without compromising security or stability of a preexisting wired LAN.

# Integrated Wired and WLAN Services using the Cisco Infrastructure and Cisco IOS Software

With the Cisco wireless-aware framework, Cisco has introduced new features specific to WLANs to their familiar IOS software. Anyone familiar with Cisco's product line is also familiar with their IOS software, which provides the ability to monitor and configure switches and routers. Cisco is now using the IOS software to allow network administrators to use a familiar tool to configure and monitor new WLAN equipment. As of the time of this writing, the new wireless features of Cisco's IOS software is only available on the Cisco Aironet line of APs and client wireless adapters. In the early part of 2004, they will be releasing updates and new wireless-aware LAN features to new and existing lines of Cisco switching and routing hardware.

The new Cisco IOS features include end-to-end delivery of WLAN services such as rogue AP detection, security, mobility, quality of service (QOS), and network management. These new features help make integration of a WLAN into an existing wired LAN a relatively painless task. Also, seamless integration between the WLAN and wired LAN make support and maintenance of the overall network infrastructure a manageable feat rather than a nightmare of mismatched equipment and services.

## CiscoWorks WLAN Solution Engine

Part of the Cisco wireless-aware LAN framework is the CiscoWorks WLAN Solution Engine (WLSE) 2.0. The WLSE is the management platform that Cisco provides for network engineers to administer and control their WLAN. It has the capability of managing hundreds to thousands of local or remote APs in a single WLAN through an easy to use Web-based user interface. The WLSE is a critical part of the Cisco wireless-aware LAN deployment and provides a number of features to help manage wireless-aware LANs.

# Wireless Domain Services for IEEE 802.1X Local Authentication Service and Fast Secure Roaming Support

Wireless domain services (WDS) is another new offering from Cisco that is part of their wireless-aware framework. WDS is basically a batch of Cisco IOS software features that help enhance client mobility in the WLAN and simplify the WLAN deployment and management. All Cisco APs in a subnet register themselves with the WDS and work together to monitor the WLAN. Some of the features offered as part of this are rogue AP detection, interference detection, and assisted site surveys. Each of these features are discussed in the next section, but first, let's cover a couple of additional feature sets provided by the WDS—fast secure roaming and IEEE 802.1X local authentication.

Fast secure roaming is a new feature that Cisco has included as part of the WDS. Typically in a WLAN, switching between wireless APs requires either a configuration change on the client side or a long delay in communication as the client is authenticated with the new AP. Fast, secure roaming eliminates this configuration change or delay by allowing Cisco wireless client adapters or Cisco compatible client adapters to quickly switch between APs on the same subnet. The delay in switching between APs has been narrowed down to less than 150ms.

Though it is not yet available, Cisco is also working on introducing the same fast secure roaming features when going between subnets. This feature will be released with or soon after Cisco's release of wireless-aware LAN features for Cisco switches and routers. Roaming is covered in more detail in Chapter 3 of this book.

Another feature of the WDS is IEEE 802.1X local authentication. With this feature, Cisco Aironet APs can be configured to act as a local Remote Authentication Dial-In User Service (RADIUS) server. Using an AP as a RADIUS server in this manner allows clients to authenticate to the WLAN even when the Cisco Secure Access Control Server (ACS) is unavailable. Using RADIUS, the end user is able to authenticate and gain access to normal network resources such as file shares or shared printers.

The RADIUS authentication features of Cisco Aironet APs can be configured and managed through the Cisco WLSE software. From this central point of management, you can configure the APs to act as RADIUS servers and manage RADIUS accounts. Keep in mind that if an AP is configured to act as a RADIUS server, it still provides functionality as an AP. These features are in no way mutually exclusive.

# Rogue AP Detection and Location

A common problem with WLAN or wired LAN implementations is that users can easily add their own APs to the network. Obviously, this creates a large security risk as anyone can connect to these APs and they are usually not configured securely. In the past, rogue APs were usually discovered by a network or security administrator roaming through the building using a utility such as NetStumbler to identify any unexpected APs. With the Cisco wireless-aware LAN framework, rogue APs can be automatically detected, located, and disabled with minimal intervention of the network administrator.

The Cisco wireless-aware framework makes finding these rogue APs easy due to its new RF scanning and monitoring features. With the wireless-aware framework, scanning is performed by authorized APs on the WLAN as well as client wireless adapters. This is a break from traditional manual scanning and actually goes far above and beyond automated scanning as it includes data input from the client wireless adapters as well as authorized APs. Because of this feature, a much wider physical range is covered in the scanning, so chances of finding rogue APs in "dead zones" is greatly improved.

All of the information coming in from the client wireless adapters as well as the WLAN's authorized APs is compiled by WDS and accessible through the WLSE. This gives a single point of reference for keeping track of all of the WLAN data including any identified rogue APs. Figure 1.15 shows the Cisco WLSE Location Manager displaying an identified rogue AP.

**Figure 1.15** Cisco WLSE Location Manager

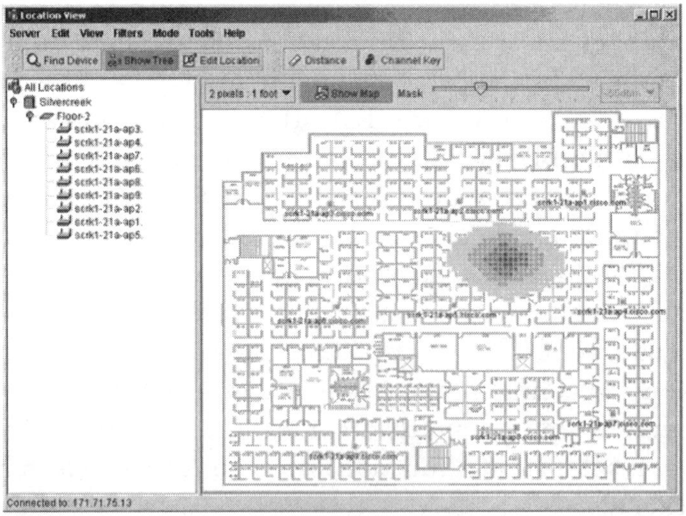

## Interference Detection to Isolate and Locate Network Interference

Another common problem in WLAN implementation is the detection and elimination of RF interference. This is typically done via an initial site survey where any pre-existing interference is detected and removed, if possible, followed by periodic follow-up surveys. Besides being time consuming, these surveys often do not catch the interference when it is actually occurring so the WLAN appears to have intermittent connectivity or latency issues in areas and a cause is unidentified. Since RF interference can be caused by something as simple as a cordless phone operating in the 2.4 GHz range, these problems can be difficult to track down.

With the Cisco wireless-aware LAN, the entire WLAN (consisting of APs and client wireless adapters) are scanning for RF interference along with their normal scanning for rogue APs. Because of this, causes for interference can be identified the moment they occur. This can help a great deal in providing a higher QOS throughout a WLAN environment. All information from the APs and client wireless adapters are sent to the WLSE where the information is centrally available for the network administrator. The WLSE can show the administrator the type of interference detected, the location of the interference, and whether it is caused by a rogue AP or some other wireless broadcast.

## Simplified WLAN Deployment Processes with Assisted Site Surveys

Site surveys are a critical part of any WLAN design. These surveys help the design team identify placement locations for APs, determine how many APs are needed, identify possible areas of RF interference, and determine the expected range of the WLAN along with a slew of other important information. With the amount of time and effort required to complete these surveys, they can be very costly and are typically outsourced to an outside contractor that has a great deal of experience and access to RF scanning equipment. This process is time consuming, mostly due to the fact that a great deal of guesswork is involved in the initial planning followed by modifications to improve performance post-implementation.

The Cisco wireless-aware LAN framework has some features that help make this part of the design phase much easier. The wireless-aware framework allows the network administrator to use tools included with the WLSE to perform their own site surveys using in-house staff, thus cutting down on cost and time. Using these tools, administrators can follow a simple 5-step process to perform a complete site survey in the WLAN environment. These five steps are:

1. A floor plan of the location to be surveyed is imported into the tool. The tool supports a variety of electronic file formats including .bmp, .jpg, and .gif. If an electronic file is unavailable, a rough building diagram can be drawn within the tool.
2. Initial AP locations are added to the diagram to provide a rough estimate of the number of APs required for the facility.
3. Cisco Aironet Series APs are installed in the facility at locations corresponding to their diagram placements.
4. The installed Cisco Aironet Series APs are set to a site survey mode known as "AP Scan Mode," where they all assume the same channel and transmit at maximum power. In this mode, the APs detect the presence of one another and automatically select transmit power, frequency selection, and other settings to fully cover the facility area.
5. Finally, the AP RF settings are fine-tuned in the "Client Walkabout" mode. In this mode, an individual walks the facility areas where coverage is needed, including the perimeter, with a client device that is sending continual RF measurements back to the Cisco Aironet Series APs.

These steps are all that are required to perform a site survey using the Cisco wireless-aware LAN framework. All future WLAN management and monitoring can be done through the WLSE based on current information from the constant WLAN RF monitoring data stream. Additional APs can be added or existing APs can be moved based on this information.

## Streamlined WLAN Management and Operations Support

Besides the features of the WLSE centralized management tool already discussed, there are some additional features that can help with management of the WLAN and the devices on the WLAN. With large WLAN environments, upgrading the firmware on all APs in the environment can be a very daunting task. The WLSE provides the ability to automatically send firmware updates out to local or remote APs, mitigating a great deal of time and effort normally associated with this task. Some additional features provided by the WLSE in this area are:

- Proactive fault and performance monitoring based on user-defined thresholds
- Centralized autoconfiguration of newly deployed APs
- Configuration archive of Cisco APs
- Security policy and fault/performance monitoring of the Cisco WLAN infrastructure
- Integration with existing network management infrastructure (SOAP/XML interface, Simple Network
- Management Protocol (SNMP) traps and Syslog messages)
- XML API for data export
- Integration with CiscoWorks LMS
- Centralized mass conversion of Cisco Aironet 1200 Series AP VxWorks operating system configuration files into Cisco IOS Software configuration files using an expanded version of the Cisco Aironet Conversion Tool for Cisco IOS Software

Use of these tools helps network administrators handle most of the manual tasks associated with a WLAN implementation easily and quickly with a very small learning curve. This cuts down on additional staff overhead dedicated to the WLAN and helps provide a higher QOS to the end users.

Tools are also included for performing troubleshooting and WLAN diagnostics. The WLSE provides a number of reports to give network engineers a good view of the health of the WLAN environment. Everything from displaying the RF interference detected (mentioned previously) to client tracking and performance monitoring is available through the WLSE. Also, the WLSE includes tools for fault monitoring and can send Simple Network Management Protocol (SNMP) alerts or syslog messages as needed to keep engineers abreast of any errors occurring throughout the system. The WLSE also incorporates some "self-healing" features to help minimize downtime based on the monitoring information that is available.

# Seamless Delivery of Enhanced Network Security Solutions

Security has always been a major issue with traditional wired LANs, and is even more critical with WLANs, as the physical medium is easily accessible by intruders. The Cisco wireless-aware LAN framework takes the issue of overall LAN security in mind and offers some excellent features to help you secure your network.

We have already gone over the IEEE 802.1X authentication features using your AP as a RADIUS server. Some additional features are security policy monitoring, centralized security settings, monitoring and notifications of the IEEE 802.1X security servers, client device response time monitoring, and IEEE 802.1 1i AES encryption support. Table 1.3 shows these various features and Cisco's definition of this support for the features under the wireless-aware LAN framework.

**Table 1.3** Wireless-aware LAN Network Security Solutions Support

| Security Feature | Support Description |
|---|---|
| Security Policy Monitoring | Monitoring of security policies for predefined Cisco Wireless Security Suite parameters across all APs is included. Alerts are generated for violations in areas such as Service Set Identifiers (SSID), broadcasts, 802.1X EAP settings, and WEP. Alerts can be delivered via e-mail, Syslog or SNMP trap notifications. |
| Centralized Security Settings | Parameters such as 802.1X EAP, WEP and Wi-Fi Protected Access (WPA) are ensured through centralized WLAN management of all local and remote AP settings. |
| Monitoring of the 802.1X EAP RADIUS or AAA server | The RADIUS or AAA server providing support for Cisco LEAP and Protected-EAP (PEAP) is monitored and the availability of Cisco Secure ACS and Committed Access Rate (CAR) EAP servers is verified. |
| Notification of 802.1X EAP RADIUS or AAA server management thresholds | Notifications of user-defined security thresholds are managed via e-mail, Syslog, and SNMP trap notifications. |
| Client device response time monitoring | The client device response time is monitored by simulating a client device via CiscoWorks WLSE. |

Continued

**Table 1.3** Wireless-aware LAN Network Security Solutions Support

| Security Feature | Support Description |
|---|---|
| IEEE 802.11i AES encryption support | Future support for IEEE 802.11i AES encryption is planned. |

# Wireless-aware Design Considerations

When designing a wireless-aware LAN, most of the design considerations are similar to those covering WLANs in general, as described earlier in this chapter. Attenuation, RF interference, application considerations, and structural considerations all need to be kept in mind when designing a wireless-aware LAN. As much thought, planning, and design work should go into a wireless-aware LAN design as in a standard WLAN design. Any addition to an existing infrastructure should always have its potential benefits weighed against its potential risk and costs to determine if it should be done at all.

Earlier sections of this chapter went over the major benefits of implementing a wireless-aware LAN over a standard WLAN, but from a design perspective there are a few important benefits to keep in mind. First, design time and costs are greatly reduced due to the assisted site survey feature. Secondly, ongoing maintenance and management tasks are simplified by using the WLSE console to centrally manage the wireless-aware LAN. Thirdly, the Cisco wireless-aware LAN is designed to easily integrate with an existing Cisco network infrastructure. And finally, many security issues inherent with the implementation of a WLAN are mitigated by the tools and features included as part of the Cisco wireless-aware LAN framework.

One other consideration to bear in mind when designing a Cisco wireless-aware LAN is compatibility. This type of WLAN is designed specifically to work with Cisco equipment. While equipment from other vendors may be able to be integrated into the wireless-aware LAN, the most benefit will be reaped when all the equipment used is from Cisco or is Cisco-compatible. This may require changes to existing LAN infrastructure hardware and is important to keep in mind.

# Summary

This chapter went over a great deal of information related to WLANs and wireless-aware LANs. It started with the WLAN itself and how it works and then discussed the IEEE 802.11 series of standards and went over the packet design used for WLANs. Next, it covered some of the many benefits offered by WLANs. These benefits range from their added convenience to the end user to their overcoming some of the limitations and costs associated with traditional wired LANs. This chapter also went over some design considerations for WLANS such as attenuation, RF interference, and application, structural, security, and network management considerations. Lastly, it covered the three WLAN modes of operation: ad-hoc, AP, and hybrid.

This chapter also discussed what a wireless-aware LAN is and how it differs from a standard WLAN. It covered a great deal of information about the benefits of a wireless-aware LAN ranging from its easy integration with traditional wired LANs to new security features offered as part of the wireless-aware LAN framework. It discussed the WLSE centralized management console, fast secure roaming, rogue AP or interference detection, 802.1X authentication support, and assisted site surveys. This bundle of features provides a great deal of benefit to the implementation of the wireless-aware LAN and its integration into an existing environment.

Lastly, this chapter covered design consideration specific to wireless-aware LANs. Obviously, many of the design considerations associated with WLANs in general still apply, but there are other considerations to keep in mind. These considerations are primarily the added benefits provided as part of the wireless-aware LAN framework, but you must also keep compatibility in mind as a major design consideration.

This chapter should serve as an introduction to WLANs as well as Cisco wireless-aware LANs and provide some good basic knowledge as a foundation for the rest of this book. The WLAN and wireless-aware LAN benefits and design information discussed here will be expanded on throughout the remainder of the book and much more detail will be provided around some of these key areas.

# Solutions Fast Track

## What is a WLAN?

- ☑ A WLAN is simply a LAN that uses radio waves as its physical medium.
- ☑ Wired LANs typically cost a great deal to implement and have specific distance and mobility limitations.
- ☑ WLANs are based on the IEEE 802.11 series of standards.
- ☑ The design of a WLAN data packet differs substantially from a normal Ethernet data packet.

## WLAN Benefits

- ☑ WLANs offer greater convenience and mobility to end users.
- ☑ WLANs can be cheaper to implement than full wired LAN implementations due to the high cost of cabling and network equipment.
- ☑ Cable limitations inherent with wired LAN implementations are no longer an issue with WLANs.
- ☑ WLANs can be integrated with an existing wired LAN infrastructure. This integrated LAN is known as a hybrid LAN.

## WLAN Design Considerations

- ☑ Since WLANs rely on radio waves for transmissions, attenuation and RF interference become major issues in a WLAN implementation.
- ☑ There are application considerations that should be kept in mind when designing a WLAN, including the application's ability to run properly over the WLAN.
- ☑ The building design and construction is critical to a WLAN implementation, as variances in building design and the compounds used to build it can alter the performance of a WLAN.

- ☑ Security and network management are also important design considerations for a WLAN and should be part of the initial WLAN design planning.

## WLAN Modes of Operation

- ☑ WLANs can operate in two different modes based on the needs of the network.
- ☑ Ad-hoc mode is a simple peer-to-peer mode of operation in which all systems on the WLAN communicate with each other.
- ☑ AP mode differs in that each system on the WLAN communicates with centralized APs rather than all the other systems on the network.
- ☑ These two modes can be combined as a hybrid mode, but this is not a recommended design.

## What is a Wireless-aware LAN?

- ☑ A wireless-aware LAN is Cisco's new framework for integrating WLANs with an existing LAN infrastructure.
- ☑ Many features are provided with the wireless-aware framework to help eliminate a lot of the issues inherent with a standard WLAN implementation.
- ☑ The wireless-aware framework makes it relatively easy to add WLAN support to your network infrastructure with a minimum of additional work.
- ☑ A lot of the additional workload associated with implementing a WLAN is mitigated with the wireless-aware LAN framework, cutting down on additional workload for network engineers.

## Wireless-aware LAN Benefits

- ☑ Wireless-aware LANs provide a great deal of benefits over standard WLANs.
- ☑ The Cisco wireless-aware LAN framework provides for simple integration with existing LAN infrastructure.

- ☑ Features such as 802.1X authentication support, rogue AP detection, and other security features by default make wireless-aware LANs much more secure than normal WLANs.
- ☑ Additional centralized management, interference detection, and assisted site survey features make it much easier to implement a wireless-aware LAN over a traditional WLAN.

## Wireless-aware Design Considerations

- ☑ Most of the design considerations for WLAN implementations also apply to wireless-aware LAN implementations.
- ☑ A lot of the risk and additional workload generated by implementing a WLAN is mitigated by the wireless-aware LAN framework.
- ☑ Compatibility with third-party vendors is an important consideration to keep in mind when designing a wireless-aware LAN.
- ☑ The Cisco wireless-aware LAN framework is designed specifically to work with Cisco or Cisco-compatible hardware; therefore, the greatest benefit will be reaped when adhering to this standard.

# Frequently Asked Questions

The following Frequently Asked Questions, answered by the authors of this book, are designed to both measure your understanding of the concepts presented in this chapter and to assist you with real-life implementation of these concepts. To have your questions about this chapter answered by the author, browse to **www.syngress.com/solutions** and click on the **"Ask the Author"** form. You will also gain access to thousands of other FAQs at ITFAQnet.com.

**Q:** The company I work for is considering implementing a WLAN. Some users already have APs at their desk that they have added on their own. Should we implement the WLAN, or just let the users who really need it add their own equipment?

**A:** In a situation like this, you should definitely implement a WLAN. The fact that your users are already implementing their own indicates that there is a need. For good security practices, any extension to your existing LAN should be carefully designed and implemented by capable network engineers, not thrown together as needed by the end users.

**Q:** Cisco equipment is certainly not the cheapest on the market. Is it worth the additional cost to implement a wireless-aware LAN rather than a standard WLAN?

**A:** It depends on your environment. If you are looking at implementing a WLAN in an environment where you will have a small number of APs and do not really need centralized management, then a standard WLAN may be best. If you are working in a large environment with over 10 APs, I highly recommend implementing a wireless-aware LAN. The cost savings on installation and maintenance will typically make up for the additional equipment costs.

**Q:** My management team has heard a lot about security risks with WLANs. Is this a valid concern with wireless-aware LANs?

**A:** Yes, security should always be a concern when implementing anything new. While the wireless-aware LAN framework does help mitigate some security issues, keep in mind that your LAN traffic is still being broadcast over simple radio waves. There is always some risk there and you should implement the highest level of security and encryption possible.

**Q:** I have set up a Cisco wireless-aware LAN and the WLSE is showing that there a two rogue APs on the WLAN. Why is this a problem?

**A:** The problem with rogue APs is that they are not in your control, therefore you cannot ensure that they adhere to the security practices and polices of your company. These rogue APs are often an open door to your WLAN and are not secured properly. Any rogue APs on a WLAN should be immediately removed.

**Q:** With the Cisco wireless-aware LAN assisted site survey feature, should I have to hire outside consultants to help us design our WLAN?

**A:** For small to mid-size WLANs, no. If you are dealing with somewhere in the range of 500+ APs ranging across a large number of sites then I still recommend having a professional site survey done just to ensure that your deployment will go as smoothly as possible.

**Q:** I have a great deal of non-Cisco equipment on my LAN. I am considering implementing a wireless-aware LAN, but I do not want to lose my investment in third-party equipment. Should I just implement a normal WLAN?

**A:** It still might be a good idea to consider implementing a Cisco wireless-aware LAN with the Cisco Aironet APs. There are a lot of features offered by the wireless-aware LAN that do not require Cisco switching and routing hardware and are worth implementing.

Chapter 2

# Designing Wireless-Aware LANs

**Solutions in this Chapter:**

- Radio Frequency (RF) Basics
- Laws, Regulations, and Environmental Considerations
- Institute of Electrical and Electronic Engineers (IEEE) 802.11 Standards
- Planning for RF Deployment

☑ Summary

☑ Solutions Fast Track

☑ Frequently Asked Questions

# Introduction

This chapter covers the physical medium of wireless local area networks (WLANs) (the electromagnetic (EM) frequency spectrum) and ways in which it can and must be manipulated in order to have an effective WLAN. This coverage ranges from theory to practice while exploring conducting site surveys in advance of WLAN setup.

# Radio Frequency (RF) Basics

This section covers the scientific principles that make wireless communication possible. Radio components and their associated frequency range is the primary mode of transmission in the wireless industry.

*Radio* is the wireless transmission and reception of electric impulses or signals by means of EM waves. EM waves are present at all frequencies; however, currently only a small part of this total spectrum can be utilized to transmit communication signals. This small subset of frequencies is commonly referred to as the *RF spectrum* and ranges from approximately 9 kilohertz (KHz) to 300 gigahertz (GHz). When designing a wireless network, understanding the science behind wireless communication allows you to recognize potential complications such as signal-to-noise (S/N) ratio, attenuation and multipath scattering, and channel spacing.

This section also explores the basic science of EM waves, and how and why radio signals are modulated onto carrier waves. It explores antenna design, the relationship between wave propagation technologies and signal power, and what elements make up a wireless network.

## Transmitting Radio Signals over EM Waves

The German scientist Heinrich Hertz demonstrated in 1887 that electrical energy could be transmitted through space by way of EM waves. Even though Professor Hertz was the first to demonstrate this phenomenon, he did not grasp the impact of his discovery. It took the work of Nikola Tesla and Guglielmo Marconi, who were inspired to use Hertz's discovery to transmit signals—thus, the first radio was born.

Electric fields are induced by the separation of positive and negative charges. Moving charges (electric current flow) induce electric fields. Changing electric fields induces magnetic fields. Therefore, alternating current flow—such as current flow into and out of an antenna—induces an oscillating electric field, which then induces an oscillating magnetic field. An EM wave is the propagation of electrical energy caused by oscillating electric fields inducing oscillating magnetic fields, which then induce further oscillating electric fields, which then induce further oscillating magnetic fields, and so on.

# Anatomy of a Waveform

Radios transmit and receive signals over vast distances in the form of EM waves, at a particular frequency level that differentiates them from other EM waves in the frequency spectrum, such as infrared (IR) or X-rays (discussed later in this chapter). The *sinusoidal waveform* is the most common waveform and is used to represent all types of waves. Figure 2.1 shows the key properties of a sinusoidal wave.

**Figure 2.1** Sinusoidal Wave

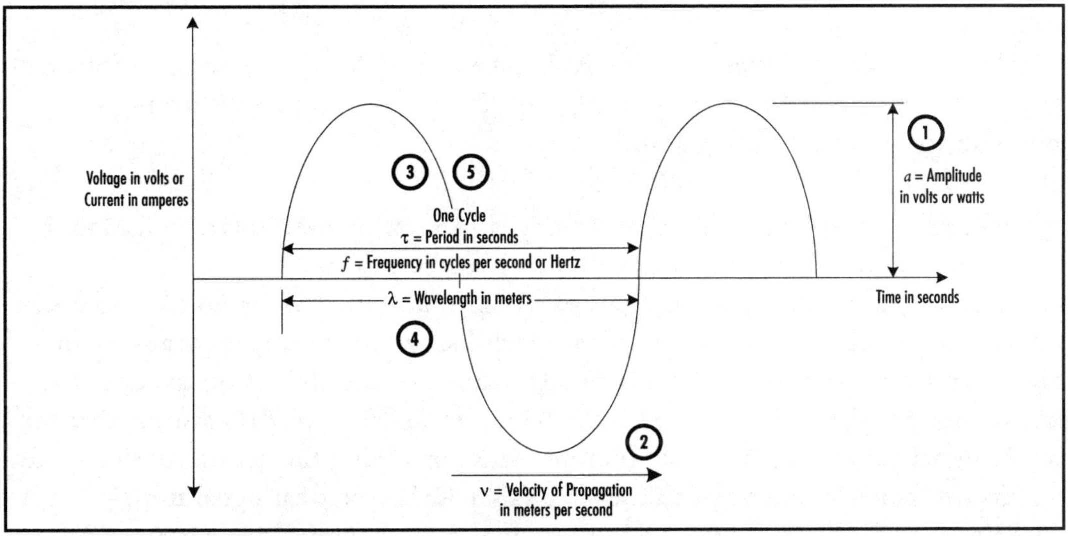


A *cycle* is the smallest portion of a waveform that, if repeated, represents the entire waveform. Waveforms can be described as having the following properties:

1. *a* = Amplitude

   The measurement of a waveform above a center reference. With EM waves this is usually measured in volts or watts.

2. *v* = Velocity of Propagation

   The velocity of propagation of a wave is the velocity that a wave travels through a medium, and is usually measured in meters per second.

3. $\tau$ = Period

   The period of a wave is the time it takes for one cycle to pass a fixed point and is usually measured in seconds. It is designated by the Greek letter tau ($\tau$).

4. $\lambda$ = Wavelength

   The wavelength of a wave is the distance that the wave will propagate in one cycle and is usually measured in meters and designated by the Greek letter lambda ($\lambda$).

5. $f$ = Frequency

   The frequency of a wave is the rate at which individual cycles pass a given point and is usually measured in cycles per second or Hertz (Hz), named after Heinrich Hertz, who discovered EM waves.

All of these properties except amplitude are related by the following formula:

$$f = 1/\tau = v/\lambda$$

The velocity of propagation for EM waves is relatively constant, and for practical purposes is equal to the speed of light ($3.00 \times 10^8$ m/s). Substituting this constant for velocity yields the following:

$$f = (3.00 \times 10^8 \text{ m/s}) / \lambda$$

Therefore, frequency ($f$) and wavelength ($\lambda$) can be used interchangeably by using the preceding formula to convert one to the other.

Some of the most common types of signals transmitted via radio are audio signals such as voice or music. True audio signals like voice are usually near-random signals and very hard to graph and conceptualize. For this discussion, assume that the input signal being transmitted is a 1 KHz sinusoidal wave. Also assume that the audio signal has already hit a microphone, thus converting the acoustical signal into an electric signal. If you want to transmit this 1 KHz electrical signal from point A to point B using a conductive medium such as copper wires, you only need to connect a pair of wires to each endpoint—one wire is the point of reference or *ground*, and the other wire carries the alternating signal voltage from point A to point B. However, transmitting this signal without wires is more complicated: the signal needs to be transmitted without interference.

## *Modulating a Radio Signal*

To transmit a 1 KHz wave without wires, it must first be modulated onto a *carrier wave* with a frequency many times higher than the input signal. There are several reasons why the desired signal must be modulated onto a much higher carrier signal.

The first reason is for better transmission. Most of the radio signals transmitted are low-frequency signals. These signals do not propagate well as EM waves. Therefore, *modulation* is used to increase the frequency to allow more effective transmission.

The second reason is to allow multiple signals to be transmitted at the same time without interference. Your voice is in the same frequency range as my voice. Assume we both try to use a two-way radio to talk to our remote friends simultaneously. If we did not modulate our signals onto different carrier waves, our signals would get mixed together and it would be impossible to distinguish on the remote end. However, if we modulated our signals onto carrier waves of different frequencies, our signals would not interfere with one another and we could talk simultaneously. Each specific carrier frequency is called a *channel* (discussed later in this chapter).

A third reason to modulate signals onto high frequency carrier waves is due to restrictions on antenna size. The length of an antenna is based on the length of the wave it was designed to transmit or receive. The simplest antennas are a fraction of the wavelength, usually one-half or one-quarter of the wavelength. Since lower frequencies have longer wavelengths, the antennas designed for low frequencies are bigger. For example, 60 Hz is at the very lowest range of human hearing. The wavelength of a 60 Hz wave traveling at the speed of light is 3107 miles, or the distance from Boston, MA to San Francisco, CA. Therefore, a one-half wavelength dipole antenna would be approximately 1,500 miles long—not a feasible length for an antenna. Antenna design is discussed in more depth later in this chapter.

When talking about modulation, we are talking about a minimum of two waves, the *signal* and the *carrier*. Certain properties of the carrier waveform are modified (modulated) to represent the signal waveform. The signal wave is also called the *modulating wave* because it is the wave that modifies (or modulates) the carrier wave. The modulating wave can be anything from analog audio to a computer-generated digital square wave. The carrier wave is called the *modulated wave* because it is the wave that is being changed by the modulating or signal wave. Almost all carrier waves are a periodic sinusoidal wave with a frequency many times higher than the frequency of the modulating wave.

There are many types of modulation. Some were developed to carry analog waveforms, however, since the invention of the computer, many types of modulation have been developed to carry digital waveforms. The following sections discuss a few widely used types of modulation.

## Analog Modulation Schemes

There are two analog modulation schemes that are widely used and familiar to anyone who has ever tuned a modern audio broadcast radio. These two forms of modulation are:

- Amplitude Modulation (AM)
- Frequency Modulation (FM)

AM is the modulation of the amplitude of the carrier wave. Figure 2.2 illustrates the AM modulated signal.

**Figure 2.2** Amplitude Modulation (AM)

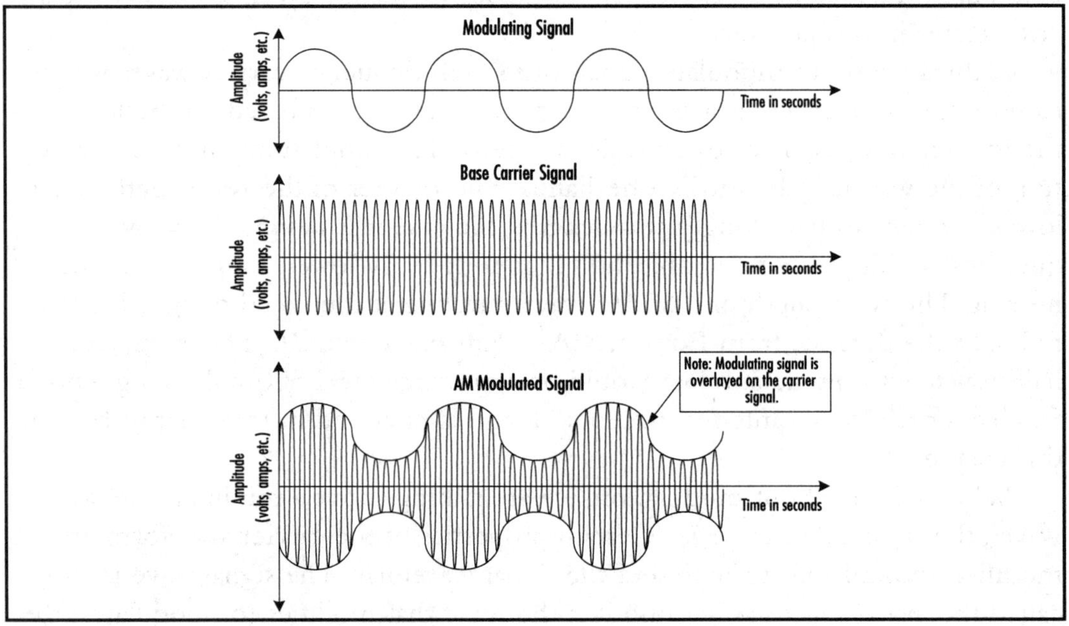


FM is the modulation of the frequency of the carrier wave to represent the frequency. Figure 2.3 illustrates the FM modulated signal.

**Figure 2.3** Frequency Modulation (FM)

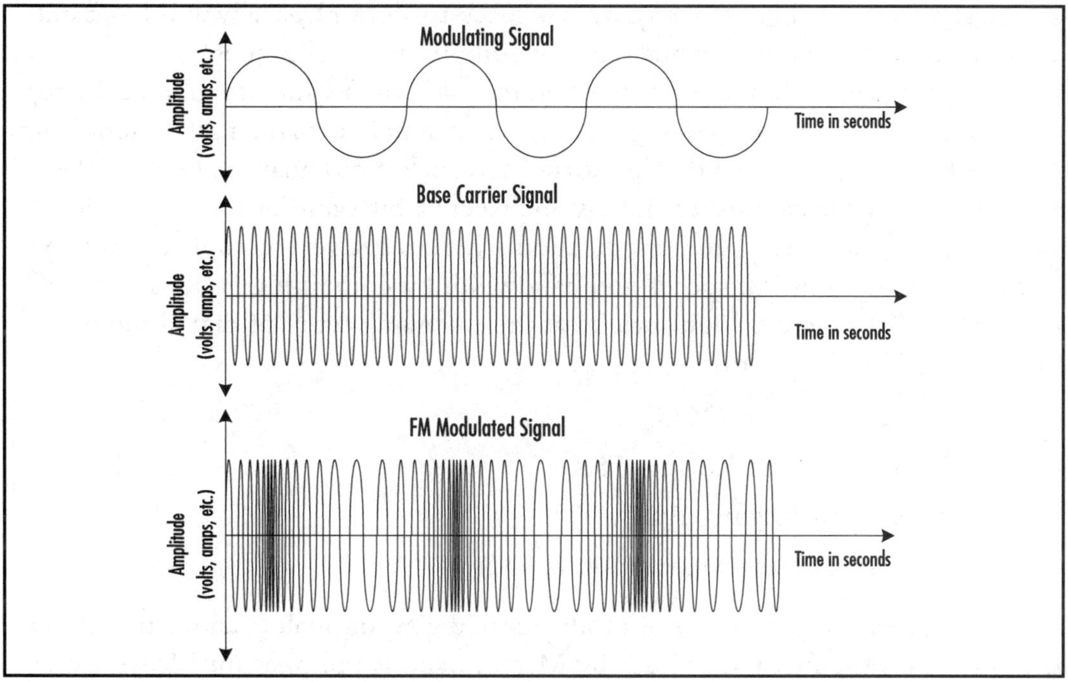


## *Digital Modulation Schemes*

Most sources of information that are transmitted are analog signals. Human speech, music, video, and pictures are all analog by nature. However, because computers use binary to store and process information, analog sources of information must be digitized. This means that the signal is represented by a code of 1s and 0s that the computer uses to recreate the original signal as closely as possible. The error introduced when the signal is digitized is called a *quantization error*. If the digitizing encoding technology is designed well, the resulting signal so closely resembles the original signal that the differences are imperceptible to humans (such as music stored on CDs).

As computers continue to take a more active role in capturing, storing, transmitting, and modifying these signals, more information can successfully be digitized to satisfy the growing demand to transmit information over the air. This results in a growing need for modulation schemes that are designed to carry digital information.

One advantage of digital signals is the increased ease of compression. Most analog signals, once digitized, require less space to store physically and less bandwidth to transmit due to various types of compression techniques.

Digital signals are commonly referred to as *bit streams* and are graphically represented as a square wave (see Figure 2.4). In its simplest form, digital modulation is easier to conceptualize and to perform than analog modulation, because there are only two signal states to distinguish between: a bit value of one and a bit value of zero. However, digital modulation schemes get very complex as we try to maximize transmission speeds and bandwidth by combining various types of modulation. This section looks briefly at the following types of digital modulation in their simplest forms:

1. On/Off Keying (OOK)
2. Frequency Shift Keying (FSK)
3. Phase Shift Keying (PSK)
4. Pulse Amplitude Modulation (PAM)

OOK is the simplest form of modulation, digital or analog, and is the modulation used on the first radios built by Marconi and is the basis for Morse Code. OOK simply involves making (on) or breaking (off) the connection between the carrier signal's oscillator and the antenna in order to represent the digital signal. Figure 2.4 illustrates OOK modulation.

**Figure 2.4** OOK Modulation

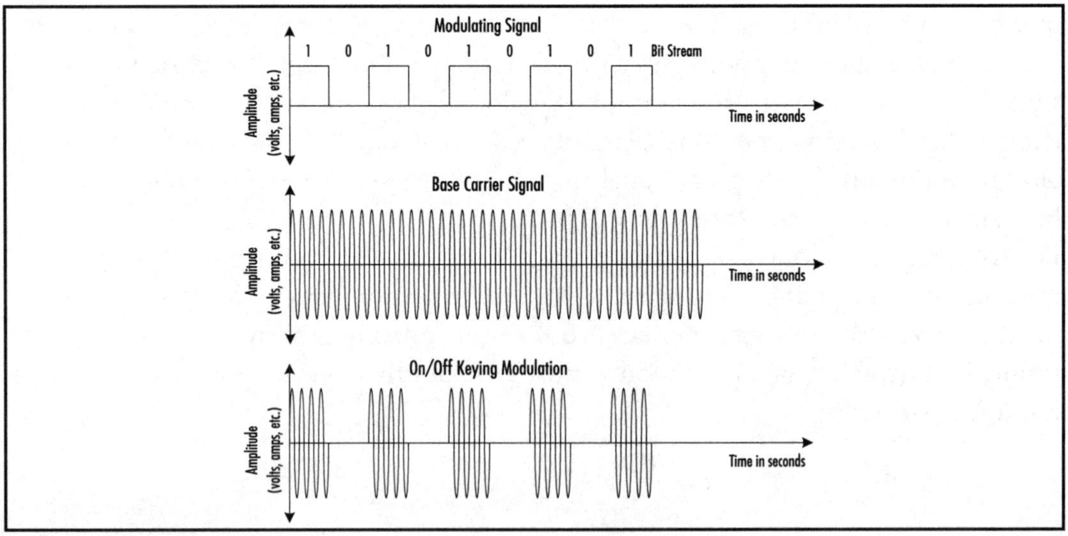

FSK is similar to OOK, but instead of alternating between the carrier frequency (on) to no frequency (off), FSK alternates from the carrier wave frequency to the carrier wave frequency plus an *offset* frequency. The detection of this frequency change yields the transmitted digital signal. Figure 2.5 illustrates FSK modulation.

**Figure 2.5** FSK Modulation

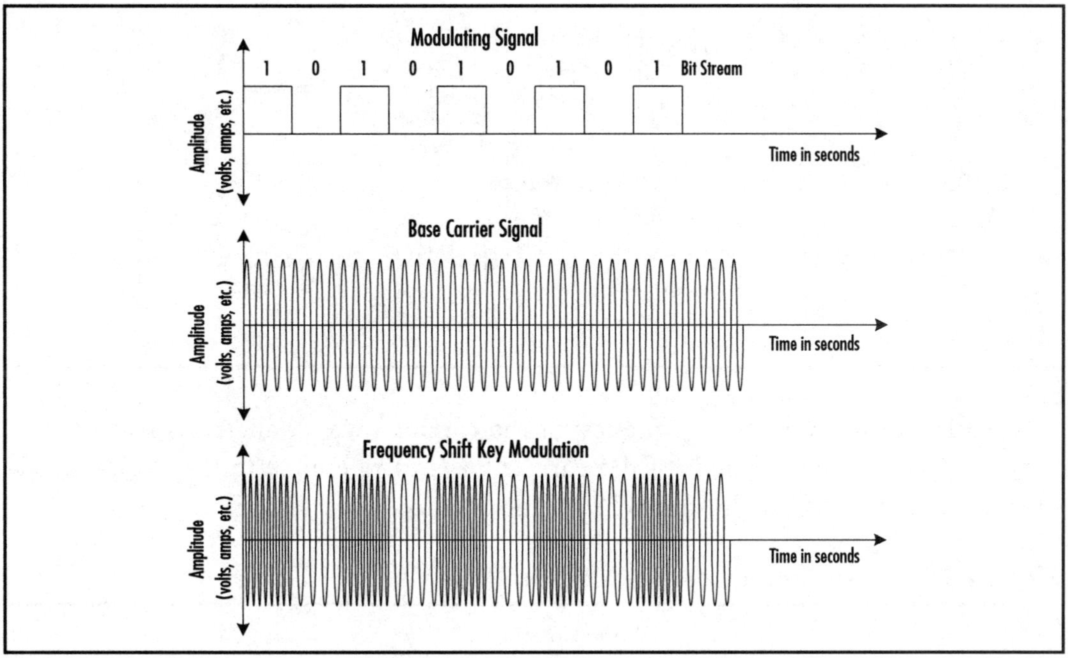


PSK differs from OOK and FSK in that it does not change the frequency of the carrier wave. PSK changes the phase of the carrier wave in reference to the digital modulating wave. The detection of these phase shifts yields the transmitted digital signal. In its simplest form, PSK shifts the phase by one-half of a wavelength, or 180 degrees. Figure 2.6 illustrates PSK modulation.

**Figure 2.6** PSK Modulation

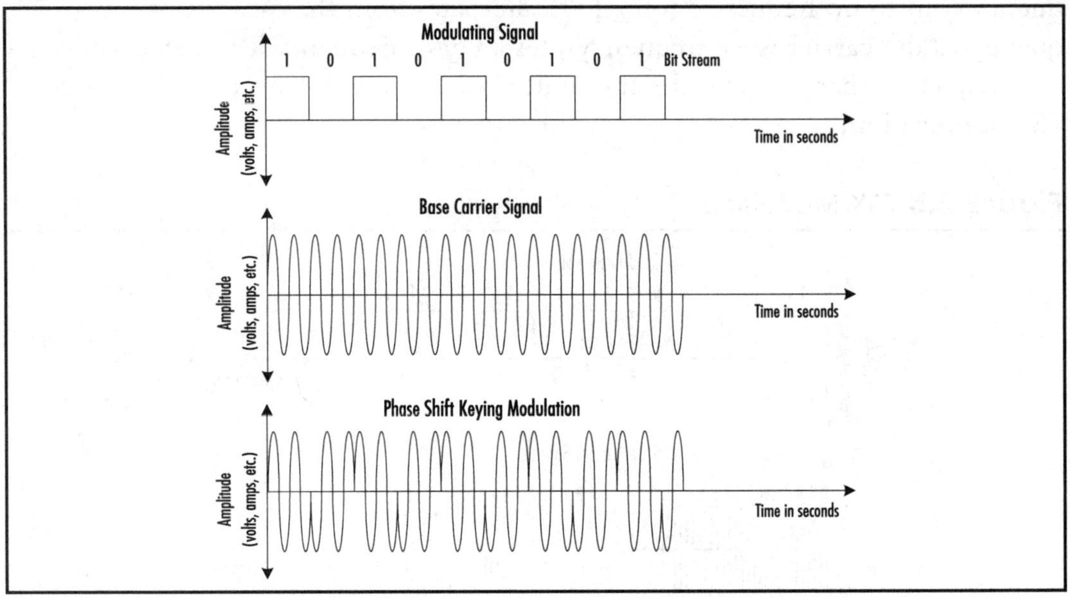


PAM does not vary the frequency of the carrier wave at all. As its name implies, it varies the *amplitude* of the carrier wave in reference to the digital modulating wave. Figure 2.7 illustrates PAM modulation.

**Figure 2.7** Pulse Amplitude Modulation (PAM)

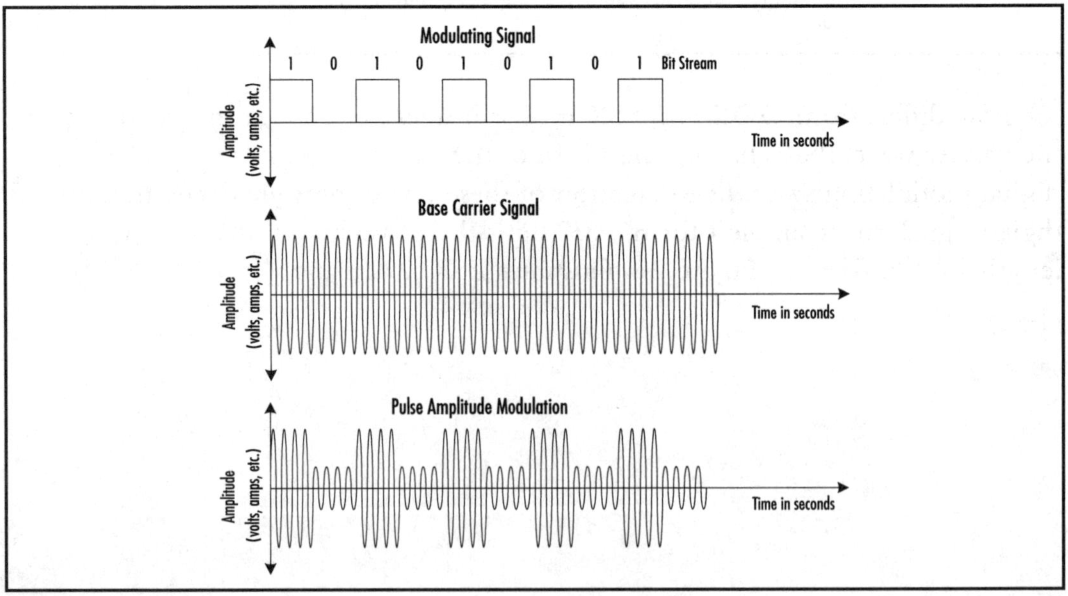

Knowledge of how modulation is accomplished and of the different types is important because it applies to how computer systems communicate. FSK is the modulation technique used in frequency hopping spread spectrum (FHSS) and direct sequence spread spectrum (DSSS), used in wireless technology standards such as 802.11. In addition, modulation is also a factor in both mobile and optical wireless communications.

# Propagating a Strong Radio Signal

In order for wireless communications to function, the signal must have a path from the transmitter and the receiver and arrive with enough power left in the signal for the receiver to comprehend what is being sent. There are many factors that affect how a signal propagates from the transmitter to the receiver. Some of the factors that affect propagation affect low frequency signals differently from high frequency signals. This section looks at several factors that impact the propagation of EM waves, and compares the benefits of low frequencies and higher frequencies.

## Understanding Signal Power and S/N Ratio

One of the principle requirements for wireless communication is that the transmitted EM wave must reach the receiver with ample power to allow the receiver to distinguish the wave from the background noise. An analogy can be made to human hearing: when someone is talking to you, they must talk loud enough for your ears to pick up the sound, and it has to be loud enough and clear enough for your brain to be able to recognize and translate the sounds into individual words.

Just as your ears and brain require a minimum volume and clarity level to be able to discern what is being said, a radio receiver also requires a minimum power level of received signal in order to discern and recreate the transmitted modulating wave. Signal strength for EM waves is usually measured in watts, or more specifically, a logarithmic ratio of the signal strength divided by 1 milliwatt. This logarithmic ratio is called *decibels above 1 milliwatt* (dBm).

Another common property used to describe signal strength is the S/N ratio. The S/N ratio does not describe the absolute power in the signal, but instead describes the power of the signal in comparison to the power of the background noise. The higher the S/N ratio, the better or more powerful the signal. Looking back at the hearing analogy: someone talking to you in a quiet room would be able to whisper and you would still be able to hear; however, if there was a lot of background noise, say at a rock concert, that person would have to yell in order for you to hear. The same concept applies to RF wireless communication. Since the S/N ratio accounts for the level of background noise, it is a very valuable and widely used indicator of signal strength.

Different modulation and encoding technologies require different minimum S/N ratios to function. Most digital modulation schemes require a lower S/N ratio than analog modulation schemes, because the receiver of a digitally modulated carrier wave only has to distinguish between certain levels that represent a logical 1 and a logical 0. Even in the presence of a lot of noise, the receiver is able to distinguish between the predefined threshold levels and then regenerate the digital square wave. In contrast, the receiver of an analog-modulated signal has an infinite number of levels that must be distinguished and maintained. It cannot assume that the received signal was supposed to be a 1 or a 0 and regenerate the signal—it must receive the signal, demodulate it, and pass the resulting representation of the original signal to the next processing device, such as an amplifier. Therefore, any noise added to an analog signal during propagation will alter the original signal. When the power of the modulated RF signal is several times greater than the power of the background noise, the added noise will not be noticed or can be reduced by filtering. On the other hand, when the noise is nearly as powerful as the signal, the resulting demodulated signal will be noticeably different than the original modulating signal. This is commonly called *static*.

The S/N ratio is an important aspect of network design. There are engineering rules established by vendors, specific to their equipment, to provide a set of guidelines for your design. Depending on the geographic span of your design, the S/N ratio may warrant devices to amplify or regenerate the transmitted signal.

*Attenuation*, discussed in the following section, is another important consideration in wireless networking. This will dictate the acceptable span between antennas. The engineering rules mentioned earlier also include the attenuation parameters acceptable by specific equipment.

## Attenuation

Anyone who has tried to listen to a radio while driving in rural areas knows that signals get weaker the farther you get from the source. This weakening of a signal is known as *attenuation*. There are several factors that cause attenuation, but to see how distance alone can cause attenuation first consider the example of propagation in free space.

Unlike audio waves (which are pressure waves and must have a medium to propagate), EM waves do not require a medium to propagate and can travel through the vacuum of space. In free space there are no other factors that cause resistance to the signal, yet there is still attenuation because the signal density diverges. Figure 2.8 demonstrates this phenomenon using light.

**Figure 2.8** Attenuation by Diversion of Rays from an Omnidirectional Source

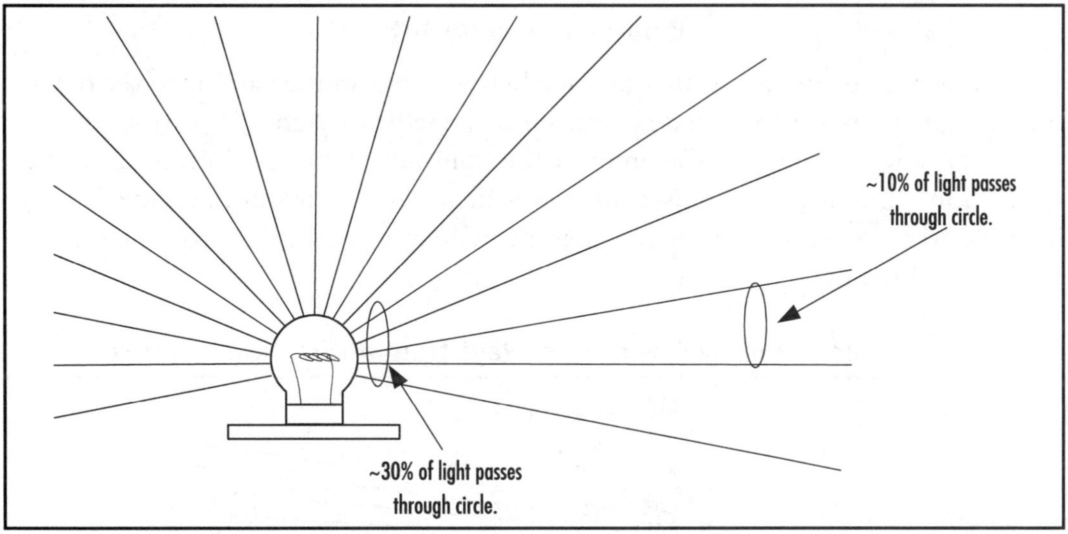


Assume that each ray of light represents an equal amount of the total light energy transmitted. You can see that the rays diverge as the distance from the source increases. The result is a decrease in light intensity. Visible light waves are high frequency EM waves; therefore, this analogy also applies to EM waves in the RF spectrum. Since the waves are propagating in all directions, it is impossible to collect them all back at the receiver. Thus the receiver receives only a small portion of the energy transmitted, and this amount of energy received continually decreases as the distance increases and the "rays" diverge further. The affects of distance on the strength of EM waves in free space are given by the following equation:

$$P \text{ (proportional to) } 1/r^2$$

where **P** is power and **r** is the distance from the source to the receiver. The inverse square relationship means that when the distance doubles ($r \times 2$), the power received is reduced by a factor of four ($2^2 = 4$).

Passing through objects further attenuates EM waves. The amount of attenuation depends on the frequency of the wave and the thickness and composition of the object through which the wave is passing. Some objects, like mountains, attenuate 100 percent of the signal, thus blocking communication. This general attenuation equation gets worse when obstacles such as rain, buildings, mountains, and so on are placed in the path of the signal. The resulting affect for terrestrial EM propagation can be estimated by the equation:

$$P \text{ (proportional to) } 1/r^3$$

In this approximation, as the distance between transmitter and receiver doubles ($r \times 2$), the power received is reduced by a factor of eight ($2^3 = 8$).

One way to minimize the amount the transmitted energy diverges is to use a directional antenna that focuses the waves in a specific direction. Figure 2.9 illustrates the previous example but uses an analogy of a flashlight to represent a directional antenna.

**Figure 2.9** Attenuation by Diversion of Rays from a Directional Source

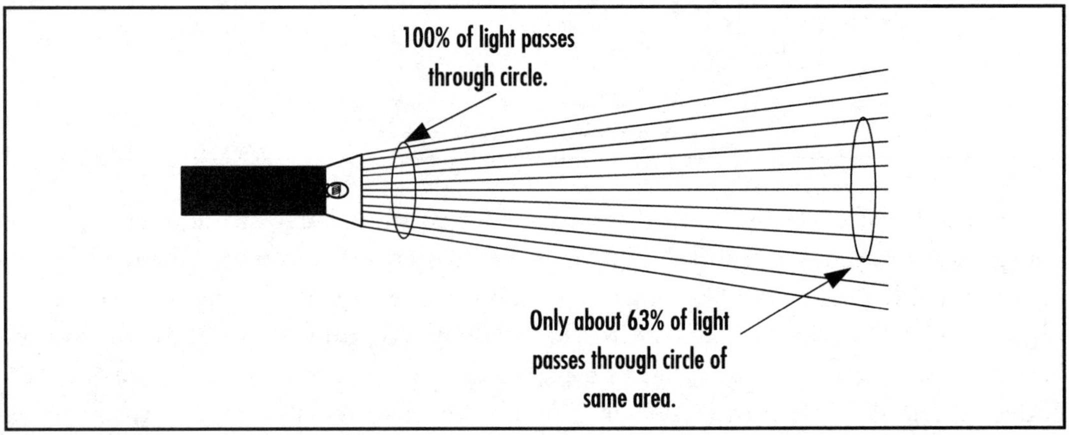


Assuming the omnidirectional light fixture in Figure 2.8 and the flashlight in Figure 2.9 both transmitted the same amount of energy, it is easy to see that the flashlight gets a stronger signal to a receiver that is the same distance away. Directional antennas are discussed in more depth later in this chapter.

EM waves do not penetrate the earth well. Therefore, for most land-based to land-based communications, the distance of the horizon is the ultimate constraint to the distance a signal can propagate. By elevating the transmitter and receiver on mountains and/or towers, however, you extend the horizon. Figure 2.10 illustrates how towers can extend the horizon.

**Figure 2.10** How Towers Can Extend Transmission Distance

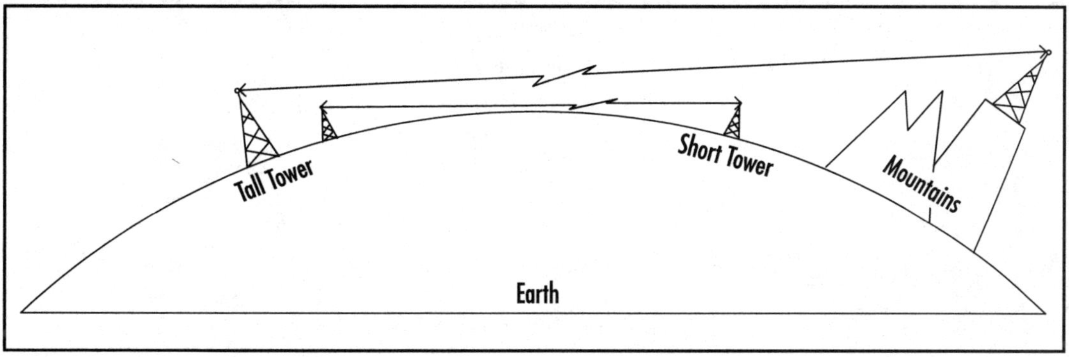


## *Rain Attenuation*

Rain attenuation is the attenuation to a signal due to precipitation. This affects high frequency waves more than low frequency waves because high frequency waves do not penetrate water as well.

The phenomenon of rain attenuation has been used to the advantage of some systems, for example, weather radar. Water droplets in the air that signify rain or clouds reflect and attenuate the high frequency radar signal differently than the surrounding air. This allows the radar system to paint a picture of the moisture.

# Bouncing

EM waves can pass through some objects, and can also be reflected off of objects. In many cases, part of the signal's energy attempts to penetrate the object, while the rest of the energy of the signal is reflected. (Imagine looking into a pool of water—some of the light passes through the water and is reflected off the bottom of the pool, allowing you to see the bottom. At the same time, if you readjust your focus you can see a reflection of yourself. This means that some of the light is penetrating and some of the light is reflecting.) Reflecting a signal is sometimes referred to as *bouncing* a signal and/or *scattering* a signal.

Bouncing can degrade the performance of some systems and enhance the performance of others. Both technology and physical conditions play a factor in whether a specific application makes use of or is hindered by bouncing. For example, AM broadcast radio signals can be bounced off of the upper layers of the earth's atmosphere. Figure 2.11 illustrates how this can extend the distance a signal can be transmitted to well beyond the horizon.

**Figure 2.11** How Signal Bouncing Can Extend Transmission Distance

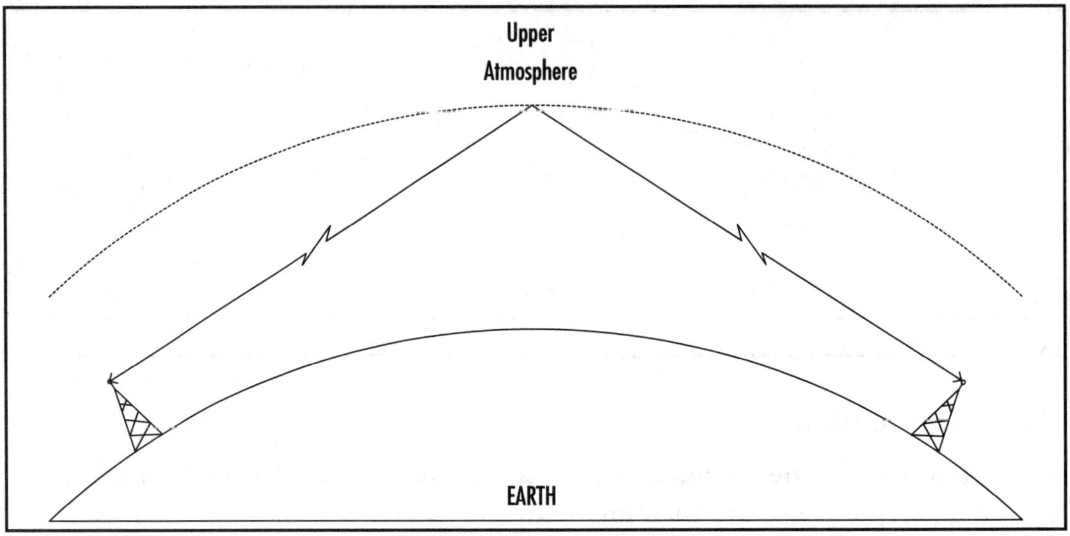


Many applications that use low frequency waves can use the layers of the atmosphere as a passive reflector, thus enhancing their distance performance; however, higher frequency waves do not bounce off the atmospheric layers well. High frequency waves penetrate through the atmospheric layers and into space without reflecting. This makes them well suited for communications with satellites. Satellites can be used as active reflectors by receiving and then retransmitting the signal to broadcast signals beyond the horizon; multiple satellites can be linked together to relay a signal completely around the world.

A specific example of using bouncing to enhance propagation is a fixed wireless telephone link this author saw in the mountains of Colorado. A house had been built up in a canyon, well beyond where the phone lines ended, but the residents wanted to have access to a phone. With a little monetary encouragement, the phone company set up a fixed wireless link to the house. However, one of the problems with choosing a wireless solution was that, obviously, the signals could not penetrate through the mountainous walls of the canyon, and the canyon was L-shaped with the house situated *around the corner* from where the last telephone access pedestal was. The fix was creative, simple, and inexpensive, albeit a little crude: a transceiver (a transmitter/receiver combination) and a directional antenna were placed at the mouth of the canyon near the last telephone access pedestal. The antenna was pointed up the canyon, aimed at a large granite rock on the far side of the L-corner of the canyon. A similar antenna was placed at the house and pointed at the same rock face. The rock actually became a reflector to bounce the signal around the corner of the canyon.

Not all results of bouncing are positive, however. One prevalent type of bouncing that adversely affects most mobile communications is called *multipath scattering*. Multipath scattering is where a signal reaches a receiver from multiple paths due to part of the signal bouncing off of various objects. If these signals arrive at the receiver out of phase, they can cancel each other. If the signals arrive in phase but are not synchronized, you can get echo signals. Echoes are probably most apparent on weak broadcast television signals where you see a main picture with a fuzzy picture just off to one side of every object in the picture. Figure 2.12 illustrates how multipath scattering occurs and demonstrates how it can cancel the signal.

**Figure 2.12** Illustration of Multipath Scattering

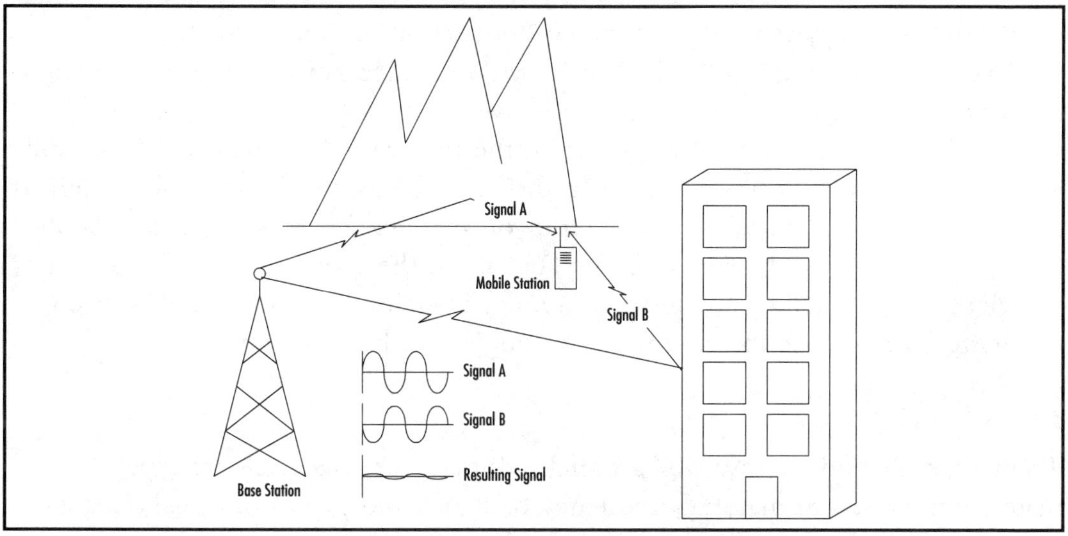


One technology that makes use of multipath is Code Division Multiple Accessing (CDMA) such as Sprint's PCS phones. CDMA uses a device called a *rake receiver* to receive multiple signals and then to align them in phase so that they all amplify each other.

## Refracting

Another property that affects the path of propagation is *refraction*, that is, the bending of a wave. Just as the lens of an eyeglass bends the light waves, suspended particles and water droplets in the atmosphere can bend radio waves. A signal can refract and bend with the curve of the earth, to a certain extent.

The *absolute horizon* is a straight line from the transmitter or receiver, and is tangential to the earth's surface. As discussed earlier, for most cases, if both the transmitter and receiver are not above this line, communication will not work. However, if the signal bends to follow the curve of the earth, it can reach receivers that are beyond the absolute horizon (a signal will refract with the curve of the earth only to a certain point). This distance is called the *apparent horizon*.

## Line of Sight

The straight clear path from the transmitter to the receiver is called the *line of sight*. All signals propagate best when they have a line of sight path, but as a general rule, high frequency signals require a line of sight signal *more* than low frequency signals. IR transmission is particularly sensitive to obstructions in line of sight.

Usually the term line of sight is from the reference point of the wave being discussed. For example, assume a certain wave can pass easily through a wood wall. In this case, line of sight might mean the receiver is on one side of the wall and the transmitter is on the other. On the other hand, assume the line of sight is for a mini-dish satellite television system; you may be able to see the sky plainly through the leaves of the tree looming above, but the signal to your dish may be sufficiently attenuated by the leaves to render it useless. It is important to keep this frame of reference in mind when talking about line of sight.

## Penetration

The factors that affect how well a signal will penetrate materials are type of material, thickness of material, frequency of signal, and power of signal. This is most apparent in the application of medical x-rays. Extremely high frequency x-ray waves penetrate through the softer tissues of the body but are blocked by bone. This causes the x-ray-sensitive film to be etched with a picture that represents the bone structure.

This plays a large factor in terrestrial-based communications because of natural obstacles such as mountains, hill, and trees, and man-made obstacles like buildings. Table 2.1 gives a comparison to the penetrating power of high- and low-frequency radio waves to various materials.

**Table 2.1** Penetration Levels of Different Frequency for Various Materials

| Material | Level of Penetration Low Frequency | Mid Frequency | High Frequency |
|---|---|---|---|
| Vacuum | Good | Good | Good |
| Air | Good/Bouncing | Good | Good/rain attenuation |
| Water | Fair | Poor | Extremely Poor |
| Earth | Poor | Poor | Poor |

Note that Table 2.1 does not include *metal* among the materials. Metal is a special case for penetration of EM waves. First consider how an antenna works. At the transmitting end, a cable is hooked to an antenna. The signal travels down the cable and into the antenna. This generates an oscillating voltage potential in the antenna, which in turn generates an oscillating electric field between the antenna and the ground plane. The oscillating electric field creates an oscillating magnetic field and the magnetic fields create additional electric fields, and the wave propagates away from the antenna without wires. At the receiving end, the magnetic and electric fields induce an oscillating voltage potential in the receiving antenna. This voltage signal is carried away from the antenna by a cable.

Now consider a piece of metal that does *not* have a conductive path to ground and is acting as an antenna. The magnetic and electrical waves induce a voltage signal in the piece of metal. Since the antenna is not connected to ground, and there is no cable to drain the signal as in the previous example, there becomes a standing wave in the antenna. This standing wave will regenerate an electric field and the wave will propagate away from the antenna again. Therefore, as soon as a signal is received, it is retransmitted. Since part of the signal is retransmitted from each side of the piece of metal, it looks like part of the signal is being reflected and part of the signal penetrated the metal.

When this piece of metal *is* connected to ground, the EM wave hits the metal and induces a voltage in the metal. This voltage is dissipated to the ground and the signal is lost. In this case, the signal does not penetrate the metal.

The last case of penetration of metal is based on a principle by a scientist named Michael Faraday and deals with metal enclosures. You may have observed that your wireless phone does not work well in elevators. The reason is not simply that the signal cannot penetrate metal. Assume you have a metal sphere. As an EM wave hits the leading edge of the sphere, it acts like an antenna and the wave induces a voltage in the metal. This voltage travels around the outside of the sphere to the trailing edge where the wave repropagates into the air; therefore, the signal does not penetrate to the interior of the sphere.

These properties can be discouraging, especially since metal is so prevalent as a construction material. However, these same properties can make metal very useful as a shield from EM waves. Coaxial cable uses an outer shield of metal mesh to protect the signal that is propagating down the center conductor of the cable—the metal mesh shield acts as an antenna to receive the unwanted EM waves that would interfere with its main signal. This metal shielding is connected to ground and as the EM wave induces a voltage in the mesh, it is dissipated to ground. Now imagine expanding that mesh shielding to surround an entire computer or an entire room. This type of shielding is important for many devices such as sensitive medical equipment, aircraft computerized flight control systems, and microwave ovens.

# Understanding the Wireless Elements

What are the wireless components in a network? Primarily, there are just two components, the antenna and the wireless device. That may seem oversimplified, but remember, a wireless network typically is not a stand-alone network. For example, a local area network (LAN) is composed of access points, antennas, and wireless PC cards. The only connection that is truly "wireless" is between the antenna and the PC card. The access point is connected to the wireline network infrastructure, and then wired to the antenna.

There are numerous types of antennas, and each type is optimized for a particular environment or application. It is important to understand the distinguishing characteristics when choosing the appropriate antenna for your wireless network design. This section also discusses mobile wireless base stations, mobile stations, and access points.

## Generic Radio Components

Generally, how is a signal manipulated during transmission? The various types of modulation and attenuation were covered earlier in this chapter; illustrated in this section are the radio components that show the process of the signal from origin to destination. Keep in mind that these are general principles.

All radios share a basic conceptual design. In this generic design many of the specific components are commonly simplified into a "black box" schematic. Figure 2.13 highlights the generic "black box" radio components.

**Figure 2.13** Generic Radio Components

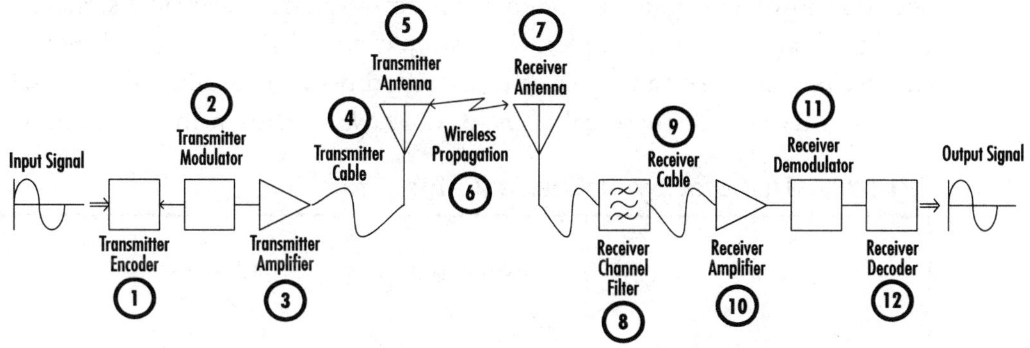


Each of the boxes in this figure represent entire subsystems of the radio; each is very complex, but simplified here as a box representing their function. The function of each of these boxes, following the system from the origin of the signal to the output of the receiver, can be described briefly as follows:

1. **Transmitter Encoder** The input signal enters the encoder. In this context, "encoder" is a very generic term. This could be a microphone that encodes analog sound waves into analog electrical waves or it could be a complex CDMA analog-to-digital converter. The main thing to keep in mind is that the original signal usually is modified or encoded before it is input into the modulator.
2. **Transmitter Modulator** As explained earlier in this chapter, the modulator box performs the modulation of the carrier wave.
3. **Transmitter Amplifier** After the signal is modulated, it is amplified so that it can be radiated with enough power to reach the receiving antenna. You may have heard radio stations boasting something like, "5,000 watts of rock and roll coming your way!" This is the result of a large amplifier.
4. **Transmitter Cable** Up to this point, most of the components would be located in one physical device. Sometimes even the antenna is integrated into the same box. However, the antenna is often mounted a dis-

tance away from the rest of the radio, for example outside the building for better transmission. That creates the problem of getting the signal from the amplifier to the antenna. It might seem like a simple wire connection; however, cables designed for high frequency electrical signals are significantly more complex than simple wire. They are called *transmission lines* because the signal is transmitted down the cable. The most common design for these cables is *coaxial cable*, as shown in Figure 2.14.

**Figure 2.14** Cutaway of Coaxial Cable

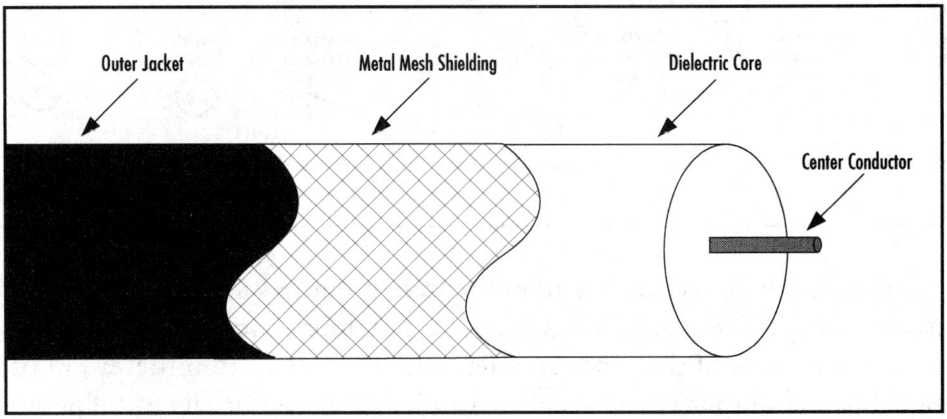


As mentioned earlier, the outer metal mesh of coaxial cable acts as a shield that shunts unwanted interference to ground. Since the shielding is tied to ground, it acts as a baseline reference. The center conductor carries the signal in reference to the outer, grounded shield. The spacing between the center conductor and the outer shield is very important and must remain constant; the dielectric material between the center conductor and the outer shield maintains this distance. Kinking coaxial cable can permanently damage the cable and can attenuate the signal, so it is important to be careful with the cable. Since the spacing is so critical to the function of a coaxial cable, splicing two cables is not as simple as connecting the inner conductor and the outer shields together. Splicing coaxial cable usually requires a splicer that looks like two connectors linked end-to-end and is crimped onto the cable with a special tool.

5. **Transmitter Antenna** The purpose of an antenna is to convert electrical signal to radio waves and vice versa. The antenna is one of the simplest subsystems of a radio because most antennas are passive devices,

yet a tremendous amount of engineering goes into antenna design. Antennas are discussed in more depth in the next subsection.

6. **Wireless Propagation** The oscillating voltage potential in the antenna generates an oscillating electric field between the antenna and the ground plane. The oscillating electric field creates an oscillating magnetic field, the magnetic fields create additional electric fields, and the wave propagates away from the antenna.
7. **Receiver Antenna** Similar to the transmitter antenna, the receiver antenna converts the radio waves back into an electrical signal.
8. **Receiver Channel Filter** Even though antennas are designed and tuned for a specific frequency, they will still receive EM energy from the entire spectrum. Most electrical components are designed to work at a specific range of frequencies and they do not deal well with frequencies outside this range. For this reason the received signal is filtered to allow only the intended frequencies to pass to the subsequent receiver components.
9. **Receiver Cable** Same as the transmitter cable. On the transmitter end, the signal is amplified before leaving the main circuitry of the transmitter and entering the cable; therefore, cable loss is not a problem on the transmitter side. However, on the receiver, the weak signal can be drastically affected by loss due to the length of the cable and the quality of the cable.
10. **Receiver Amplifier** The received signal is usually very weak and must be amplified before it is processed by the more complex receiver components. Some receiver designs place this amplifier or add an additional amplifier before the cable and very near the antenna to boost the received signal so that cable loss does not kill the signal as it travels from the antenna to the main receiver.
11. **Receiver Demodulator** Separates the original encoded modulating signal from the carrier signal. On a clean signal, the output of the receiver demodulator should closely represent the input of the transmitter demodulator.
12. **Receiver Decoder** Decodes the demodulated signal to get a representation of the original input signal.

As noise increases and/or the received signal strength decreases, the output of the receiver resembles the transmitted signal less until the point where it can no longer be recognized as the same signal.

# Laws, Regulations, and Environmental Considerations

As the number of wireless devices dramatically increases, it is not difficult to see why there is a need for regulatory agencies. This section describes the regulatory agencies responsible for the development and operation of wireless systems. Although the Web sites provided by these agencies are the best source for the most current information, covered here are some general operational and technical rules and regulations associated with domestic wireless broadband data networks. Remember, many rules and restrictions are based entirely on regulations proposed by federal, state, and local agencies in an evolving business and regulatory climate, so it is recommended to reference the regulations committees before designing a wireless network.

## Regulatory Agencies

The International Telecommunication Union (ITU) (www.itu.int/ITU-R), headquartered in Geneva, Switzerland, is an international organization that coordinates standards and regulations, and promotes the efficient use of the RF spectrum. The ITU established an international Table of Frequency Allocations that designates specific bands of frequencies for specific uses within different regions of the globe.

The ITU Radiocommunication Sector (ITU-R) was created in 1993 and comprises what once were the International Radio Consultative Committee (CCIR) and the International Telephone Consultative Committee (CCIF). The ITU-R is responsible for all of the ITU's work in the field of radio communications.

In the U.S., the Federal Communications Commission (FCC) and the National Telecommunications and Information Administration (NTIA) regulate the use of the RF spectrum, through the management of frequency allocations and the designation of specific bands of frequencies. The two agencies determine which portions of the spectrum are reserved for federal use, for non-federal use, and for shared use within the United States.

The FCC (www.fcc.gov) is an independent federal regulatory agency accountable directly to the United States Congress, established by the

Communications Act of 1934. The FCC allocates frequency spectrums for commercial use including specific types of fixed, mobile, and broadcasting services such as broadcast television services, cellular telephone, paging, personal communications services, public safety, and other commercial and private radio services. Because the primary directive of the FCC is to govern the efficient use of the EM spectrum, it generally leaves the development of standards (such as the 802.11b standard) to the industries. Recently, the FCC has been associated with its additional objective of developing a domestic telecommunications infrastructure that provides services on national and global levels.

The NTIA, an agency of the U.S. Department of Commerce, is the Executive Branch's principal representative in domestic and international telecommunications and information technology issues, and frequently works with other Executive Branch agencies to develop and present the Administration's position on these issues. NTIA works to encourage innovation and competition, promote job creation, and provide consumers with more choices and the best quality telecommunications products and services at lower prices.

### The Need to Know

With the exception of a few unlicensed frequency bands, all transmitters in the private, commercial, and broadcast services require an FCC license prior to operation. An FCC fine for unlicensed operation varies from radio service to radio service, but penalties in the broadcast and commercial bands can be severe—for example, $5000 per day plus potential imprisonment. Unauthorized private systems are also subject to fines. Licensing and other legal requirements for operating any radio transmitter change periodically; consult the FCC regulations for the latest information.

## Regulations for Low Power, Unlicensed Transmitters

Three changes in FCC regulation contributed to the fast-paced growth of wireless communications. In 1985 the U.S. released the Industrial, Scientific, and Medical (ISM) frequency bands, provided that certain technical restrictions on transmitter power and modulation are met. The deregulation of this frequency spectrum eliminates the need for users and organizations to perform costly and time-consuming frequency planning to coordinate radio installations that avoid interference with existing radio systems. Successful introduction of new RF spectrum-dependent systems into congested frequency bands for worldwide use is difficult; however, requirements imposed by the regulations on unlicensed wireless networking equipment are relatively simple. First, the signal strength is limited, usually to less than 1 watt. Second, the signal must be transmitted using one of two spread-spectrum methods in which the signal is spread out over a certain range of frequencies, or hops, among a certain minimum number of narrow slots each second. Many devices operating in the ISM bands use these types of modulation; however the FCC does permit narrowband modulation techniques to be used, so all devices have to contend with interference from other unlicensed devices operating in the same frequency range.

The most specific restrictions imposed upon the ISM bands by the FCC is that the U.S. RF systems must implement spread spectrum technology. Microwave systems are considered very low power systems and must operate at 500 milliwatts or less, but because of the interference avoidance characteristics of spread spectrum technology, these devices are permitted greater output power than transmitters not using spread spectrum technologies that operate in the same band. Again, although these restrictions may not generally apply to the consumer, for rules concerning eligibility for licensing, frequencies available, permissible communications and classes and number of stations, and any special requirements, the FCC remains the sole authority. If implementing a spread spectrum system outside the U.S., it is important to investigate the local regulations (as mentioned earlier, the regulatory bodies associated with the ITU are considered an excellent point of reference). Since licensing is not required when operating under Part 47 Section 15.247 of the FCC Rules and Regulations, implementation of spread spectrum systems is simple and cost-effective.

# Environmental Considerations

There is some concern in relation to health problems caused by RF radiation. Extensive research has been done on this subject and there is a great deal of literature regarding the biological effects of RF radiation. This section covers some of the important details and results of this research.

High intensities of RF radiation can be harmful due to the ability of RF energy to quickly heat up biological tissue. Using the same principles employed by a microwave oven to cook food, exposure to high densities of RF radiation can result in the heating of tissue in the human body and cause an increase in body temperature. This can cause damage to the tissue primarily because the body is unable to quickly dissipate this excessive heat. There are a lot of factors involved in this effect including the frequency of the radiation, the size of the area exposed, the duration of exposure, and the efficiency of heat dissipation. These biological effects resulting from the heating of human body tissue are known as "thermal" effects.

Research is still being done on "non-thermal" effects of RF radiation. Some of these are the possibilities of changes in the immune system, neurological effects, behavioral effects, and a "calcium efflux" effect. Experiments relating to these non-thermal effects are still being conducted and so far results show that these only occur with very high exposure to high densities of RF radiation. Many contradictory results have been reported from these experiments; therefore a great deal more testing needs to be done.

So what effect does this have on your wireless network? The RF radiation created in a wireless network is minimal. The radio waves used for wireless transmission are typically done with very low power within office buildings and homes. The only area where high levels of RF radiation would be found is with long distance microwave radio networks. Even in this situation where more power is being used for the transmission, the possibility of any negative effects of the radiation are minimal. It is typically not suggested to stand right in front of the antenna when transmissions are occurring, but otherwise the radiation being emitted will cause no harm. In one manufacturer's experiments, a person would have to put the broadcasting antenna for a wireless radio *on the surface of the eye* to get any effect whatsoever. That puts the radiation levels generated into perspective.

The Royal Society of Canada has also done some research on the subject in order to ensure that wireless devices' safety guidelines are accurate and up-to-date. Their research paper (available at www.rsc.ca/english/RFreport.pdf) also indicates that the radiation intensity emitted by wireless devices is not powerful enough to cause adverse health effects.

# IEEE 802.11 Standards

As in all 802.x standards, the 802.11 specification covers the operation of the Media Access Control (MAC) and physical layers. As seen in Figure 2.15, 802.11 defines a MAC sublayer, MAC services and protocols, and three physical (PHY) layers.

**Figure 2.15** 802.11 Frame Format

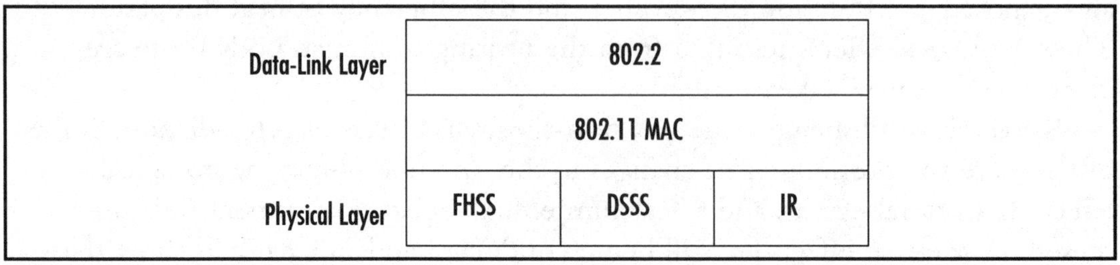


The three physical layer options for 802.11 are IR baseband PHY and two RF PHYs. Due to line-of-sight limitations, very little development has occurred with the IR PHY. The RF physical layer is composed of FHSS and DSSS in the 2.4 GHz band. All three physical layers operate at either 1 or 2 Mbps. The majority of 802.11 implementations utilize the DSSS method.

FHSS works by sending bursts of data over numerous frequencies. As the name implies, it hops between frequencies. Typically, the devices use up to four frequencies simultaneously to send information and only for a short period of time before hopping to new frequencies. The devices using FHSS agree upon the frequencies being used. In fact, due to the short time period of frequency use and device agreement of these frequencies, many autonomous networks can coexist in the same physical space.

DSSS functions by dividing the data into several pieces and simultaneously sending the pieces on as many different frequencies as possible, unlike FHSS, which sends on a limited number of frequencies. This process allows for greater transmission rates than FHSS, but is vulnerable to greater occurrences of interference. This is because the data is spanning a larger portion of the spectrum at any given time than FHSS. In essence, DHSS floods the spectrum all at one time, whereas FHSS selectively transmits over certain frequencies.

The third physical layer outlined in the 802.11 frame format is IR. This type of communication is done using an IR transceiver and is based on modulating the light emitted by a laser diode or a light emitting diode (LED). IR communications only work over a short distance and require that there is no physical interference between the transmitter and the receiver. Since this technology does not have the ability to work through walls and so on, it is not very popular for network communications.

Designing & Planning…

## Additional Initiatives of the 802 Standards Committee

**802.1 LAN/MAN Bridging and Management** 802.1 is the base standard for LAN/Metropolitan area network (MAN) bridging, LAN architecture, LAN management, and protocol layers above the MAC and logical link control (LLC) layers. Some examples include 802.1q, the standard for virtual LANs, and 802.1d, the Spanning Tree Protocol.

**802.2 Logical Link Control** Since LLC is now a part of all 802 standards, this working group is currently in hibernation (inactive) with no ongoing projects.

**802.3 CSMA/CD Access Method (Ethernet)** 802.3 defines that an Ethernet network can operate at 10 Mbps, 100 Mbps, 1 Gbps, or 10 Gbps. It also defines that Category 5 twisted pair cabling and fiber-optic cabling are valid cable types. This group identifies how to make vendors' equipment interoperate despite the various speeds and cable types.

**802.4 Token-Passing Bus** This working group is in hibernation with no ongoing projects.

**802.5 Token Ring** Token Ring networks operate at 4 Mps or 16 Mbps. Currently, there are working groups proposing 100 Mb Token Ring (802.5t) and Gigabit Token Ring (802.5v). Examples of other 802.5 specs are 802.5c, Dual Ring Wrapping, and 802.5j, fiber-optic station attachment.

**802.6 Metropolitan Area Network (MAN)** Since MANs are created and managed with current internetworking standards, the 802.6 working group is in hibernation.

**802.7 Broadband LAN** In 1989, this working group recommended practices for broadband LANs, which were reaffirmed in 1997. This group is inactive with no ongoing projects. The maintenance effort for 802.7 is now supported by 802.14.

Continued

**802.8 Fiber Optics** Many of this working group's recommended practices for fiber optics get wrapped into other standards at the Physical layer.

**802.9 Isochronous Services LAN (ISLAN)** Isochronous Services refer to processes where data must be delivered within certain time constraints. Streaming media and voice calls are examples of traffic that requires an isochronous transport system.

**802.10 Standard for Interoperable LAN Security (SILS)** This working group provided some standards for data security in the form of 802.10a, Security Architecture Framework, and 802.10c, Key Management. This working group is currently in hibernation with no ongoing projects.

**802.11 Wireless LAN (WLAN)** This working group is developing standards for Wireless data delivery in the 2.4 GHz and 5.1 GHz radio spectrum.

**802.12 Demand Priority Access Method** This working group provided two Physical layer and repeater specifications for the development of 100 Mbps demand priority MACs. Although they were accepted as International Organization for Standardization (ISO) standards and patents were received for their operation, widespread acceptance was overshadowed by Ethernet. 802.12 is currently in the process of being withdrawn.

**802.13** This standard was intentionally left blank.

**802.14 Cable-TV Based Broadband Comm Network** This working group developed specifications for the Physical and MAC layers for cable televisions and cable modems. Believing their work to be done, this working group has no ongoing projects.

**802.15 Wireless Personal Area Network (WPAN)** The vision of Personal Area Networks is to create a wireless interconnection between portable and mobile computing devices such as PCs, peripherals, cell phones, Personal Digital Assistants (PDAs), pagers, and consumer electronics, allowing these devices to communicate and interoperate with one another without interfering with other wireless communications.

**Continued**

**802.16 Broadband Wireless Access** The goal of the 802.16 working group is to develop standards for fixed broadband wireless access systems. These standards are key to solving "last-mile" local-loop issues. 802.16 is similar to 802.11a in that it uses unlicensed frequencies in the unlicensed national information infrastructure (U-NII) spectrum. 802.16 is different from 802.11a in that quality of service for voice, video, and data issues are being addressed from the start in order to present a standard that will support true wireless network backhauling.

# Does the 802.11 Standard Guarantee Compatibility across Different Vendors?

The primary reason WLANs were not widely accepted when they first became available was the lack of standardization. It is logical to question whether vendors would accept a nonproprietary operating standard, since vendors compete to make unique and distinguishing products. Although 802.11 standardized the PHY, MAC, the frequencies to send and receive on, transmission rates and more, it did not absolutely guarantee that differing vendors' products would be 100 percent compatible. In fact, some vendors built in backward-compatibility features into their 802.11 products in order to support their legacy customers. Other vendors have introduced proprietary extensions (for example, bit-rate adaptation and stronger encryption) to their 802.11 offerings.

To ensure that consumers can build interoperating 802.11 wireless networks, an organization called the Wireless Ethernet Compatibility Alliance (WECA) tests and certifies 802.11 devices. Their symbol of approval means that the consumer can be assured that the particular device has passed a thorough test of interoperations with devices from other vendors. This is important when considering devices to be implemented into your existing network, because if the devices cannot communicate, it complicates the management of the network. It is also important when building a new network because you may be limited to a single vendor.

Since the first 802.11 standard was approved in 1997, there have been several initiatives to make improvements. As seen in the following sections, there is an evolution unfolding with the 802.11 standard. The introduction of the standard came with 802.11b. Then along came 802.11a, which provides up to five times the bandwidth capacity of 802.11b. Accompanying the ever-growing demand for multimedia services is the development of 802.11e. Each task group, outlined next, is endeavoring to speed up the 802.11 standard, making it globally accessible, while not having to reinvent the MAC layer of 802.11:

- **The 802.11d working group** is concentrating on the development of 802.11 WLAN equipment to operate in markets not served by the current standard (the current 802.11 standard defines WLAN operation in only a few countries).
- **The 802.11f working group** is developing an Inter-Access Point Protocol due to the current limitation prohibiting roaming between access points made by different vendors. This protocol would allow wireless devices to roam across access points made by competing vendors.
- **The 802.11g working group** is working on furthering higher data rates in the 2.4 GHz radio band.
- **The 802.11h working group** is busy developing spectrum and power management extensions for the IEEE 802.11a standard for use in Europe.

## DSSS

As mentioned earlier, DSSS functions by dividing the data into several pieces and simultaneously sending the pieces on as many different frequencies as possible, unlike FHSS, which sends a limited number of frequencies. This process allows for greater transmission rates than FHSS, but is vulnerable to greater occurrences of interference. This is because the data is spanning a larger portion of the spectrum at any given time than FHSS. In essence, DSSS floods the spectrum all at once, whereas FHSS selectively transmits over certain frequencies.

# IEEE 802.11b Direct Sequence Channels

Ignoring the FHSS and IR physical mediums, the 802.11b PHY uses DSSS to broadcast in any one of 14 center-frequency channels in the 2.4 GHz ISM radio band. As Table 2.2 shows, North America allows 11 channels; Europe allows 13, the most channels allowed. Japan has only one channel reserved for 802.11, at 2.483 GHz.

**Table 2.2** 802.11b Channels and Participating Countries

| Channel Number | Frequency (GHz) | North America | Europe | Spain | France | Japan |
|---|---|---|---|---|---|---|
| 1 | 2.412 | X | X | | | |
| 2 | 2.417 | X | X | | | |
| 3 | 2.422 | X | X | | | |
| 4 | 2.427 | X | X | | | |
| 5 | 2.432 | X | X | | | |
| 6 | 2.437 | X | X | | | |
| 7 | 2.442 | X | X | | | |
| 8 | 2.447 | X | X | | | |
| 9 | 2.452 | X | X | | | |
| 10 | 2.457 | X | X | X | X | |
| 11 | 2.462 | X | X | X | X | |
| 12 | 2.467 | | X | | X | |
| 13 | 2.472 | | X | | X | |
| 14 | 2.483 | | | | | X |

There are many different devices competing for space in the 2.4 GHz radio spectrum. Unfortunately, most of the devices that cause interference are especially common in the home environment, such as microwaves and cordless phones. As you can imagine, the viability of an 802.11b network depends on how many of these products are near the network devices.

One of the more recent entrants to the 802.11b airspace comes in the form of the emerging Bluetooth wireless standard. Though designed for short-range transmissions, Bluetooth devices utilize FHSS to communicate with each other. Cycling through thousands of frequencies a second, Bluetooth looks as if it poses the greatest chance of creating interference for 802.11. Further research will determine exactly what, if any, interference Bluetooth will cause to 802.11b networks. Many companies are concerned with saturating the 2.4 GHz spectrums, and are taking steps to ensure that their devices "play nicely" with others in this arena.

These forms of interference will directly impact the home user who wishes to set up a wireless LAN, especially if neighbors operate interfering devices. Only time will tell if 802.11b will be able to stand up against these adversaries and hold on to the marketplace.

## IEEE 802.11a OFDM Physical Layer

IEEE published the 802.11a standard prior to publishing the 802.11b standard. This sounds logical, but if you look at the wireless LAN marketplace, devices using the IEEE 802.11b standard came out long before devices using the IEEE 802.11a standard. The reason for this is the ease of implementation of the standards in wireless devices. The use of DSSS in the IEEE 802.11b standard was covered previously, and it should be noted that hardware utilizing DSSS is fairly easy to design and manufacture.

The IEEE 802.11a standard specifies the use of orthogonal frequency division multiplexing (OFDM) for its Physical layer. This technology is a little more difficult to design and build into wireless devices; therefore devices using the IEEE 802.11b standard hit the market first. OFDM basically squeezes multiple modulated carriers tightly together, which reduces the required bandwidth. Additionally, it keeps the modulated signals orthogonal so one signal does not interfere with the others.

## IEEE 802.11a Channels

Due to the overwhelming demand for more bandwidth and the growing number of technologies operating in the 2.4 GHz band, the 802.11a standard was created for WLAN use in North America as an upgrade from the 802.11b standard. 802.11a provides 25 to 54 Mbps bandwidth in the 5 GHz spectrum (the U-NII spectrum). Since the 5 GHz band is currently mostly clear, the chance of interference is reduced. However, that could change since it is still an unlicensed portion of the spectrum. 802.11a is still designed mainly for the large enterprise organization primarily due to cost.

As shown in Table 2.3, three 5 GHz spectrums have been defined for use with 802.11a. Each of these three center-frequency bands covers 100 MHz.

**Table 2.3** 802.11a Channels Usable in the 5-GHz U-NII Radio Spectrum

| Regulatory Area | Frequency Band | Channel Number | Center Frequencies |
|---|---|---|---|
| USA | U-NII Lower Band | 36 | 5.180 GHz |
| | 5.15 – 5.25 GHz | 40 | 5.200 GHz |
| | | 44 | 5.220 GHz |
| | | 48 | 5.240 GHz |

Continued

**Table 2.3** 802.11a Channels Usable in the 5-GHz U-NII Radio Spectrum

| Regulatory Area | Frequency Band | Channel Number | Center Frequencies |
|---|---|---|---|
| USA | U-NII Middle Band | 52 | 5.260 GHz |
| | 5.25 – 5.35 GHz | 56 | 5.280 GHz |
| | | 60 | 5.300 GHz |
| | | 64 | 5.320 GHz |
| USA | U-NII Upper Band | 149 | 5.745 GHz |
| | 5.725 - 5.825 GHz | 153 | 5.765 GHz |
| | | 157 | 5.785 GHz |
| | | 161 | 5.805 GHz |

# Planning for RF Deployment

The most important part of a WLAN implementation is deployment planning. A poorly planned deployment is much worse than no deployment at all. If appropriate planning is not done, your wireless LAN deployment can be a complete failure due to its inability to function or its inability to properly serve your users' needs.

This section covers the most important aspect of planning for your RF deployment, starting with the coverage of your WLAN and how to plan coverage appropriately. It then discusses the WLAN data rates with the different 802.11x standard and the variances caused in throughput due to client density. Various antenna options are also covered along with how to choose an antenna based on your needs.

Site surveys are critical to both the initial planning stages of your RF deployment and the ongoing maintenance of the WLAN. These are covered in detail and a sample site survey form is provided to get you started. Finally, interference detection and how this step of planning can help you to successfully deploy your WLAN is covered.

## WLAN Coverage

Each access point and antenna combination produces a single area of coverage. Each of these single areas is referred to as a *cell*. Multiple overlapping cells are used to provide wireless coverage for areas larger than a single cell alone can produce. This is *cellular architecture*.

DSSS wireless LANs have 11 total channels that can be used for RF transmission. Each channel is 22 MHz wide, and all channels combined equal the entire spectrum that can be used for 802.11b wireless LANs. When designing wireless LANs, multiple channels become an issue only when overlapping coverage (multiple cells) is required, which is usually the case in most designs.

When two access points have overlapping coverage (they have a cellular architecture), each access point must use a different channel so that the client can distinguish the difference between the RF for each access point. The only three channels that do not overlap concurrently are channels 1, 6, and 11.

Rate requirements also factor into the cellular architecture of a wireless coverage area because the distance from an access point affects the data rate. The data rate decreases as the coverage area increases until of course, you have no coverage at all.

Depending on the coverage rate required for a given area, you may need more or fewer access points to fulfill the coverage requirements. As seen in Figure 2.16, it requires many more access points configured at the 11 Mbps rate to cover a specific area than to cover the same area with access points configured at the 2 Mbps rate. Bandwidth is sacrificed for distance.

**Figure 2.16** Cellular Architecture

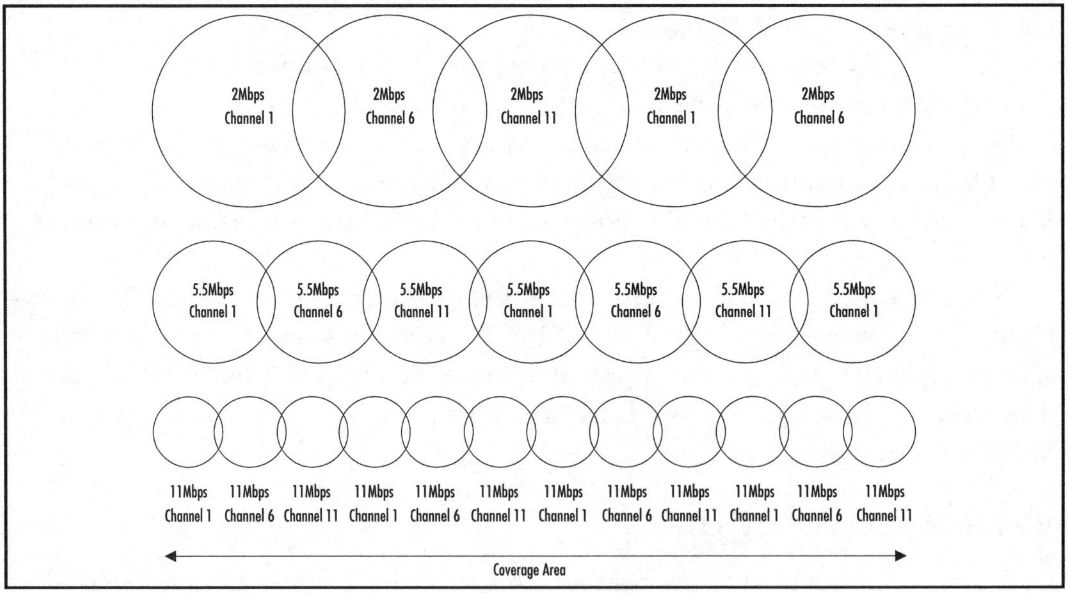

You can extend the coverage for a client by using an access point in repeater mode to extend the coverage of an existing access point. The repeater access point does not have a wired connection to the network. Instead, the client associates to the wired, root access point *through* the access point acting as a repeater. This solution can provide additional coverage when the wiring infrastructure is not available for another access point. There are limitations to the use of repeaters to extend coverage. You cannot continuously add access points in repeater mode to gain extremely long coverage areas. Repeater cells need 50 percent overlap with a wired access point cell. Each repeater loses approximately half its normal coverage distance as they are added farther away from the wired root access point. For example, you may have about 200 feet of 11 Mbps coverage from a wired root access point. You add one repeater, and you gain an additional 100 feet of 11 Mbps coverage. The next repeater you add will give you approximately 50 feet of additional coverage, and so on. Eventually, adding an access point in repeater mode will give you only a little bit of additional coverage at 11 Mbps, and that repeater access point will cost more than the value of the coverage it adds.

Generally, within buildings, the availability of Ethernet connections is fairly predominant. Repeaters are typically used to extend access points from the building edge to the surrounding outdoor portions or additional rooms of a building as a temporary measure. For example, the owner of a retail store may use access points in repeater mode to extend coverage into the parking lot of their facility during an outdoor tent sale.

## *Designing Seamless Roaming*

When a client travels throughout a wireless cell (one access point coverage area), it is called *roaming*. The smooth transition from one wireless cell to another (one access point coverage area to another) is called *seamless roaming*. A large factor when designing a wireless system is determining whether clients require seamless roaming. All devices that require seamless roaming must be turned on when moving from location to location. Seamless roaming is not required for devices that are turned off before being moved. Several factors are involved when designing and surveying for a wireless LAN that requires seamless roaming. You must ensure overlapping coverage for the entire roaming path, you must maintain a consistent client IP address, and you must stay on the same subnetwork.

If a client is going to travel between more than one access point, coverage must exist for the entire path. The Internet Protocol (IP) address must be consistent within the coverage area because the client will acquire its IP address at its starting point and use that address throughout the path. If the IP subnet for each access point is located on different switches, and the switch is a Layer 3 switch or is separated from the other switch by a Layer 3 device such as a router, consider using switch trunking such as Inter-Switch Link (ISL) or 802.1Q to span the virtual local area networks (VLANs) to ensure that a single broadcast domain is used for all access points.

As discussed previously, multipath distortion can be a problem in RF. The transmission between a client and an access point travels on a RF. Those signals interact with the surrounding environment and can be deflected while in transit to the access point. Under optimal conditions, the RF signals arrive at their destination in the same order in which they were sent. There is a good possibility that some of the RF signals will reflect off surrounding objects and arrive out of phase at the antenna, which causes the signals to cancel each other out and result in inoperability. This scenario is very similar to Transmission Control Protocol (TCP) packets arriving out of order within a wired LAN infrastructure.

If the RF between the two devices is strong, it can sometimes give the misperception of good connectivity. Even if the signal strength is good, the signal quality might be poor, thus causing traffic performance to suffer. By adding a second antenna to the access point, you can increase the area in which signals are received and thus minimize, if not eliminate, the "dead path" and increase the signal quality and performance. Using antennas with access points is called *antenna diversity*. The access point chooses the best antenna and uses that antenna to receive signals. Only one antenna at a time is active; the active one is selected on a per-client basis for the optimal signal. It applies only to a specific client. The access point can jump back and forth between the antennas when talking to different clients.

Cisco wireless network cards (client interfaces) can also use antenna diversity because they have a diversified antenna built in to them or have diversified antennas available externally. Whether using an access point or a client card, you can turn off the diversity through the configuration menu of both devices. Antenna diversity is used to overcome multipath issues, not to increase the coverage area of an access point.

When configuring access points for cellular coverage, the amount of overlap required to allow a mobile client to seamlessly roam throughout the coverage area is approximately 15 percent. The amount of this overlap depends upon how mobile the users will be. In a highly mobile environment, the amount of overlap required to allow the users to seamlessly roam may be higher than would normally be required. Your site survey will allow you to determine how much overlap will be necessary in your environment. Be sure to test this roaming capability in your environment and especially in highly mobile user environments by performing the roaming yourself as closely as possible to the actual use the wireless system will see when in production.

# WLAN Data Rates

The data rates of your WLAN will vary based on the strength of the wireless signal, the IEEE 802.11x standard used, and the wireless load. The following section covers wireless load (client density), but here we go over the data rates as they relate to the IEEE standards and the problems with signal strength.

Table 2.4 shows the data rates for each of the IEEE 802.11x standards.

**Table 2.4** IEEE 802.x Standard Data Rates

| | 802.11 | 802.11a | 802.11b | 802.11g |
|---|---|---|---|---|
| Raw Data Rate | 2 MBit/sec | 54 MBit/sec | 11 MBit/sec | 54 MBit/sec |
| Frequency | 2.4 GHz | 5 GHz | 2.4 GHz | 2.4 GHz |
| Transmission Technology | DSSS/FHSS | OFDM | DSSS/CCK | OFDM/DSSS |

As seen in Table 2.4, both IEEE 802.11a and 802.11g allow data rates of up to 54 MBit/sec and IEEE 802.11b only goes up to 11 MBit/sec. Keep in mind that these are maximum data rates. Typical throughput varies, but should be ~27 Mbit/sec with IEEE 802.11a, ~ 4–5 Mbit/sec with IEEE 802.11b, and ~20–25 MBit/sec with IEEE 802.11g. This is due to the overhead required for any network as well as various environmental considerations.

The strength of your wireless signal can have a great impact on your actual throughput. Basically, with low signal strength, there is a greater possibility of interference in the signal or dropped data. All of the transmission technologies auto-correct for this at the Physical layer, but the time and additional work required for this correction does cause the overall data rate to drop. Therefore, when planning your WLAN deployment, always make sure that your access points are distributed so that the clients have the strongest signal possible.

# Client Density and Throughput

Throughput and data rates do not equal each other in either a wireless or wired network. Typical data throughput on an Ethernet network is about 60 percent of the nominal capacity. Therefore, a 10 Mbps Ethernet network under typical load would yield a throughput of 6 Mbps. Factors that affect nominal throughput on a network include overhead, number of users, operating systems, and so forth. A wireless link has slightly greater overhead associated with it than a wired link and therefore typically yields approximately 5.5 Mbps of throughput on an 11 Mbps network (based on the use of Cisco wireless gear).

The load on an access point (the total number of potential clients) should be considered in any design. For design purposes, an 11 Mbps wireless network roughly equates to a 10 Mbps Ethernet network in terms of the number of users the wireless LAN can handle. One potential problem with wireless LANs is that the number of clients is very dynamic because the freedom of a wireless system allows any number of people to converge within an area. The actual number of clients is limited by a table within the access point; Cisco access points have a table that will theoretically allow 2,048 clients. Although this is the maximum, it is not practical. Keep in mind that wireless LANs are a shared infrastructure, and the more clients on the access point, the less overall available bandwidth there is for each individual user. Therefore, the distribution of the clients among more access points in congested areas may be required. The number of clients that can be handled by a single access point is a function of the applications supported, the data rate of the access point, and the desired performance of the application by the customer. The number of users per access point should not exceed 20 to 30 to maintain acceptable performance levels in the average environment. This number will of course be higher or lower depending on the applications, operating systems, type of use, and simultaneous usage of the users on the network.

Cisco access points give you the ability to load balance users across access points to further enhance performance in your wireless network. By placing two access points in the same coverage area you have the ability to "balance" the load on any single access point in a coverage area. This increases performance in your wireless system.

Expectations are everything in a wireless implementation. If the correct expectation is set and accepted, your wireless implementation will be a great success, and you will reap the benefits of that success. One of the most important expectations to set will be that of throughput in your wireless system. If your environment has, for example, FastEthernet already in place, users are accustomed to those speeds on their network and will notice the difference in performance from the wired 100 Mbps network to the wireless 11 Mbps network. If you set the expectation that throughput will not be at the accustomed speed, but will be more like a 10 Mbps network, usually there is no problem when the wireless network "goes live" in regards to performance. I like to compare speeds to T-1 wired circuits, especially where Internet access over wireless will take place. When a customer hears that they will have the same speed as their wired network Internet access and that their wireless connection equates to about six T-1s, the expectation is set and the client walks away feeling confident in their decision to go wireless.

# Antenna Options

By definition, an antenna is a conductive device used to transmit and/or receive radio waves. As mentioned earlier, antennas are passive devices and can be the simplest components in a wireless system. However, there is a tremendous amount of engineering and complex math that goes into designing antennas to meet certain needs.

Some antennas are designed to broadcast a signal in all directions, known as *omnidirectional* antennas; other antennas are designed to focus their beam in a specific direction, known as *directional* antennas. All of the antennas described in this section are used in wireless networking; however, each type is optimized in certain environments. Selecting the right antenna is crucial in designing a wireless system.

## Omnidirectional Antennas

Omnidirectional antennas propagate or receive signals in all directions. These types of antennas are useful in point-to-multipoint scenarios like a radio station, and for mobile devices that are constantly changing their aspect to their peer antenna.

### *Half-Wavelength Dipole (Half λ Dipole)*

The half-wavelength dipole antenna is one of the simplest antennas in design and construction. It consists of two conductors positioned end-to-end with a small gap between them. This gap is usually filled with a dielectric such as air, plastic, silicon, or rubber. The total length of the two conductors is one-half of the wavelength of the wave that they are designed to send or receive. If the antenna is designed for a range of frequencies such as the FM broadcast radio band, the length is usually half the wavelength of the center frequency of the range. Figure 2.17 illustrates a half-wavelength dipole antenna and demonstrates its omnidirectional propagation pattern.

**Figure 2.17** Dipole Antenna and Associated Omnidirectional Beam Pattern

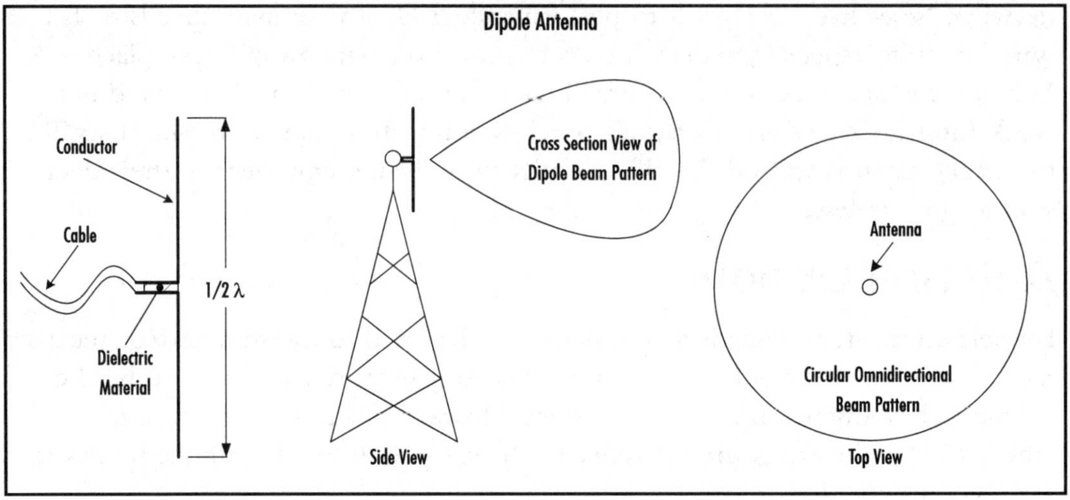


When the antenna is not near half the wavelength, the performance of the antenna is drastically affected due to a mismatch in characteristic impedances and standing wave ratios.

### *Quarter-Wavelength Monopole (Quarter λ Dipole)*

The quarter-wavelength dipole is a special version of the half-wavelength dipole. It consists of one side of a half-wavelength dipole mounted above a ground plane, such as the roof of a car. The ground plane acts as a reflector to simulate the second arm of a half-wavelength dipole. Quarter-wavelength dipoles do not have as high of a gain as half-wavelength dipoles, but they are close.

Quarter-wavelength dipoles are probably the most recognized form of antennas. They are found on almost all cars and "boom boxes" for AM and FM broadcast radio reception. They are also found on most handheld transceivers such as cellular phones, wireless phones, and two-way radios.

## Directional Antennas

Directional antennas can take the power coming from a transceiver and magnify the effect of the radiated signal by focusing most of the radiated power in one or two general directions. This focused radiated energy is referred to as a *beam*. Directional antennas fall into two general categories: *parabolic* and *phased array*.

Parabolic antennas function similar to a flashlight in that they use a dish-shaped reflector to concentrate the signal into a tight beam.

It is hard to visualize how phased arrays work because they involve very complex EM theory and complex mathematics. The concept behind phased array antennas is the same as the concept of multipath scattering and how waves can amplify or interfere with one another based on their phase. Multiple antennas work together to amplify some waves and cancel others so that the overall sum of radiated energy is in the form of a focused beam.

## *Yagi Array Antennas*

Yagi antennas are named after their inventor, Dr. Hidetsugu Yagi. Yagi antennas consist of three or more dipole antennas, called *elements*, mounted on a common boom. All of the elements work together as a phased array to direct the radiated energy into a focused beam. This gives Yagi antennas much higher gains than a half-wavelength dipole. Figure 2.18 illustrates the design and construction of a Yagi antenna.

**Figure 2.18** Yagi Phased Array Antenna and Associated Directional Beam Pattern

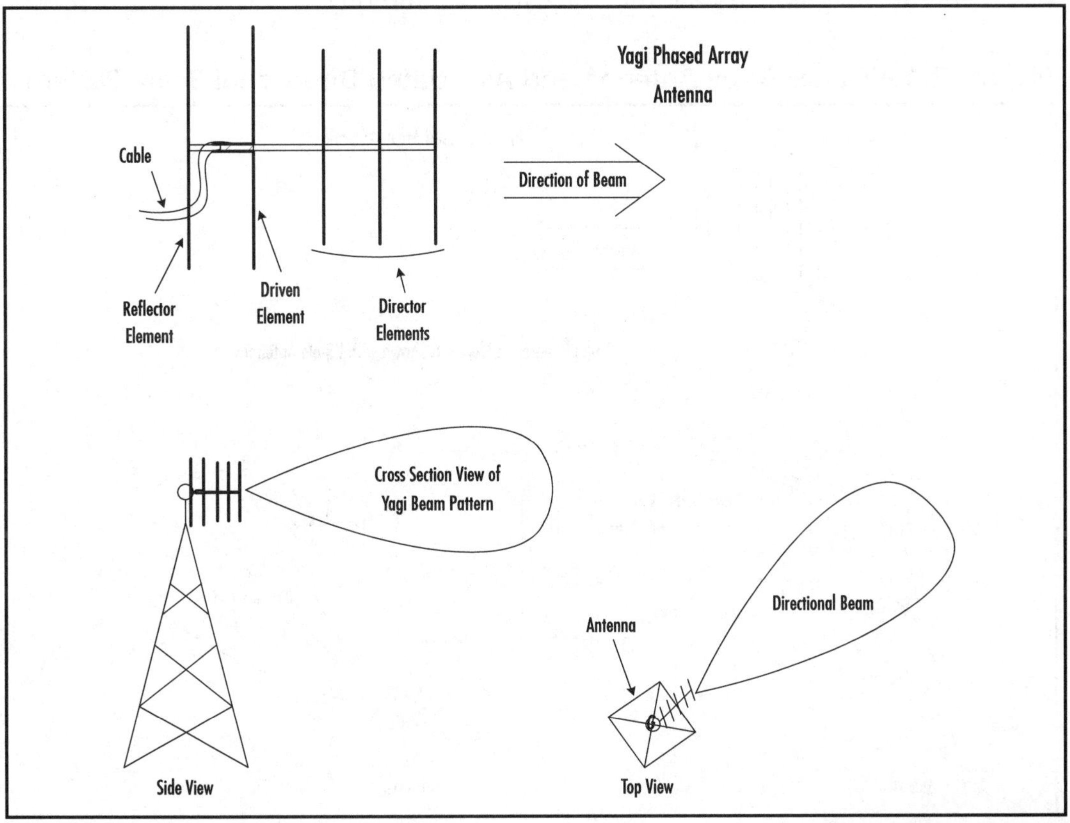

The elements are longest at the rear and gradually get shorter towards the front. The rear element is called the *reflector*. Immediately in front of the reflector is the *driven element*. In front of the driven element are one or more *director* elements.

The driven element is the only active element on a Yagi antenna and is the only element that connects to the transceiver via a cable. The remaining elements are known as parasitic elements because they feed off of the radiated power from the driven element. As described earlier, if a piece of metal receives a signal and it is not drained from the metal, it will be re-radiated from the metal. This is how the parasitic elements of a Yagi work.

Broadcast television antennas are examples of Yagi type antennas.

## *Planar Array Antennas*

Planar array antennas are similar in concept to Yagi antennas except all elements, both active and parasitic, lie in the same plane. This results in a flat antenna that can be mounted flat on a wall, yet still have the properties and gain of a directional antenna. Figure 2.19 shows a planar array antenna.

**Figure 2.19** Planar Array Antenna and Associated Directional Beam Pattern

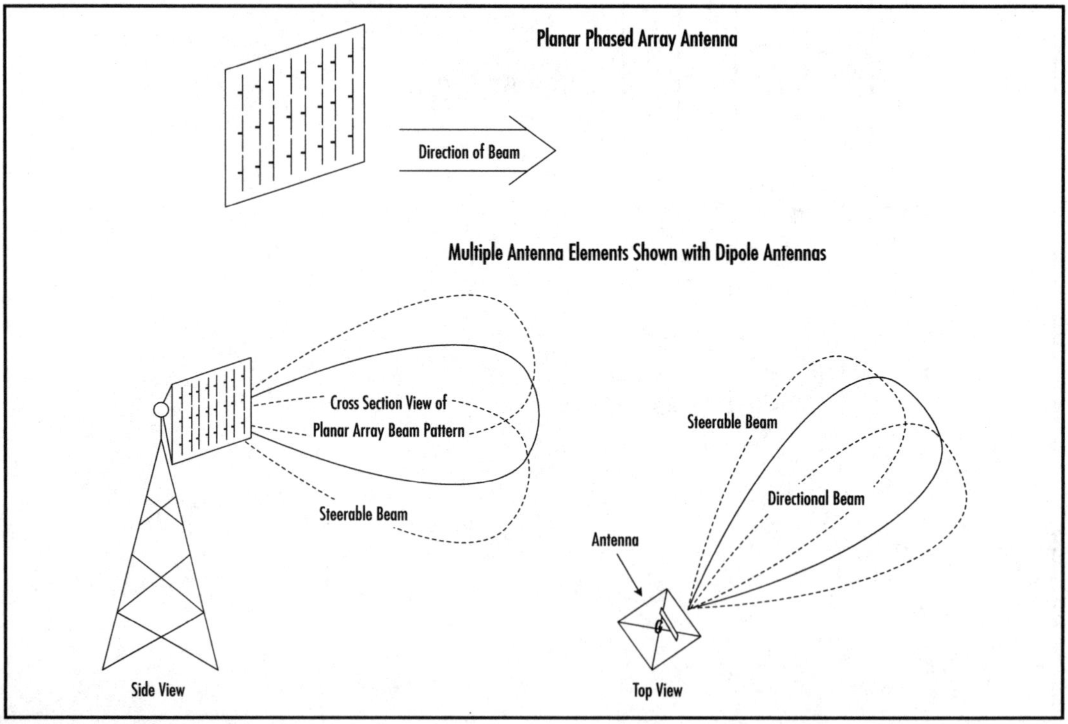

Planar arrays can have more than one active element. By changing the phase and power of the signal to hit specific active elements, the beam can be steered without the antenna physically moving. A useful application of this is on military tracking radars. A computer can adjust the input signals to the various elements of a planar array and steer the beam faster than the whole array could be moved physically, thus allowing for tracking of multiple fast-moving targets.

## *Sectorized Array Antennas*

Sectorized array antennas are a type of phased array antenna designed to split up a circular coverage area into sectors to help in channel allocation and reuse. Most sectorized antennas have a beam width of about 120 degrees that allows them to divide a circle into three sectors. Sectorized antennas are commonly used in wireless phone applications and can be seen on wireless phone towers all over the U.S. Figure 2.20 illustrates how sectorized antennas work to divide a coverage area.

**Figure 2.20** Sectorized Array Antenna and Illustration of Sectorization

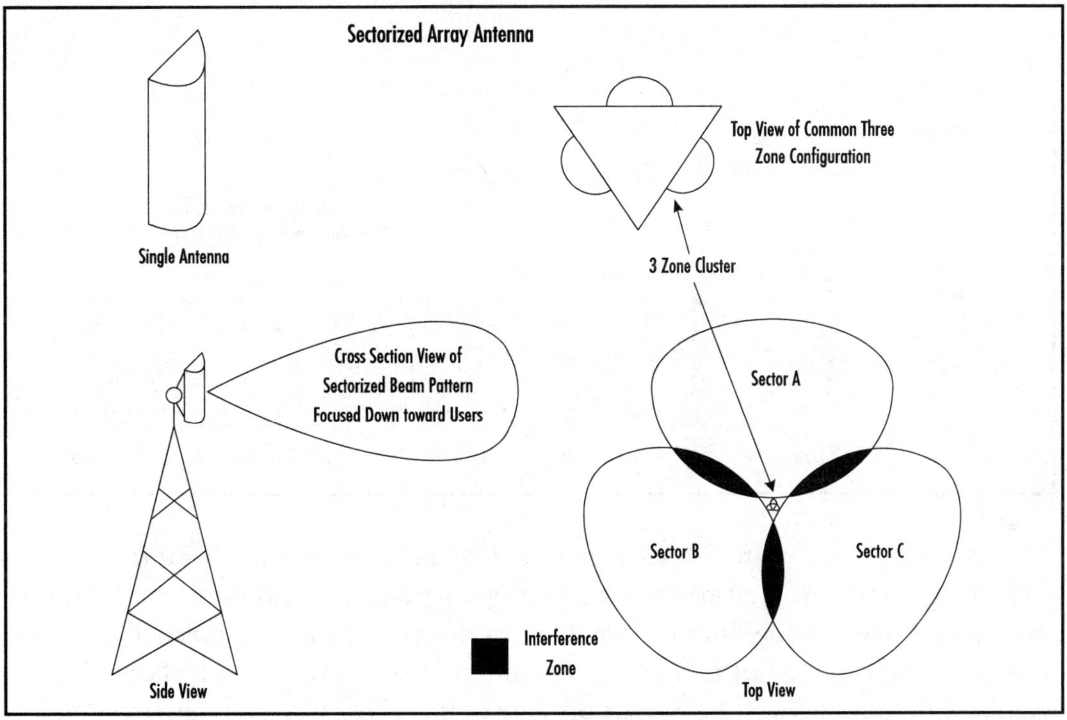

It is important to plan your zones carefully to minimize interference zones or to make interference zones reside in regions with no users.

## *Parabolic Antennas*

The most common example of parabolic antennas is satellite dishes. Parabolic antennas have an emitter that is mounted so that it is aimed into a bowl-shaped reflector. Just as in a common flashlight, the reflector acts to focus the signal from the emitter into a very tight beam. On the receiving end, the dish reflector increases the area of the antenna, collecting more of the transmitted signal and focusing that signal back onto the receiver. Figure 2.21 illustrates how parabolic antennas work.

**Figure 2.21** Parabolic Antenna and Focused Beam Pattern

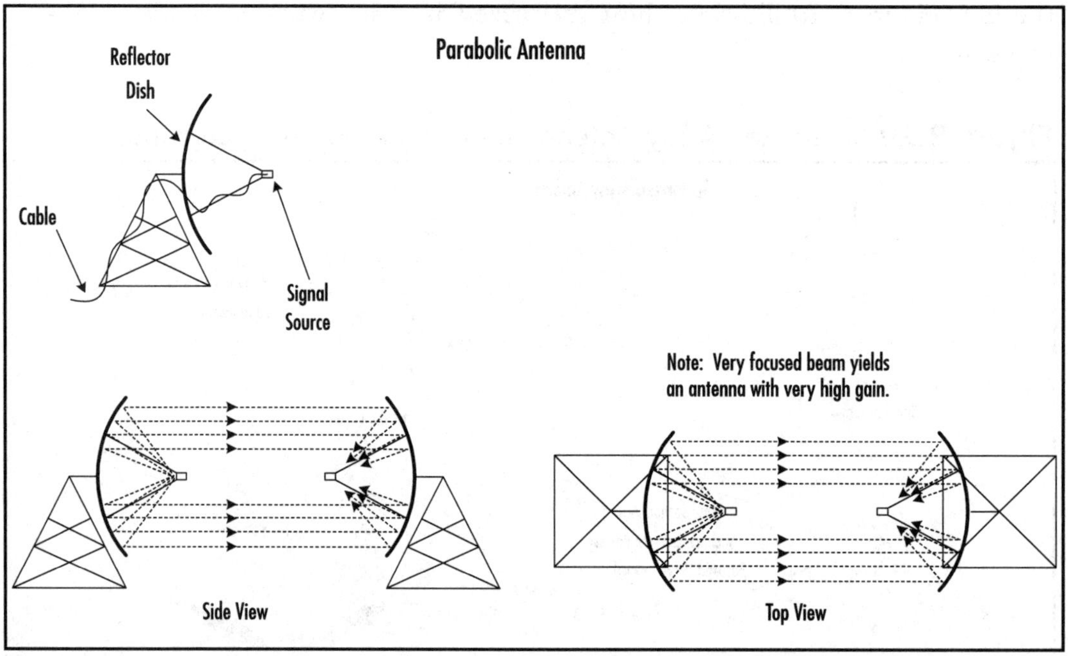


Parabolic antennas are used for terrestrial-to-stellar communication (ground-to-satellite) and for terrestrial-to-terrestrial point-to-point communication. Microwave long-distance telephone links use parabolic and cone antennas to carry phone conversations from one point to another. The number of microwave telephone links is rapidly diminishing with the advent of fiber-optic cables; however, terrestrial point-to-point links using parabolic antennas could see new life in creating cheap alternatives to leased lines for short enterprise network connections.

## Interference Detection

Interference detection is the process of identifying any possible interference with your RF deployment before and after the implementation of the WLAN. Detecting possible sources of interference prior to your RF deployment can help you in choose the correct WLAN deployment for your environment as well as help you eliminate these interference sources.

Other devices operating within the frequency range of the IEEE 802.11x standard that you are using are always considered possible sources of interference. If you are aware of any of these devices such as cordless phones, microwave ovens, and Bluetooth devices for IEEE 802.11b and 802.11g, it is best to eliminate them in advance or change to the IEEE 802.11a standard.

Whether or not you are aware of any devices operating in your RF frequency band, it is still a good idea to scan for them prior to your implementation. Scanning is covered briefly in the next section on conducting site surveys, but you should be aware of the methods available to do the scanning.

The first way to perform a scan for interference is to use a *spectrum analyzer*. This device allows you to scan for any RF activity in a particular frequency range or in all ranges supported by the device. There are many spectral analyzers on the market, but one common factor among them all is their high cost. These devices are very expensive and are not generally a good investment if you are not planning on performing scans on a regular basis across many sites. The good news is that they can usually be rented so that a purchase of a spectral analyzer is not a required part of your RF deployment.

The other method of scanning for interference makes use of the new Cisco wireless-aware LAN infrastructure. Using the software provided by Cisco in conjunction with Cisco wireless access points and client adapters, you can perform an *assisted site survey*. This type of site survey uses the access points and wireless client adapters to scan for interference in a range of the devices and report back to a central management console. This type of interference scan works very well but requires an up-front investment in the Cisco wireless hardware.

## Conducting Site Surveys

Wireless site surveys are critical to the successful implementation of a wireless LAN. There are as many ways to perform a wireless site survey as there are differing environments and businesses. As the person performing the survey, you must be able to creatively address all of the unique issues of the business and/or environment while staying within the best practices of the wireless technology. This is not always an easy task. As seen in previous sections, wireless may not be the answer to your business needs, or the cost to work around some business issue may outweigh the practicality of the installation. You have to be knowledgeable on the wireless equipment you are installing and the wired equipment with which you may be interfacing, the physical environment, the application considerations and the structural environment.

A site survey performed by you or a certified wireless professional, will verify the feasibility of the initial (rough) design in the face of obstacles such as wired connectivity limitations, radio hazards, and application requirements. The survey will help you determine the number of access points needed throughout a facility to provide the desired coverage, and it will determine the placement of those access points, detailing the necessary information for placement. Point-to-point surveys will also provide you with detailed information for placement of both the bridge units and antennas as well as determine feasibility of the link desired. With the information gathered from a site survey, a site survey report can be generated to assist you in "selling" the productivity and return-on-investment benefits of wireless LANs to the decision makers in your company. For those who already understand the benefits of wireless, the successful wireless site survey will allow you to properly install the wireless LAN and have efficient, reliable wireless access.

In your site survey, whether interior or exterior, try to identify potential problems up front and discuss how these issues will be handled. This can potentially save you a lot of time and trouble during the installation. You do not want to discover these issues during the installation or the "go live" period. A faulty or incomplete wireless site survey can be detrimental to your business for a very long time.

This chapter has covered many design considerations. Keeping those design considerations in mind, we now discuss what is needed to prepare and perform a site survey and the differences in performing exterior versus interior surveys. Best practices for surveying are integrated into this section, in addition to examples of creative approaches to specific issues.

## Preparation

Preparation for your wireless site survey will provide the important information needed to perform the survey and will guide you in your design process. In general, find out as much about the facility and business environment as you can before performing your survey. You will have enough to do without worrying about whether you are allowed in an area without escort or interviewing to find out what possible RF interference is in or around the coverage areas. To this end, the pre-site survey questionnaire was created. The following section contains a sample questionnaire form; as you use this form, you may find it useful to change or add sections to tailor it to your specific requirements.

## *Sample Pre-Site Survey Form*

Filling out a pre-site survey form helps your preparation prior to arrival at your site and helps ensure that you design a WLAN that meets your needs and requirements.

### Sample Pre-Site Survey Form

| | | | |
|---|---|---|---|
| Customer: | | Contact: | |
| Site Address: | | | |
| Department: | | E-mail: | |
| Phone: | | Fax: | |
| Mobile: | Pager: | | |
| • **Total Number of Sites (Please fill out one form for each site)** | | | |
| Site No.: | of | No. Buildings at site: | |
| Require 802.11 Compliance: ☐YES ☐NO | | Require Redundant WLAN?: ☐YES ☐NO | |
| Hours of Operation: | | Approximate sq. footage: | |
| Hours during which site survey may be performed: | | | |
| Total No. users: | Minimum | | Maximum |
| Packet Size: | Average | | Maximum |
| PC and/or Terminal type(s): | | | |
| • **RF Coverage** | | | |
| ☐ Complete Inside | ☐ Selective Inside (Provide diagram of facility indicating desired coverage) | | |
| ☐ Complete Outside | ☐ Selective Outside (Provide diagram of facility indicating coverage) | | |
| Using WEP Encryption: ☐YES ☐NO | | | |
| Desired Link Speed(s): ☐ 11Mb | ☐ 5.5Mb | ☐ 2Mb | ☐ 1Mb |
| Utilize Rate Shifting: ☐YES ☐NO (Please indicate where desired link speed will be needed) | | | |
| • **Existing Network** | | | |
| Existing Network Topology: ☐Ethernet | | ☐Token Ring | ☐Other |
| All sites use the same topology? ☐YES ☐NO | | | |
| Explain: | | | |
| | | | |
| Application type(s) to be used on WLAN: | | ☐Client/Server ☐Emulation | ☐Other |
| If emulation, what type? | | | |
| If Other, explain | | | |
| | | | |
| Protocol(s): | | | |
| | | | |
| Host Environment: | | | |
| | | | |
| WAN Connectivity: | | | |
| | | | |
| LAN Connectivity: (Check all that apply) | ☐10Mbps | ☐100Mbps | ☐Hubs |
| | ☐Switches | ☐Routers | ☐Bridges |
| Media: (Check all that apply) | ☐Coax | ☐Copper | ☐Fiber |
| | ☐Cat 5 | ☐Cat 7 | ☐Gigabit |
| Existing Wireless LAN? ☐YES | ☐NO | | |
| If Yes, explain: | | | |
| | | | |
| | | | |
| | | | |

## Sample Pre-Site Survey Form Continued

- **Site Information**

Ceiling Height: ☐8-10ft. ☐10-20ft ☐20-30ft. ☐30+ft.

Is lift available? ☐YES ☐NO If No, will customer provide? ☐YES ☐NO

In-Building Roof Access? ☐YES ☐NO

Racking/Shelving? ☐YES ☐NO If Yes, please describe construction:

Clearance above storage level: ☐<4ft. ☐4-8ft. ☐8+ft.

Any Hazardous areas? ☐YES (Please indicate Hazardous areas on diagram) ☐NO

If Yes, please describe:

Temperature range inside facility: Temperature range outside facility (if applicable):

Any Freezers? ☐YES (Please indicate freezer area(s) on diagram) ☐NO

Temperature (Check all that apply) ☐30°+ ☐15°-30° ☐0°-15° ☐-5°–10° ☐<-10°

Special Safety Rquirements: ☐Steel Toed Boots ☐Hard Hat ☐Safety Glasses ☐Other:

Other Non-WLAN RF Equipment installed at facility?
☐YES (Please indicate on diagram where installed) ☐NO

If Yes, please describe and provide frequency used for the device.
Device: Frequency:
Device: Frequency:
Device: Frequency:

Materials Stored at Site:

Do stock levels fluctuate? ☐YES ☐NO

Explain:

Current Stock Levels: ☐High ☐Average ☐Low

Any Conveyors? ☐YES (Please indicate location on site diagram) ☐NO

Sample Pre-Site Survey Form Continued

| | | |
|---|---|---|
| Available Power: | ☐110VAC | ☐220VAC |
| | | |
| Does site experience power problems? | ☐YES | ☐NO |
| | | |
| If Yes, please describe: | | |
| | | |
| | | |
| Site Construction (Please provide a brief description of each) | | |
| Building: | | |
| Ceiling: | | |
| Walls: | | |
| Floor: | | |
| | | |

# Other Preparations

In order to perform a successful and complete wireless site survey, the following items and/or services should be made available where applicable.

- In-building surveys require blueprints, CAD drawings, or if those are not available, fire escape drawings of the facility. These should detail the location of office spaces, cubicles, and equipment to scale.
- Point-to-point surveys require topographical maps of the area to include all the facilities involved in the survey.
- Provide a dedicated escort, if required, to allow full access to the facilities being surveyed.
- Provide facility identification or badges, if necessary for full access.
- Provide any facility guidelines or restrictions concerning equipment mounting.
- If the facility is a union facility, please provide a declaration of operating equipment limitations, if any.
- Provide information regarding asbestos construction, if applicable.

- If the facility has been designated a historical site, provide any limitations or specifications for modifying the interior or exterior of the building.
- Be prepared to provide an Occupational Safety and Health Administration (OSHA) representative where regulations prohibit non-OSHA certified personnel from access or installation.
- Be prepared to provide qualified personnel to access facility rooftops for the duration of any point-to-point survey.

If a customer or client fills out the pre-site survey instead of you, do not be surprised if many of the more technical questions go unanswered. Remember, your customer is not likely to know what Wired Equivalent Protocol (WEP), rate shifting, or what LAN or WAN is, so it will be difficult for them to answer those questions. Be prepared to explain to the customer, in layman's terms, what some of the various technical questions are asking for. You will be in a much better position to survey and produce a survey report if you already have these questions answered.

It is not uncommon for a client to request 11 Mbps coverage in all areas of the questionnaire. However, after the survey, when they understand how many access points are required to provide that coverage, the customer usually decides that 5.5 or 2 Mbps is sufficient for several of the areas. If you do not follow-up the questionnaire with a meeting or phone call to discuss the answers you received, you may find yourself spending additional time to resurvey your coverage areas for the lower speeds. By asking why your customer needs 11 Mbps coverage everywhere, for example, you can ensure that you will survey an area only once.

Depending on your local laws and regulations, you may have other inquiries that you will want to add to this questionnaire. In the U.S., such things as safety requirements and regulations may impede or hinder your site survey, so you should be aware of these regulations and plan accordingly. OSHA will sometimes require you, through the business, to complete paperwork verifying your understanding and compliance of specific or unique work safety requirements in a given area. An example is surveys performed in a biohazard or chemical manufacturing area, requiring body suits and/or masks.

During your installation, fire codes for office, manufacturing, healthcare, and other facilities will need to be adhered to. Prior to your survey, you will need to find out if the facility contains firewalls used in building construction. In building construction, a firewall keeps a fire from spreading from one building or part of a building to another. Firewalls are used to divide overall structures into allowable areas permitted by building code. Building codes vary by locality and country. Firewalls can be difficult to spot if you are not looking for them, thus they are on the list of questions to be answered prior to performing the survey. Because firewalls typically extend to the roof of most buildings, it is important to determine during the survey, whether you will need to penetrate a firewall for any reason. During the survey, it is unlikely you will have to penetrate the firewall, but you may discover this need for your implementation. In the event that you must penetrate a firewall, procedures exist in your locality for this. In the U.S., most of these procedures must be compliant with the National Electric Code (NEC). You can usually obtain copies of the NEC from local electrical suppliers.

In addition to firewalls, another typical fire code-related question commonly encountered is whether *plenum cable* is required for this facility. Plenum cable is cable that is coated with a fire-retardant coating (usually Teflon) so that in case of a fire, it does not give off toxic gasses and smoke as it burns. Twisted-pair and coaxial cable are made in plenum versions. In building construction, the plenum is the space that is used for air circulation in heating and air conditioning systems, typically between the structural ceiling and the suspended ceiling or under a raised floor. The plenum space is typically used to house the communication cables for the building's computer and telephone network(s). This impacts your installation and design only if you must have either your antenna cabling or your data cabling running through plenum areas of a facility to get to your access points and/or bridges, and the facility requires plenum cabling. Knowledge of plenum cabling requirements in your facility prior to the survey will help you determine how to cable your wireless gear for your design and implementation.

You do not want to cause undue strife by performing certain functions during the site survey that union personnel are required to perform, such as running a man-lift to get to the ceiling of a facility. The questionnaire will help you determine what coordination with other parties, if any, will be necessary to complete the survey.

Another preparation to consider is the need for permits to work in and on historical sites. Due to the nature of these sites, you must use extreme care to ensure that you make little or no changes to the facility or site. You should consider the differences required in surveying these sites as opposed to installing. Permission from one or more authorities may be needed to allow the installation, and even the survey, to occur.

Based on the preliminary information obtained from your questionnaire, you should be able to form one or more rough designs that may accomplish the desired coverage. You will use these rough designs to test during your site survey. The last section of your questionnaire contains some very important requests. The first of these is the request to have blueprints, CAD drawings, or fire escape drawings of the facility and any obstacles, such as offices, cubicles, and equipment, to be diagramed to scale. This is important for your preparation because without it you cannot create any rough designs for anticipated coverage. To do this while surveying takes a significant amount of time and will force you to start your testing with no predetermined designs to guide you through the creative aspect of the survey.

The other extremely important request in the last section of the questionnaire pertains to escorts, badges, IDs, and, in general, access to facilities or areas that you need to survey or areas you must go through to get to the survey areas. Verify that you are going to be surveying these areas by speaking to the manager or security company in charge of the survey areas ahead of time. If an escort is needed, confirm that they are at the location prior to going to the site. They may have called in sick or an emergency is preventing them from being there, in which case the area or facility may not have someone else to escort you. In the case of badges and IDs, this is usually a little easier provided that your company does not require "special" badges to get you into certain areas. If this is the case, you should request the appropriate badge(s) as far ahead of time as possible and confirm that the badge is ready before you go to the area or facility requiring it.

One more thought on preparation: Understand that surveys are typically less intrusive than the actual implementations of a wireless system. As you prepare to survey, always keep in mind what it will take to actually install your system and how this will affect production of services or goods in the particular environment. Is the business a 24/7 shop? Does the business have a weekly or monthly business cycle that is critical? When and where can you install the wireless system so that it will have minimal impact on the business? These are but a few of the keys to the success of your implementation.

## Infrastructure Awareness

In order to properly perform a site survey, you should make yourself aware of the environment you are surveying. What types of media and infrastructure devices are used in this environment? Will bleed-through of radio waves affect any sensitive equipment in the area you are in? What about floors above and below you in multi-floor facilities? Will you receive interference from or transmit interference to any devices? Is there adequate power and network connectivity for your access points? These questions are a good starting point for making yourself aware of the environment in which you are surveying and in which you intend to install a wireless system.

The next section describes the common infrastructure items you should be aware of while preparing and performing a wireless site survey.

## *What Types of Network Media are Used?*

Be aware of the media types that encompass the network. Most networks likely use some type of copper cabling, and many use fiber-optic cabling for backbone or longer runs. As you are surveying a facility and deciding on locations for your access points or bridges, you should also be looking for ways to connect them to the network.

The most frequently used cabling in today's networks are Category 5 (Cat5) or Category 5e (Cat5e) unshielded twisted pair (UTP). which consists of eight strands of solid copper wires, grouped into four pairs. Each pair is twisted, with a specific number of twists per inch, to create magnetic "shielding" when current is applied. This helps UTP to avoid cross-talk or interference from the other strands.

In Category 5e UTP cabling, the four already twisted pairs of cable (as in Cat5) are twisted together inside the cable sheath to provide additional (though not complete) shielding from outside interference. This is why it is called Category 5 enhanced. Cat5 and Cat5e are typically terminated with RJ-45 connectors or punched down to a patch panel or wall jack. In an Ethernet topology, Cat5 or Cat5e can be run a maximum of 100 meters or 328 feet.

The typical sheathing on UTP cabling is made of polyvinyl chloride (PVC). PVC can melt, smoke, and give off toxic fumes in the event of a fire. As discussed previously, the alternative to PVC sheathing is plenum cabling. Upon first observation, plenum is exactly the same as normal UTP, except the cable is much stiffer and harder to work with. In addition, the cable is marked with a code, for example "CMP," which indicates a plenum rated, unshielded cable.

The following are some common sheath (jacket) types and their ratings:

- **Teflon** Plenum rated
- **FR-PVC** Flame-retardant PVC
- **PVC** Non-plenum PVC

The following are some common copper wire types for UTP:

- **CM** Communications cables
- **CMP** Communications plenum cable
- **CMR** Communications riser cables that passes the UL1666 test
- **CM** Communications cables that passes the UL15812 test

When calculating cable runs for either your wired network connection or your antenna connection, always measure what the actual cable path will be. Do not measure a straight line from end to end because bends or turns in the cable path can add significantly to the cable length. In your design, this could mean cable runs that are too long for your topology causing loss of data or bad performance and will also throw off any estimates you may obtain from wiring contractors or unnecessarily increase the cost of the implementation if you are installing the wiring.

Do not run cabling at an angle; instead, try to run cabling as straight as possible using 90-degree turns where necessary to avoid obstacles. Avoid running cable on top of ballasts in fluorescent light fixtures, because this will wreak havoc by interfering in your data transmission through the cable.

Always calculate for service loops at both ends of your cable run. Service loops give you or your wiring contractor some "play" in the cable in the event the cable has to run around some unforeseen object or in case the cable must be terminated numerous times. Service loops are usually an additional 10 to 15 percent of the estimated cable length. For example, on a 200-foot run, an additional 20 feet to act as a service loop is typical. This would work out to 10 feet on each end of the cable.

If your design calls for fiber connectivity to the access points or bridges, you will require a media transceiver because Cisco radios only interface the wired media via an RJ-45 connection.

## *What Operating Systems, Protocols, and Drivers are Used?*

Different operating systems can use different protocols on the LAN. These protocols have different overhead and bandwidth requirements. Some are more efficient than others. Ask about the operating systems that are being used for the clients and servers and find out specifically what protocols are being used in the current LAN and which ones will be required to traverse the wireless LAN. The reason for this is to determine what protocols, if any, can be filtered from accessing the wireless LAN and thus increase performance.

Some drivers are not yet available or supported from Cisco. For example, Cisco provides drivers for Macintosh operating system 9.*x* only. In addition, Cisco PCI client adapters are not supported for use with Apple computers.

## *What Hubs are Used?*

Hubs may be 10, 100, or 10/100 hubs. The Cisco access points have 10/100 auto-sensing ports, and they will try to work on either port, but whenever possible you should try to connect via a 100 Mbps–capable port.

## *What Switches are Used?*

Access points communicate with each other on Layer 2 of the OSI model. Access points communicate with each other only if they are going to have clients roaming from one access point to another. In such a situation those two access points would need to be on the same broadcast domain. If for some reason two access points that are going to have clients roaming between them cannot be on the same broadcast domain, the Layer 3 devices must be configured to pass required packets for the access points to communicate with each other.

The client determines if it will change access points; the prerequisite is that the client has an encryption key for the new access point and a matching service set ID (SSID). The client uses three criteria to make this determination: signal strength, packet error rate, and access point load.

Switches have the capability for each port to be seen as a VLAN. VLANs may be grouped together to form larger VLANs. Switches are designed for wired networks with stationary users. They were not designed to handle mobile users. If the switch sees each port as a VLAN, and there are access points on each port, the switch is not set up to handle users moving from one VLAN to another.

Cisco access points are set up to work with these switch features. When a client roams from AP1 to AP2, AP2 sends a multicast packet with the source address of the roaming client. This packet is sent by the access point on behalf of the client and updates the switch's addressable memory. AP1 can then forward any packets that it has for the client to AP2.

Your application may not be set up to handle a switched network. The application may send out broadcast packets. If the client is connected to an access point that is not on the same virtual LAN as the server(s), the broadcast may never reach their destination(s)

A potential solution to this problem is to group the ports with the access points connected to them with the ports the host is using to form a VLAN. This may or may not work for you depending on requirements of the host to your wired clients.

Another solution is to network all of the access points to the same hub the host uses. Cable distance limitations may make this impossible. Still another solution may be to network all of your access points together via hubs and have them connect to the same hub the host uses. This is not a viable solution if the host is remote. This solution may present problems for some people. Under the 802.3 standard, when using a switch, you should not extend beyond two hops when using a 100 Mbps network. Although the wireless link between the client and the access point is not considered a hop, a remote host may be well past the two-hop limit.

The ideal solution for switching across VLANs connected to access points is to install a router or Layer 3 switch between the VLANs to perform routing, thus providing the ability to deny or allow the appropriate traffic between VLANs.

## *What Routers are Used?*

Routers present problems similar to switches in that they stop broadcast packets and may present a problem for the application or for clients trying to use dynamic host control protocol (DHCP). Static routes configured in the router may be necessary if the users on the wireless segment intend to use a remote host.

## *What Bridges are Used?*

Bridges can also present challenges because of their tables. Most bridges used today build dynamic tables. Some facilities may need to build their tables manually, sometimes by choice or sometimes because they are using older bridges. Because most IT personnel are not eager to work with these tables, you may need to configure them in order for wireless LAN applications to work properly, especially if they will be accessing a remote host.

## *How is Power Supplied?*

Cisco access points and bridges require power to function. The 340 series access points and bridges use traditional power inputs; the 350 series access points and bridges utilize inline power. Inline power consists of sending DC power over standard Category 5 UTP cable up to 100 meters. Instead of requiring wall power, access points and bridge devices can utilize power provided from Cisco line power-enabled devices, such as Cisco Catalyst Switches and line power patch panels. You can also use a line power injector, included with the Cisco 350 series access points and bridges, to provide the inline power required from a traditional wall outlet.

During the site survey, you will need to look for methods of power for your access point and/or bridge devices for the locations in your design requirements. Take note of any areas that do not have power available and consider how you will provide the power—traditionally or via inline power.

**NOTE**

**If your design requires the wireless network to stay up in the event of a power outage, inline power is much more efficient than traditional power because you can place an uninterruptible power supply (UPS) on the switch or line power patch panel that is providing the power to the access point. Otherwise, UPS will be necessary for every one of your access point or bridge devices as well as the switches or hubs they are connected to, to maintain connectivity to the network during the power outage.**

## Preparing a Site Survey Kit

A site survey kit contains all of the equipment necessary to evaluate, test, and record the possible wireless designs and their implementation ramifications for a given site. The following sections discuss the different types of equipment you should have to perform a site survey and when and where they are used, and also provides you with sources for some of the more difficult equipment to obtain.

Although not specifically listed in the following sections, note that you will need a device, preferably a portable device, capable of running a Cisco wireless client adapter and the Cisco client software and utilities in order to perform a wireless site survey. I use a lightweight laptop computer with plenty of battery life (I actually have two batteries), using Windows 2000 as my operating system, but you can equally use Linux, Windows 95, 98, or NT to run the client software.

Many of the sections list additional or optimal tools and equipment for a specific task. Your need for this additional or optimal equipment has much to do with your role in wireless site surveys. Differentiation is made between a one-time or limited site surveyor, for example, an information systems employee at a company intending to implement wireless for themselves, and a surveyor who intends to perform multiple surveys over time, as with a wireless consultant or engineer.

**NOTE**

Your site survey kit should definitely be a portable unit. If you intend to perform multiple surveys, as is the case with wireless engineers and consultants, invest in foam equipment cases to protect your survey equipment from damage due to weather and the various bumps and bangs of moving your equipment around.

Site surveys are often requested on very short notice, and the ability to carry the case in the trunk or backseat of a car, or check it as luggage allows you to travel with your equipment and solves the problem of your kit being lost or detained during shipping.

### *Using Client Adapters in the Survey*

Cisco recommends that you survey with the wireless network adapter you intend to use in your rough design. Therefore, if you intend to have desktop computers act as wireless workstations in your design, you should use the PCI or ISA client adapter to perform your site survey. This is not always practical, especially if the intended workstations are large and bulky and are not located where they will be located under the actual implementation. A combination of creativity and skill are required in this scenario. Different client adapters can have differing types of antenna connections, giving you a variable in the spread pattern of the antennas between different client adapters. Receive sensitivity, maximum transmit power, and typical indoor and outdoor ranges vary from the Cisco 340 series to 350 series client adapters producing several other variables to consider.

The best way to handle the desktop workstation scenario is to either provide or request from your client a cart that has a desktop workstation with the client adapter card you intend to use for the implementation. This gives you the flexibility to move about during your survey and eliminates from consideration many of the variables related to using a client adapter that is not intended for the design.

Designing & Planning...

### Surveying with LEAP as a Requirement

In order to conduct a proper wireless site survey that takes into account Cisco's Lightweight Extensible Authentication Protocol (LEAP) or standards-based EAP, you must have an authenticator capable of supporting it on the network you are surveying. If you perform or expect to perform multiple wireless site surveys, you should outfit your survey laptop with Windows 2000 Professional Server and install Internet Authentication Services (IAS) configured for Remote Authentication Dial-in User Service (RADIUS) and EAP to allow you to test authentication against your design. Typically, LEAP and EAP do not add significant performance degradation, however, you should survey with this configured and tested to ensure design functionality.

Following is a list of Cisco's requirements for LEAP or EAP with their equipment:

- The minimum Cisco client adapter firmware version required for LEAP support is 4.13.
- Cisco access point firmware release 11.00 is the minimum version required to support LEAP or EAP. Release 11.00 is the first version of firmware that enables the access point to be configured as an EAP or LEAP authenticator.
- The access point requires an EAP authenticator. The RADIUS server must support the type of authentication you are using (either EAP or LEAP).

## *Using Access Points and Bridges in the Survey*

Obviously the most critical component needed to perform a wireless site survey is the radio devices themselves. You typically need only one access point for interior surveys and two bridges for exterior surveys. Be sure to have your access point or bridge console cable with you to allow you to configure the devices directly. The console cable for the access point and bridge is a straight-through cable with 9-pin male to 9-pin female connectors. Previously, these console cables did not come with the access points or bridges, but they are now being supplied. If you prefer using the browser-based configuration tool over the console, set the IP address of your devices and laptop to a network other than the subnet you are on to allow you to move from subnet to subnet without reconfiguring your laptop and radio devices each time. For example, if the network in the facility you are surveying is 192.168.0.0, set your laptop and access points for network 172.16.0.0 to prevent conflicts and allow you to go to any subnetwork without reconfiguring your laptop and access point.

Some configuration has to take place on your access points or bridges prior to surveying. You should configure the devices exactly as they will be used in a potential implementation. If your design requires a wireless network that is optimally 11 Mbps with rate shifting to 5.5 Mbps, WEP encryption, LEAP authentication, and mandatory SSIDs, configure your access point with these parameters. All of these configuration parameters are explained in detail in Chapter 5.

You should always carry more than one access point with you when surveying. This allows you to continue to survey in the event your access point fails for any reason.

### *Choosing Antennas for the Survey*

There is no single antenna that is perfect for all wireless design applications. A variety of antennas are offered by Cisco because the variety of wireless design applications possible requires them. Your choice and placement of an antenna is in many cases dictated by your customer. Your customer may not want the antenna to be visible, or it may be located in a high traffic area requiring a low profile antenna. By carrying a variety of antennas, you will be prepared for any situation.

The minimum collection of antennas should include, but not be limited to, the following:

- 2.2 dBi "Rubber Duckies" These are the rubber antennas that come with your Cisco access point or bridge
- 5.2 dBi Ceiling Mount
- 5.2 dBi Mast Mount
- 5.2 dBi Ground Plane
- 5.2 dBi Diversity Pillar Mount
- 6.0 dBi Patch
- 8.5 dBi Patch
- 13.5 dBi Yagi

If you will be performing site surveys where you are aware that you will be using an antenna that is not in your kit, carry that antenna as well. Always survey with the antenna you intend to use. Do not use a different antenna and attempt to guess what the coverage will be. The reason you are performing the site survey is to take the guesswork out of the installation.

Antenna coverage is one of the most critical factors in a wireless system deployment because it applies directly to a client's ability to roam and communicate with the wired network. A large selection of antennas is required to handle a variety of potential networks from warehouses, retail floors, outdoors, and offices.

Although you choose among a variety of antennas, only two general classes exist:

- **Omni-directional** Provide a coverage pattern that is mostly circular and is usually used for indoor implementations. The signal is strongest at the center (nearest to the antenna) and gets weaker as the signal radiates outward. Mast mount antennas are examples of omni-directional antennas.
- **Directional** Frequently installed outdoors. The coverage area is similar to a triangle, and it gets weaker as the signal extends outward. The coverage area varies from antenna to antenna, and coverage can range from 12 to 65 degrees. Yagi and solid dish antennas are examples of directional antennas.

## *Providing Battery Packs and Inverters for the Survey*

When you perform your survey, you will need to provide power for your access points and/or bridges. You will not be able to count on the site having the appropriate power in the proper locations for every survey or part of a survey performed. Therefore, you need to provide some type of portable power. Because most wireless radios, including Cisco access points and bridges, utilize AC power, you will need, in addition to a battery pack, an inverter to convert the DC power of the battery pack to AC. You should ensure that the battery pack provides you with enough power for about eight hours or a days worth of surveying. It would not do to have to reschedule your survey because your battery packs lost power in the middle of your survey.

There are several approaches to providing power for your access points for the purposes of surveying. The most common approach is to purchase commercially available battery packs and inverters. The downside to this is that you will have two pieces of extra equipment to carry with you and hook up for each access point you are surveying with, in addition to the cost of the equipment. If you are performing a survey for your company, and you are fairly sure this will be the only wireless implementation for some time, you may want to just rent or lease several battery packs and inverters for the period of the survey.

Some commercially available battery packs have inverters built in to them, thus reducing the amount of equipment to carry. A company called Statpower produces a line of mobile battery packs with inverters built in to them, called xPower. The 21-amp-hour rated xPower300 will power one Cisco 350 series access point for well over 12 hours; a larger 40-amp-hour version on wheels is also available. Keep in mind that most inverters have only two outlets, so if you plan on powering several devices, you should purchase some inexpensive five- or six-outlet power strips to plug into the inverter outlets.

Remember, you are going to be carrying this equipment around, so weight also plays a factor in choosing your battery packs. The xPower300 battery pack/inverter combination weighs about 18 pounds. Most of this weight is due to the lead in the batteries themselves. The carrying handle, however, makes this relatively easy to move about.

Another alternative for portable power is to build your own battery pack and inverter combination. This requires good knowledge of electronics, but can be affordably done and can accommodate not only the power components but also the access point or bridge in one unit.

### *Providing Tools for the Survey*

There are many other tools you should have in your site survey kit to aid you in the successful completion of your wireless site survey. Some of the tools listed in this section should be considered "needs," whereas others can be considered "nice to have, or wants. This section provides you with the extra items you will need to perform your site survey and provide an explanation of their uses.

**NOTE**

In addition to the tools and equipment needed for your site survey kit, you need some specific tools for installation. Among these are a cordless drill with a drill bit set containing both hole saw bits and dry core bits to makes holes in masonry block and brick, in sizes up to two inches. These are needed to penetrate interior and exterior walls for running antenna cabling.

You will also require a caulk gun and clear silicone caulking to seal the area around the cabling on the exterior holes. I sometimes use expanding foam sealant for this.

You may also need to provide some quarter-inch plywood to serve as backing for wall mounting access points and bridges.

---

- **Graph Paper, Ruler, Pencil and Sticky Notes** These simple, inexpensive tools will probably be the most valuable tools you can have in your site survey kit. Even if your client has provided you with scaled diagrams of the survey area(s), you may still need to write down installation and/or design notes during the survey or draw an area to scale on your graph paper that is not on the client-provided diagram. Post-It or sticky notes are also invaluable when you need to make notes on a diagram that your customer may need back (and does not want changes written on it); this allows you to copy the diagram with your notes on it.
- **Markers** Once your access point, bridge, and/or antenna placement is determined in your site survey, you will need to mark their location for ease of installation. Location markers should be very bright, resistant to dust, grease, and water, and easy to remove when necessary. They should be sturdy but temporary.
- **Surveyor's Tape** This is probably the best solution for temporary markers. It comes in a variety of colors and is inexpensive. You can tie, tape, or pin it to just about any surface. I prefer the fluorescent colored surveyor's tape. It is much easier to see when marking locations, especially in areas that are not well lighted. I usually carry two colors of this tape at a minimum. One color is used to mark the location of the access point, or bridge and the other is used as an antenna placement marker.
- **Measuring Devices** In order for you or your customer to get accurate installation costs, you will have to provide many measurements in the site survey report. These measurements need to be as accurate as possible. If you guess the Cat5 run to be 300 feet, and it turns out to be 380 feet, the cost for this portion of the installation could be more than anticipated.

  Your kit should include a measuring wheel to allow you to accurately measure cable distances. You can use more advanced measurement devices, such as laser and ultrasonic range finders, but a measuring wheel will give you the measurement detail you will need.

Vertical measurements, such as floor to ceiling distances, can be best accomplished with a simple rope marked in ten-foot increments.

Note that counting structural features such as floor tiles, ceiling tiles, or cinder blocks typically does not give accurate measurements of distance. This is because much of the time these structural elements are cut or shortened to accommodate the site architecture and are therefore unreliable for measurement.

- **Ladders, Man-lifts, and Safety Harnesses** In wireless site surveys and installations, you will frequently need to gain access to ceilings and roofs of buildings. A ladder in most buildings will get you to the ceiling. However, in warehouses where the ceilings are typically very high (20 to 30 feet) and to reach rooftops that do not have interior access, you will need some type of powered equipment such as a forklift with a personnel basket or a man-lift to give you access to these areas. This equipment can be rented or your customer may already have this type of equipment for their facility.

  In addition, you should purchase a safety harness for working in these areas. Several different types of safety harnesses are available with differing levels of protection. A full body harness and some type of compatible lifeline or lanyard, preferably self-retracting to allow ease of movement, are recommended. For insurance reasons, some organizations do not permit this type of work by anyone other than their own employees.

- **Digital Camera** A digital camera is a very useful tool in your site survey kit. It will allow you to take pictures during your survey of the coverage areas, antenna, and radio device placements, and will allow you to insert these directly into your site survey report after the survey.

- **Laser Pointers** Laser pointers are used for point-to-point wireless site surveys to determine precise line of site. Green lasers are the best because they are easier to see than the red lasers. The highest power for a green laser allowed by U.S. law is 5 milliwatts. Lasers are especially handy for aligning antennas that are more than 1 mile away. The downside to using these lasers however, is that in direct sun they are very difficult to see. They work much better during overcast or cloudy days and very well at night.

- **Global Positioning System (GPS)** Another "nice to have" item is a GPS device. Although you can use the odometer in your car to get fairly accurate distances between antennas in a point-to-point survey, a GPS will give extremely accurate readings as well as altitude. A GPS can also aid you in determining vehicle speeds in a highly mobile wireless installation and survey.
- **Spectrum Analyzer** A spectrum analyzer as used in a wireless site survey is a device that allows you to monitor a specific portion of the RF spectrum to determine what interference, if any, is present in the band of frequencies you intend to use for your wireless implementation.

  Though considered by some to be a necessary component in a wireless site survey, I consider the spectrum analyzer as a component to use if you suspect interference from other sources, such as neighboring facilities.

  Spectrum analyzers are very expensive and can range in price from $5,000 to over $30,000. This a lot of money to invest for a one-time survey; instead, you can rent them from many companies.

### *Bringing Temporary Mounting Equipment for the Survey*

Your access points and/or bridges will need to be temporarily mounted in the survey area(s). Because you will move them frequently, you should take care to mount them as securely as possible without damaging the site. You may not be installing an access point or bridge in the location you are surveying, so you do not want to unnecessarily damage a drop ceiling or I-beam by drilling holes in them. To this end, you will need a variety of tools and equipment to "soft" mount the access points.

If you are a wireless consultant or engineer, carrying both access point and bridge mounts in your survey kit is advisable to allow you to get the best ideas, during you survey, for installation mounting of these devices. You should also carry antenna mounts for this same reason.

Both wireless consultants and one-time surveyors should also have in their survey kits various alternative mounting solutions for equipment. You must again be creative. Beam clamps, C-clamps, bar clamps, tie wraps, and Velcro are common components in a good site survey kit.

During a wireless site survey, there is no bad mounting technique with the exception of a mounting technique that does not properly secure the access point/bridge, battery pack, and antenna. For safety and prudence, you should definitely double- or triple-check the temporary mounting of all your equipment during the survey. This protects your equipment from possible damage, but also eliminates the risk of injury to you or others from falling access points and antennas. As an added incentive, failure to ensure secure mounting of equipment, at a minimum can result in a loss of confidence in you (the wireless professional). The following pieces of equipment are essential for safely mounting your equipment:

- **Tools and Miscellaneous Equipment** A good socket set and driver and bit set are invaluable for your site survey kit. These are used for another piece of recommended equipment: U-bolts. U-bolts are used to attach to antennas to aid in the temporary mounting that is required for the site survey. You should have various U-bolt sizes ranging from one and one-half inches to six inches to accommodate various antennas. Another good tool for mounting antennas is a modified camera tripod for mounting exterior antennas on roofs for exterior surveys.
- **Velcro** Velcro is a good choice for strapping an access point to a beam or post as long as the only weight the Velcro must bear is the radio device itself. I typically use this for access points and antennas that do not have low-loss cable connected to them. The cabling can add substantial weight to an access point or antenna hanging from an I-beam 30 feet in the air. Velcro is typically not strong enough to hold the additional weight. Also, replace your Velcro regularly, because it tends to wear out over time and heavy usage.
- **Tape and Other Adhesive-based Products** I typically try to avoid adhesive products of any sort when soft mounting equipment. It is frequently difficult to remove the adhesive "leftovers" when moving the equipment from place to place and solvents can remove paint and other finishes as well as the adhesive. The only adhesive I carry is a removable adhesive putty-like substance. This reusable adhesive has many brand names, but it is usually blue or green in color. It is typically used to attach pictures or posters to walls without damaging the wall finish. I use only this removable adhesive to attach my location markers during site surveys. This type of adhesive product does not leave glue residue when removing the markers.
- **Tie Wraps** Tie wraps are a good, strong alternative to Velcro for soft mounting your survey equipment. The only downside to tie wrap use is

that you normally must cut the tie wraps to free your equipment. This is not too much of an obstacle because they are relatively inexpensive, especially in bulk, if you plan to perform many surveys.

- **Clamps** My personal choice for soft mounting access points, bridges, and antenna are clamps. They are easy to use and reuse, do not wear out easily, and can be low cost. If you plan to perform surveys in many different environments, you will need a range of sizes to accommodate whatever type mounting structure you may encounter. Beam clamps and C-clamps are the most inexpensive, but they typically require both hands to manipulate when mounting your equipment. I prefer the grip action bar clamps that have become increasingly popular. You can generally hold your access point or antenna in one hand and secure the clamp around them using the other hand. Grip action bar clamps are very quick to install and remove, fasten securely to even the most difficult structures, and because they usually have foam rubber grips on them, they do not damage anything. It is usually a good idea to glue or tape the rubber grips onto the clamps to prevent the grips from sliding off.

# Performing an Interior Wireless Site Survey

There are specific methodologies for performing wireless site surveys, and these methodologies differ depending on the type of survey you need to perform. There are two main styles of surveys: the interior site survey, sometimes referred to as an in-building survey, and the exterior or point-to-point/point-to-multipoint survey.

You should complete certain steps regardless of the type of site survey you are going to perform. You should make sure your equipment is operational and preconfigured prior to arriving at the site. You should ensure that your battery packs and laptop batteries are all fully charged. If your customer is providing a man-lift (from the pre-site survey questionnaire), call to be sure it is already on site, available, and can reach the ceiling of the area you are about to survey.

The interior site survey requires you to understand cellular architecture, roaming, and rate shifting. The following sections first explain these interior site survey components and conclude with the actual interior survey method.

## Considering Rate Shifting

*Rate shifting* refers to the capability of the wireless client to negotiate the data rate at which it sends and receives at any given distance from the access point. This is also referred to as *auto rate negotiation.* As an example, a client negotiates the best speed of 11 Mbps while in close proximity to an access point. As the client moves away from the access point and the distance increases, the speed (rate) is renegotiated to allow for the best possible signal quality. These rates shift down from 11 Mbps to 5.5 Mbps to 2 Mbps to 1 Mbps if the access point is configured to allow this rate shifting function. Cisco access points give you the ability to specify the rates that they will "shift to." For example, you may configure your access point to only allow rate shifting from 11 Mbps to 5.5 Mbps. This will mean that a client roaming away from an access point will renegotiate its speed to 5.5 Mbps from 11 Mbps, but will lose its association (and therefore connectivity) to the access point if they roam out of range of the 5.5 Mbps coverage instead of shifting down to 2 Mbps.

## Performing the Interior Survey

In your interior wireless site survey, you will need to determine the coverage area produced by the access point/antenna combination you chose in your rough design and intend to use in your implementation. This is done by temporarily installing your access point and antenna, then using your Cisco Aironet Client Utility (ACU) application installed on a laptop computer with a wireless network card to verify the signal rate (11, 5.5, 2, or 1 Mbps) in the area. If your customer has specified that they require 11 Mbps coverage throughout the coverage area, you move your laptop to the point where the 11 Mbps rate drops to 5.5 Mbps. This point where the rate drops is the outer edge of your 11 Mbps coverage area and should be annotated on a scaled drawing of the room or area (possibly provided by the customer); if it has not been provided, you must draw this coverage on the graph paper in your site survey kit. Typically, you will start by placing your access point/antenna combination in the corner of the room or area (see Figure 2.22, A) and survey the coverage of that access point, making a note of where the furthest point of coverage is from that access point. You will then move the access point/antenna combination to the annotated point and survey the coverage again. If you were to leave your access point in the corner for an implementation, you would waste as much as 75 percent of your coverage cell radiating an area outside the building or an area on the other side of a wall that does not require coverage. You may need to move the access point several times in order to find the best placement and coverage pattern.

Once you have established this first coverage cell, move to another corner of the facility (see Figure 2.22, B) and repeat the process until you have surveyed the entire area (see Figure 2.22, C and D). In larger facilities, you may need to repeat these steps from the entire perimeter and/or center of the facility in order to fill in "gaps" in the coverage area. You must overlap your coverage cells in order to have seamless roaming in the area.

**Figure 2.22** Survey from Corners to Middle until You Achieve the Best Coverage Area

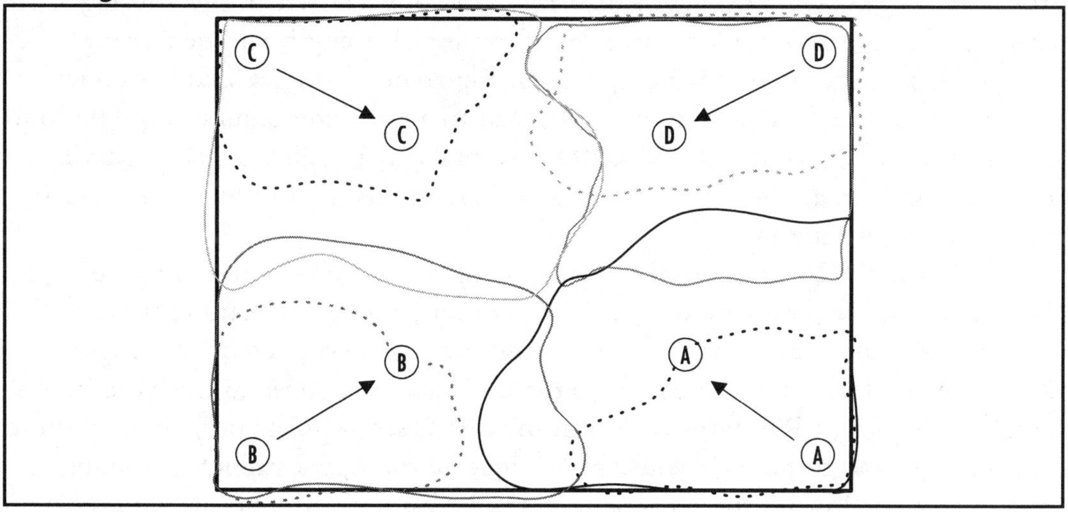


Once you have determined the best coverage for the entire area, it is extremely important for you to both annotate the access point and antenna placement on your drawing and place markers from your site survey kit at these locations. Be sure to photograph the area (with the markers in place) with your digital camera before proceeding to a new survey area.

In addition to the locations, you should take into account that only three non-overlapping channels are available and annotate the channel your access point is using in each cell of coverage. Be sure to change the channel to the channel you intend to use during the installation for each cell. For example, cell A in your rough design may be using channel 1. In the same room, cell B is supposed to use channel 6. Be sure to change your survey access point's channel to 6 before surveying cell B. This is done to ensure the accuracy of your site survey. During your survey, you may discover that you have too much overlap in some of your coverage cells. One or more access points may be providing too much coverage in an area, but without them there is too little coverage. At this point, you have a few choices. You can add more access points and use smaller antennas (lower dBi rated antennas), or you may elect to use the same number of access points, but increase the coverage by using larger antennas. Still another option is to change the power levels on one or more access points, thus changing the size of the coverage cells. You may have to use some combination of these options to properly achieve your coverage goals.

## *Using the Cisco ACU for Interior Site Surveys*

You may ask yourself, how will I determine the point at which my coverage ends during the survey? How can I tell if my signal strength and quality are acceptable in a given area? What speed am I currently using? Cisco has conveniently provided you with the tool you need to answer these questions. The tool is called the ACU, which allows you to measure signal quality, signal strength, rates, lost packets, and more. This section covers the use of the ACU in terms of your interior site survey.

Within the ACU are several screens giving you the ability to configure your wireless client for power modes, SSID, and other parameters. Among these screens is the site survey screen. You can use the site survey screen (see Figure 2.23) to help determine the best placement or coverage (overlap) for your access points. The current RF status is read from your Cisco wireless network card four times per second to provide you with a gauge of the signal strength, beacons, overall link quality, and the current access point association. The site survey screen also displays the IP address of the associated access point, the name of the access point, and the frequency channel the client is using to communicate with the access point. In addition, trends over time in graphic representations on the site survey screen indicate signal strength, beacons received, and link speed (rate).

**Figure 2.23** Passive Mode Site Survey ACU Screen

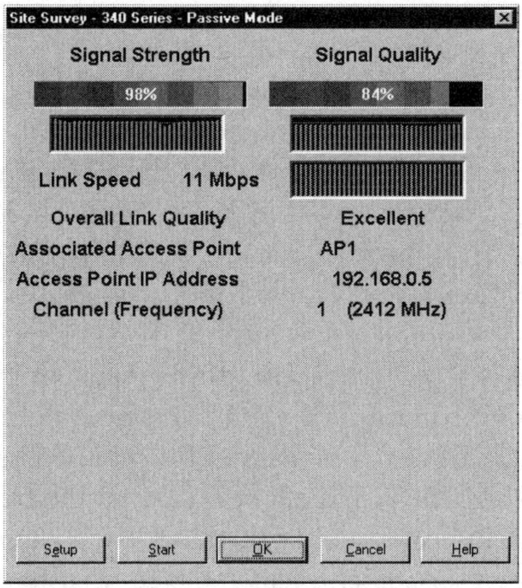


The site survey tool operates in two distinct modes: passive (the default) and active. The passive mode does not initiate any RF network traffic, it merely listens to any RF network traffic that the Cisco wireless network card hears. The active mode (see Figure 2.24) actively sends and/or receives packets to or from the associated access point, and updates the Percent Complete, Percent Successful, Lost To Target, Lost To Source, and Percent Retries accordingly.

**Figure 2.24** Active Mode Site Survey ACU Screen

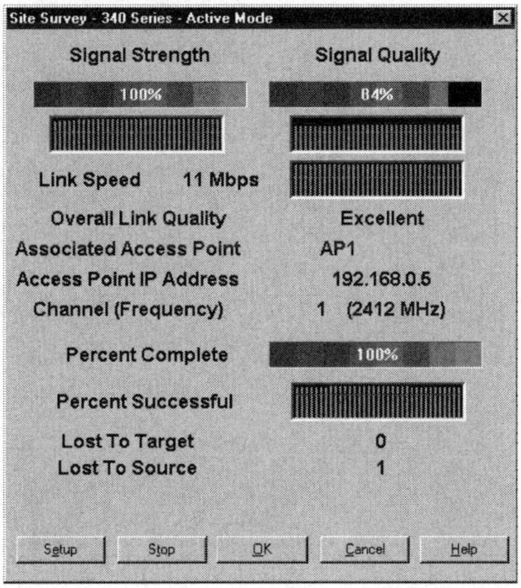

Lost To Target indicates the number of packets that were not received by the other device. Lost To Source indicates the number of packets that were lost on the way back to your wireless network card. Packets may be lost due to interference from other devices producing RF, because you are on the edge of the radio reception range, or due to multipath distortion.

To set up the Active Mode, click **Setup** at the bottom of the page. To start the Active Mode, click **Start**. To stop the Active Mode and return to Passive Mode, click **Stop** (the Start button changes to the Stop button while the Active Mode test is running). Otherwise, Active Mode will change back to Passive Mode once the Percent Complete has reached 100 percent.

Overall Link Quality is an indication of the ability of the Cisco wireless network card to successfully communicate with an access point. Ratings are Excellent, Good, Fair, and Poor. It is derived from the current signal strength and current signal quality. A result of Excellent indicates that both values are greater than 75 percent; Good indicates that both values are greater than 40 percent, but one (or both) is less than 75 percent; Fair indicates that both values are greater than 20 percent, but one (or both) is less than 40 percent; and finally, Poor indicates that one or both values is less than 20 percent. When I perform a site survey, I have found my overall link quality should be Good or Excellent for the implementation—anything less is unacceptable.

You also have the option to display the Signal Strength in dBm, the Signal Quality as Noise Level (in dBm), and the Overall Link Quality as the Signal to Noise Ratio. You can do this from the Preferences menu.

**Figure 2.25** Active Mode Setup Page

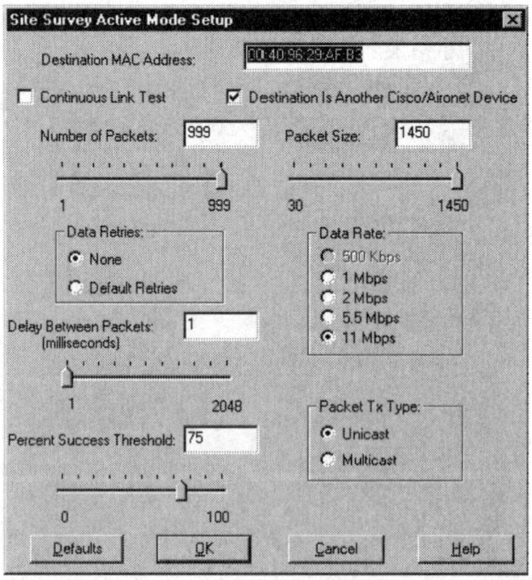

The Active Mode Site Survey Setup page (see Figure 2.25) allows you to set the parameters for the active mode. You can change the following parameters:

- **Destination MAC Address** This parameter allows you to select the access point, by MAC address, with which you will perform the active mode test. The default MAC address is the address of the access point that you are currently associated to via your wireless network card. The active mode test will not roam to other access points, allowing you to determine the size of a single cell. You should make sure the address in this field is the access point you are currently using at that moment in your survey and not another survey access point that just happens to be on in the area.
- **Number of Packets** Sets the quantity of packets that will be sent.
- **Continuous Link Test** Causes the active mode test to run repeatedly until you click **OK** or **Stop** on the Site Survey page. (The test will loop repeatedly for the number of packets that you specified.) You should choose to perform this test at least once per area you survey to provide you with continuous feedback of your RF link as you survey. It will give you information on expected loss over a period of time at any given survey point.
- **Packet Size** Sets the size of the packet to be sent. The packet size should be set to the packet size that can be expected to traverse this wireless segment when it is in production.
- **Data Retries** The number of times to retry a transmission if an acknowledgement (ACK) is not received from the destination.
- **Data Rates** Sets the bit rate at which the packet will be transmitted. No rate shifting will be performed.
- **Delay Between Packets** Sets the delay (in milliseconds) between successive transmissions.
- **Packet Tx Type** Unicast—expects an ACK back from the destination and retries can occur. Multicast—no packet retries.
- **Packet Success Threshold** The percentage of packets that are not lost. This parameter controls the red line on the "Percent Successful" histogram. Percentages greater than or equal to this value will show up as green bars; percentages below this value will show up as yellow bars on the Percent Successful histogram.

### *Watching Your Power Consumption*

When using wireless network cards, power consumption while surveying (roaming) is going to be an issue because devices within the laptop will use power, and the laptop battery has a limited life. Three modes for power are available on Cisco wireless network cards:

- **Constant Awake Mode (CAM)** CAM is best for devices when power is not an issue. This would be when AC power is available to the device, and it provides the best connectivity option, and therefore, the most available wireless infrastructure from the client perspective.
- **Power Save Mode (PSP)** Select PSP when power conservation is of the utmost importance. In this mode, the wireless network card will go to sleep after a period of inactivity and periodically wake to retrieve buffered data from the access point.
- **Fast Power Save Mode (FastPSP)** FastPSP is a combination of CAM and PSP. This is good for clients who switch between AC and DC power.

### *Setting Your SSIDs*

SSIDs are required for clients to communicate to access points. You can define three possible SSIDs on the client, although you can configure only one on the access point. The most common configuration within a corporation has the SSIDs the same for all access points. SSIDs act as a password, allowing the client to gain access into the infrastructure through the access points. The default SSID for the Cisco products is "tsunami"; it is preconfigured in the shipping product. By default, the access point is configured for "Allow Broadcast SSID to Associate =YES"; this means that clients do not have to have an SSID configured to associate to the access point. The recommended setting for SSIDs while surveying is to have the access point configured to "Allow Broadcast SSID to Associate = NO." If the access point and client SSIDs do not match, the association between the two will not happen, and access will not be granted. This will force you to configure your client with the matching SSID, but will give you a more accurate survey because most organizations require SSIDs in their wireless systems.

## *Interior Survey Problems*

Wireless site surveys are a process of trial and error. Experience is the best way to overcome many of these problems, but this may not be possible for the one-time surveyor. Most, if not all, problems encountered during your site survey are a result of unexpected design, business, or environmental issues. You may find yourself trying option after option to force a solution to a problem. This is where frustration sets in. You may find yourself working on a single thought process over and over because you do not want to start the survey over again. If you find yourself in this situation, take a break. Get a cup of coffee, go to lunch, and just get away from the problem for a bit, and more times than not you will find the solution to the problem presents itself upon your return. If it does not, you may need to wipe the slate clean and start your survey over. By starting over, you will be aware of the trouble spots in your survey and will be able to factor this knowledge in when planning the layout of your access points again. Starting the survey over again and designing the wireless LAN properly is always better than trying to force or use a solution that may not provide the best coverage and performance.

Sometimes, the location of your access points may be dictated by available network connectivity. For example, copper Ethernet cabling has a length limit of 328 feet. No matter what the problem you encounter, there is almost always a way around it.

Some business may want coverage in a large walk-in freezer in their facility. Of course they did not tell you this until you came out to perform your site survey. How will you handle this issue? The freezer is much too cold for the access point to be placed in it without expensive heated enclosures. Is this the only solution to the problem? You could mount the access point outside the freezer and install the antenna (which can withstand the cold) inside to provide the coverage required. You could even use antenna splitters to provide coverage both inside and outside the freezer for an even more cost-efficient solution to coverage. The only caveat for your antennas is to not use both antenna connections on your Cisco access point to provide this coverage (see Figure 2.26). Remember, when using antenna diversity the access point uses one antenna or the other, never both.

**Figure 2.26** Antenna Splitters

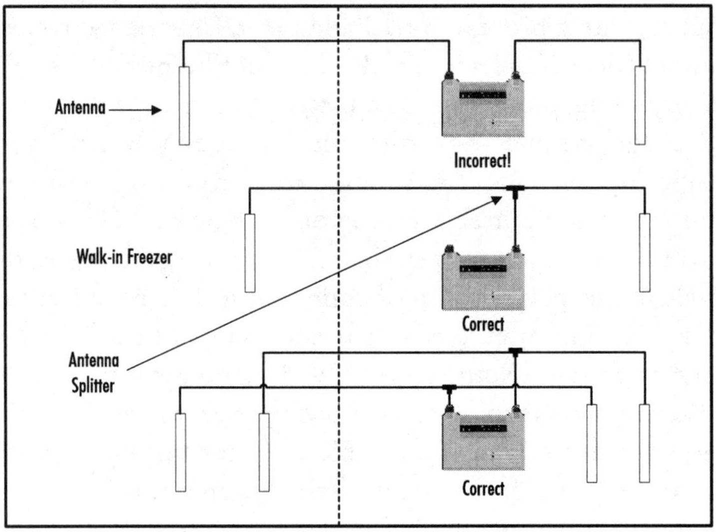


Take caution when surveying multifloor facilities because access points on different floors can cause as much interference as if they were located on the same floor. You can possibly use this bleed-through to your advantage in your design. During your site survey, you may be able to penetrate floors and ceilings with a single access point to provide coverage for floors above and below you by using a larger antenna.

# Performing an Exterior Wireless Site Survey

Exterior site surveys are very different from the interior surveys. This type of survey requires drastically fewer physical survey procedures but much more thought and analysis than interior surveys. Much of this has to do with the nature of point-to-point implementations. Unlike the interior site survey, exterior site surveys are performed without the use of the ACU. This is because clients will not directly associate with the bridges as they do with access points. The bridges simply link two or more wired LANs together to provide connectivity. However, specific steps and tools allow you to perform your survey.

Because you are usually attempting to transmit a signal over some distance, signal attenuation (loss) is of significant importance during your survey. Every component used in an exterior survey produces some loss. Cabling, connectors, splitters, the environment, and weather all affect the distance you may achieve during your site survey. When performing this type of survey, signal attenuation (loss) in your survey is of the highest importance, because the signal is not spread around as in interior wireless implementations and therefore is not as forgiving of interference and multipath issues.

Point-to-point and point-to-multipoint wireless implementations use bridges rather than access points to achieve wireless connectivity. These bridges are designed to interconnect two or more wired LANs, using narrow RF transmissions or beams over distances of up to 25 miles. As with the access points, the bridges require a wired connection to the network and either conventional power (340 Series) or inline power fed directly through the RJ-45 interface on the bridge (350 Series).

The steps for performing this type of site survey consists of the following:

1. **Link Distance Determination** You need to determine the distance of each site to be connected. This is the distance from a transmitting antenna to a receiving antenna. If these distances are long, you may wish to use the odometer in your car or a GPS to calculate this. You may already have gathered this information to create your rough design.
2. **Fresnel Zone Calculation** The next and most complicated step in the survey process, is to determine the radio line of sight for the wireless link. This is done by calculating the wireless link's Fresnel zone and possibly taking into account the curvature of the Earth (depending on link distances). If you already have your link distances and building or tower heights for your antennas, you can perform this step prior to arriving for the site survey.

Designing & Planning…

## Calculating the Fresnel Zone

Mathematics is required to calculate the size of the Fresnel zone radius at its widest point (midpoint radius). The following formula (see Figure 2.27) allows you to calculate the radius in feet of the widest point in your Fresnel zone:

**Figure 2.27** Fresnel Zone Radius Formula

$$R = 72.1 \sqrt{\frac{d_1 d_2}{F_{GHz}\,(d_1 + d_2)}}$$

Where:

- $d_1$ = the distance from the transmitting antenna (to the midpoint in the path)
- $d_2$ = the distance from the receiving antenna (to the midpoint in the path)
- F = the frequency in GHz
- R = the radius of the first Fresnel zone (at the midpoint)

For $d_1$ and $d_2$, it is usually much easier to determine the entire path length and divide that by 2 to get the distance for $d_1$ and $d_2$. Therefore, a 7-mile path length would have $d_1$ and $d_2$ values of 3.5 each.

You can also replace the 72.1 parameter with 43.3 (60 percent of 72.1) to give you your 60 percent clearance factor right off the bat, without having to calculate it later. Here is how I calculate my Fresnel zone. I have a total path distance of 7 miles. I divide my 7 miles by 2 to get my $d_1$ and $d_2$ values. In this case, it is 3.5. Multiplying 3.5 times 3.5 gives me 12.25. I then take my total path distance in miles (7) times 2.4 (my frequency in GHz) to produce a value of 16.8. I now divide 12.25 by 16.8 to get a value of .729. The square root of .729 is .854. I now multiply .854 times 72.1 to result in my midpoint Fresnel zone radius of 61.57 feet. I need 60 percent of this radius for a good link path, so I take

Continued

my radius value (61.57) times 0.6 to get my antenna height for this link, which is approximately 36 feet.

Once you have calculated the Fresnel zone's largest radius point, you must then determine what obstructions, if any, obstruct the ellipse more than 40 percent into the Fresnel zone (see Figure 2.28). If you have more than this percentage of path interference, you will experience transmission loss. Remember that the Fresnel zone should be clear of obstructions year round. Many surveyor's have been tripped up by trees in their Fresnel zone. When surveyed in the fall or winter, the trees did not have any leaves and therefore did not cause much, if any, interference with the signal. When summer arrives they have to return to correct the antenna height because the leaves in the trees in the Fresnel zone were so thick they caused noticeable interference. You should also consider a tree that may be very close to 40 percent of your zone, because it will grow and eventually hit this mark.

**Figure 2.28** Fresnel Zone Clearance

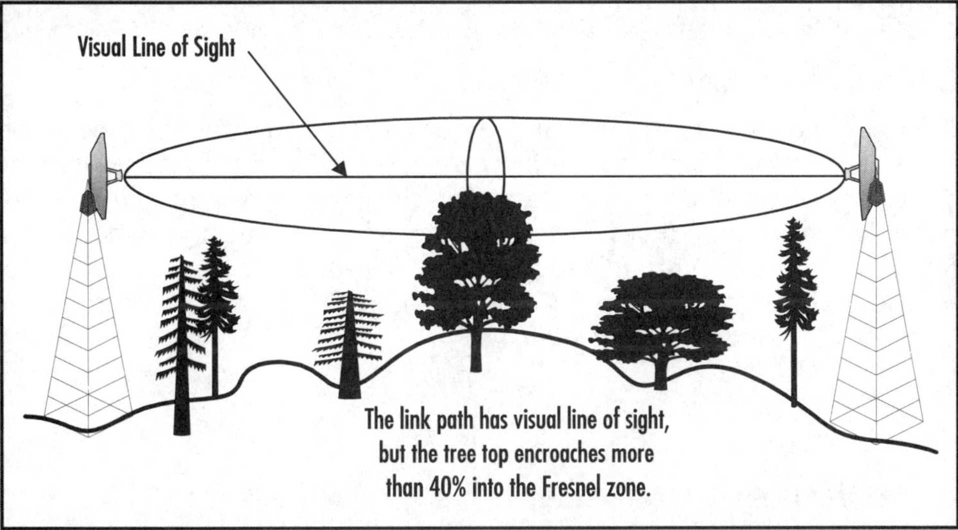


- **Link Setup and Testing** Once you have completed the previous tasks, you will set up your wireless link based on your rough design. This will entail determining antenna alignment, identifying cable lengths required, power requirements, wired connectivity for the bridge unit, and available infrastructure hardware (such as a switch for the wired connectivity to the network). The testing portion encompasses verification of connectivity and determination of optimal performance factors such as signal quality and strength and packet loss. You will use the tools built in to your Cisco bridge to acquire this information. Bridge tools and configurations for point-to-point and point-to-multipoint wireless implementations are covered in detail in Chapter 6.

- **Link Impairment Identification and Consideration** You will need to identify and consider all of the design obstacles covered earlier in the chapter (see the "Wireless Planning Considerations" section), with special attention paid to possible link impairments and taller-than-estimated trees, new building construction that may be in the path, and potential reflection points (such as flat, paved roads, pools, or other bodies of water). These considerations, if any, must then be corrected, and the solutions worked into your final design.

Designing & Planning…

## Calculating Antenna Height

Calculating antenna height simply requires you to determine the Fresnel zone radius and plan your implementation height so that your zone clears any obstructions by no less than 60 percent. When planning for paths longer than 7 miles, the curvature of the Earth might become a factor in path planning and require that the antenna be located higher off the ground. To calculate the additional height due to curvature of the Earth, use the following formula:

$$H = D^2/8$$

Where:

- H = Additional height of the antenna (in feet)
- D = Distance between antennas (in miles)

For example, we have already calculated our midpoint Fresnel zone and antenna height to be about 36 feet (for our 7-mile path). Now because we are 7 miles apart, we must consider the additional height required to compensate for the curvature of the Earth. By using the previous formula, we are able to determine that we will require approximately 6 additional feet to the antenna height to bring our total height requirement to approximately 42 feet.

# Summary

This chapter covered a lot of material on RF communications and wireless/wireless-aware LAN design. We started with the fundamentals and basics of RF communications. We talked about how radio signals are created and propagated as well as some of the problems with attenuation and interference.

It went on to cover some of the laws and regulations in place around wireless communications. and the roles of various regulating agencies and the rules in place for using low power, unlicensed transmitters. This chapter also covered some of the health risks associated with EM radiation.

This chapter next covered the IEEE 802.11x standards and how they relate to wireless communications. It discussed how the standards define the data-link and physical layers of the OSI model as well as how each of the IEEE 802.11x standards differ from the others.

Finally, planning for RF deployments was covered. This section included everything from selecting the appropriate antenna to conducting a site survey. It discussed the importance of performing site surveys and included a sample site survey form for you to use.

# Solutions Fast Track

## RF Basics

☑ Radio signals are basically the propagation of energy using EM waves.

☑ Modulating your radio signal allows signals at the same frequency to transmit simultaneously on different channels.

☑ To ensure a good wireless connection, your signal-to-noise ratio must be minimal.

☑ Attenuation is the weakening of a radio signal and can be caused by many natural phenomena.

## Laws, Regulations, and Environmental Considerations

☑ In the U.S., the FCC regulates all frequency spectrums for radio communication.

- ☑ Except for a few unlicensed bands, all radio transmitters require an FCC license.
- ☑ Low power, unlicensed transmitters can be used without a license, but there are several rules that must be adhered to.
- ☑ EM radiation can cause health problem in high doses.

## IEEE 802.11 Standards

- ☑ The IEEE 802.x standards have been developed as a common set of standards to design network communications.
- ☑ The IEEE 802.11x standard covers the data-link and physical layers of the OSI model.
- ☑ Each of the IEEE 802.11x standards has a specific number and frequency of channels that can be used.
- ☑ IEEE 802.11 defines the use of FHSS, 802.11a uses OFDM, and 802.11b uses DSSS.

## Planning for RF Deployment

- ☑ When planning your wireless coverage, always plan for overlapping areas of coverage.
- ☑ Carefully choose your 802.11x standard based on the needs for your environment and the benefits of each standard.
- ☑ Choosing the antennas that you will implement is very important.
- ☑ It is critical to perform a detailed site survey before beginning a wireless implementation.

# Frequently Asked Questions

The following Frequently Asked Questions, answered by the authors of this book, are designed to both measure your understanding of the concepts presented in this chapter and to assist you with real-life implementation of these concepts. To have your questions about this chapter answered by the author, browse to **www.syngress.com/solutions** and click on the **"Ask the Author"** form. You will also gain access to thousands of other FAQs at ITFAQnet.com.

**Q:** I am deploying a wireless-aware LAN in a hospital. Are there any special concerns I should have?

**A:** Installing any wireless communications device in a hospital is difficult. Not only do you need to be concerned about other devices interfering with your deployment, but you also have to ensure that your deployment does not interfere with the hospital's special equipment. Many pieces of test equipment can be rendered inoperable when a strong RF source is nearby.

**Q:** I want to create a wireless LAN that extends over a very large distance. Can I just boost the signal going to the antenna to increase the coverage range?

**A:** This can be done to a limited degree, but if you send too much power you will be breaking FCC regulations. Make sure you review the FCC regulations and any local ordinances to ensure that you comply with the law.

**Q:** If there is a wireless access point and antenna near a user's cubical, do they need to be notified of the health risk?

**A:** The risk is typically too low to warrant any unnecessary warnings or undue concern. If you are truly concerned about this being a problem, just approach the company's legal or Human Resources department through the appropriate channels and let them make the decision.

**Q:** I am going into the wireless LAN consulting business and am considering buying a spectrum analyzer. Is this a wise purchase?

**A:** If you are just starting out in the design business, probably not. It is usually best to rent the spectrum analyzer until you are sure that the purchase makes financial sense based on the amount that you use it.

**Q:** I am trying to decide what IEEE 802.11x standard I should use for my wireless deployment. The deployment is for a small office with no outside RF interference and they do not need a high data rate. Which should I use?

**A:** I would recommend using IEEE 802.11b for this type of installation. It is the lowest cost and will fit with the client's needs. If there were RF interference, IEEE 802.11a would be a better choice and if there was a need for high-speed access, IEEE 802.11g could be considered.

**Q:** I have finished designing my wireless-aware LAN on paper and am ready to implement. After I have the hardware in place, what should I do?

**A:** After the implementation is complete, go through and make sure that the wireless-aware LAN works in all necessary areas and that it provides for a good data rate. Also, perform another site survey so that you have a final overview of the environment for future reference.

# Chapter 3

# WLAN Roaming

## Solutions in this Chapter:

- Cisco L2 Roaming Solutions
- Cisco Solutions to Speed the L2 Roaming Process
- Cisco L3 Roaming Solutions
- WLAN Design Considerations

☑ Summary

☑ Solutions Fast Track

☑ Frequently Asked Questions

# Introduction

Unless you are setting up a WLAN in your home or very small business, your wireless network will include more than one AP. Because each AP provides RF coverage to a limited area, you will need many of them to provide complete wireless connectivity in the office building, airport, or warehouse. Even if your intent is to provide connectivity for users in only one particular area to create a "hotspot," deploying multiple APs configured for different RF channels will increase effective radio bandwidth and the number of simultaneous users your network can support. In the business environment, you can encounter WLANs with hundreds of APs. Examples of such WLANs include organizations that occupy an entire Manhattan skyscraper or have a large campus with multiple adjacent buildings. Usually if a company decides to deploy a WLAN, it opts for complete wireless connectivity throughout the organization's real estate to provide complete mobility for its workforce.

The process of a wireless client moving from one wireless cell to another wireless cell is called *roaming*. As we will learn later, the user might roam even if he or she does not physically move. If, after the association with the new AP, the client stays connected to the same IP subnet or virtual local area network (VLAN) as before, we call this *L2 roaming*. If after the roam the client ends up on a different IP subnet or VLAN, we call it *L3 roaming*. In a large, well-designed WLAN, the user's client device will usually be within the range of multiple APs and it will have to make a choice between them. This chapter discusses the behavior of the wireless clients when multiple APs are present on the network, design challenges that the mobile clients pose to the WLAN designer, and the software features that Cisco offers in its APs and client software to respond to these challenges.

At this point, the process and parameters associated with roaming are mostly not defined in the 802.11 standards developed by the Institute of Electrical and Electronics Engineers (IEEE), which leads standardization of the WLAN technologies. Each vendor has its own proprietary solutions to speed roaming. The exact algorithms and parameters used by Cisco devices are available only to development and support engineers who have access to source code and may change between different software versions and different hardware platforms. This chapter is an attempt to gather in one place all publicly available information about Cisco implementation of this process. This chapters' frame captures and analysis of the behavior of Cisco wireless devices were based on the Aironet

1231G APs with Internetwork Operation System (IOS) code version 12.2(13)JA3, Cisco Aironet 352 802.11b wireless client adapter with firmware version 5.30.17, Cisco Aironet CB21AG 802.11a/b/g wireless client adaptor with driver version 1.0.0.305, and Cisco Wireless Voice over Internet Protocol (WVoIP) phone 7920 with firmware version 3.3-01-06. Since we do not have access to the Cisco source code, some information in this chapter may be imprecise, but the reader should get an overall understanding of the roaming processes that take place on Cisco WLAN that are mostly invisible to the ordinary user.

# Cisco L2 Roaming Solutions

Let's start our discussion with L2 roaming because the L3 roaming process is a superset of L2 roaming. Figure 3.1 represents WLAN users who are moving between coverage cells provided by four APs (AP1 through AP4). All APs in this diagram are connected to two different L2 access switches (A and B) that are in turn aggregated by the distribution L2 Switch C. In this case, all APs are connected to the same VLAN X that is spanning all switches A, B, and C and that represents one IP subnet.

As the user travels to the unknown destination located on the right (probably to an important meeting in the conference room), he will roam multiple times—first between AP1 and AP2, then between AP2 and AP3, and finally between AP3 and AP4. After each roaming episode, the user, who keeps the same IP address, will stay within the same subnet boundary, so he does not have to worry about losing IP connectivity to the network. But the user's Media Access Control (MAC) address will move, first from the port 1 on Switch A to port 16 on Switch A and then to the ports 1 and 16 on Switch B. On the distribution Switch C, the user's MAC address will move from port 3/1 to port 3/2 (we assume Switch C is a modular switch). If nothing special is done to immediately update the forwarding databases on all these switches, the MAC layer frames from the router to the user may be misdirected.

**Figure 3.1** L2 Roaming Example

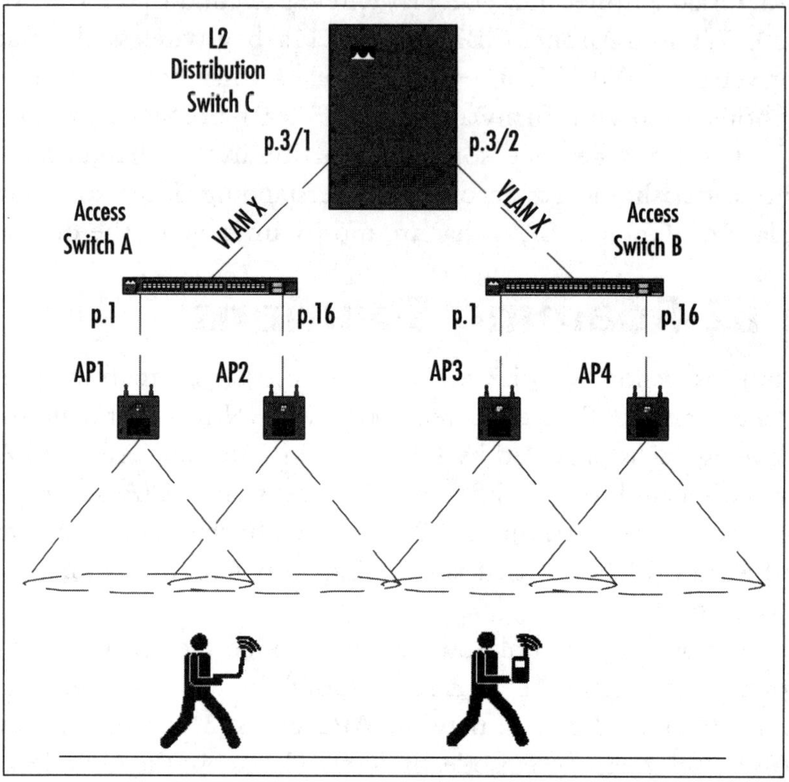


Cisco APs use the following solution to resolve this problem. After the user successfully associates and authenticates with the new AP, this device immediately sends out a multicast packet with the source MAC address of the client. This packet will update the forwarding databases (CAM table, in Cisco-speak) on all upstream switches. In addition, the new AP will send out a multicast packet using its own source MAC address to inform all APs on the VLAN that the client is now associated with. This will force all APs on this VLAN to update their association tables with the new information. These messages are part of the Cisco proprietary Inter-Access Point Protocol (IAPP). They also follow the recommendations of the recently ratified IEEE 802.11i standard that attempts to standardize the IAPP procedures from different vendors and provide interoperability that does not exist in this space today.

So far we have discussed what happens after the client roams. But why did the client decide to roam, and how did he decide where to roam? To understand that, we need to look inside a few different 802.11 management frames that APs and wireless clients use to communicate with each other.

# Beacon Frames

According to the IEEE 802.11 standard, every compliant AP periodically sends out management frames called *beacon frames*. The time interval between two consecutive beacon frames is called the *beacon interval*. The purpose of beacon frames is to advertise an AP's presence, its capabilities, and some configuration and security information to the client devices. Figure 3.2 shows a beacon frame from the Cisco AP as it can be seen on the WLAN Protocol Analyzer. (All captures in this chapter were produced using AiroPeek NX 2.0.2 software from WildPackets, Inc.).

**Figure 3.2** Cisco AP Beacon Frame

```
Packet Info
   Flags:                  0x00
   Status:                 0x01
   Packet Length:          134
   Timestamp:              16:47:29.972994000 XX/YY/ZZZZ
   Data Rate:              2    1.0 Mbps
   Channel:                1   2412 MHz
   Signal Level:           60%
   Signal dBm:             -53
   Noise Level:            0%
802.11 MAC Header
   Version:                0
   Type:                   00   Management
   Subtype:                1000   Beacon
Frame Control Flags:       %00000000
                            0... .... Non-strict order
                            .0.. .... WEP Not Enabled
                            ..0. .... No More Data
                            ...0 .... Power Management - active mode
                            .... 0... This is not a Re-Transmission
                            .... .0.. Last or Unfragmented Frame
                            .... ..0. Not an Exit from the Distribution System
                            .... ...0 Not to the Distribution System
```

**Continued**

**Figure 3.2** Cisco AP Beacon Frame

```
  Duration:              0  Microseconds
  Destination:           FF:FF:FF:FF:FF:FF  Ethernet Broadcast
  Source:                00:0F:23:D1:C9:70
  BSSID:                 00:0F:23:D1:C9:70
  Seq. Number:           1405
  Frag. Number:          0
802.11 Management - Beacon
  Timestamp:             2398413198  Microseconds
  Beacon Interval:       100
  Capability Info:       %0000000000110001
                           x....... ........ Reserved
                           .x...... ........ Reserved
                           ..0..... ........ DSSS-OFDM is Not Allowed
                           ...x.... ........ Reserved
                           ....0... ........ Robust Security Network
Disabled
                           .....0.. ........ G Mode Short Slot Time [20
microseconds]
                           ......x. ........ Reserved
                           .......x ........ Reserved
                           ........ 0....... Channel Agility Not Used
                           ........ .0...... PBCC Not Allowed
                           ........ ..1..... Short Preamble
                           ........ ...1.... Privacy Enabled
                           ........ ....0... CF Poll Not Requested
                           ........ .....0.. CF Not Pollable
                           ........ ......0. Not an IBSS Type Network
                           ........ .......1 ESS Type Network
SSID
  Element ID:            0  SSID
  Length:                8
  SSID:                  TestWLAN
Supported Rates
  Element ID:            1  Supported Rates
  Length:                8
```

Continued

**Figure 3.2** Cisco AP Beacon Frame

```
  Supported Rate:        1.0   (BSS Basic Rate)
  Supported Rate:        2.0   (BSS Basic Rate)
  Supported Rate:        5.5   (BSS Basic Rate)
  Supported Rate:        6.0   (Not BSS Basic Rate)
  Supported Rate:        9.0   (Not BSS Basic Rate)
  Supported Rate:        11.0  (BSS Basic Rate)
  Supported Rate:        12.0  (Not BSS Basic Rate)
  Supported Rate:        18.0  (Not BSS Basic Rate)
Direct Sequence Parameter Set
  Element ID:            3   Direct Sequence Parameter Set
  Length:                1
  Channel:               1
Traffic Indication Map
  Element ID:            5   Traffic Indication Map
  Length:                4
  DTIM Count:            0
  DTIM Period:           2
  Traffic Ind.:          0
  Bitmap Offset:         0
  Part Virt Bmap:        0x00
ERP Information
  Element ID:            42  ERP Information
  Length:                1
  ERP Flags:             %00000010
                          x... .... Reserved
                          .x.. .... Reserved
                          ..x. .... Reserved
                          ...x .... Reserved
                          .... x... Reserved
                          .... .0.. Not Barker Preamble Mode
                          .... ..1. Use Protection
                          .... ...0 Non-ERP Not Present
Extended Supported Rates
  Element ID:            50  Extended Supported Rates
  Length:                4
```

Continued

**Figure 3.2** Cisco AP Beacon Frame

```
  Supported Rate:        24.0  (Not BSS Basic Rate)
  Supported Rate:        36.0  (Not BSS Basic Rate)
  Supported Rate:        48.0  (Not BSS Basic Rate)
  Supported Rate:        54.0  (Not BSS Basic Rate)
Cisco Proprietary
  Element ID:            133  Cisco Proprietary
  Length:                30
  OUI:                   0x00-0x00-0x84
  Value:                 0x120700FF031100
  AP Name:               ap1............
  Number of clients:     1
  Value:                 0x000025
WPA
  Element ID:            221  WPA
  Length:                22
  WPA Value:
  .@........"...AT  00 40 96 04 00 0B 06 A5 00 00 22 A3 00 00 41 54
  ..aC..            00 00 61 43 00 00
FCS - Frame Check Sequence
  FCS:                   0x28C9FEBF
```

The beacon frame consists of the 802.11 MAC Header and multiple fields called *information elements* (IEs). Each of the IEs are numbered and contain sub-fields. Some of the IEs are standard; others are vendor proprietary. We will not go into detail here about every piece of information that can be present in the beacon frames, but instead we concentrate on the fields relevant to the roaming decision. (These fields on the figure are highlighted in boldface type).

The *Frame Info* portion is derived from the beacon frame by the client adapter software. As you can see, the client adapter software can measure RF signal strength of the received frame and the transmission bit rate. Cisco AP sends beacon frames at the lowest bit rate that is set to *require* on the AP Radio Interfaces | Radio X | Settings Web configuration screen (or to *basic* with the *speed* CLI configuration command for radio interface). If you lock your 802.11b AP into 11Mb/sec rate by setting 11Mb/s speed to *require/basic* and the rest of the speeds to *No*, the AP will start sending beacon frames at 11Mb/s. This will effectively make the AP coverage cell smaller and the borders of the cell sharper.

**WARNING**

With the current versions of IOS software [12.2(13)JA3 and below on 1200 AP with 802.11g radio], if, in addition to DSS rates, you configure any of the OFDM rates to *require/basic*, the interoperability with the 802.11b-only clients will be lost. You may arrive at this configuration while manually adjusting rates or by clicking the Best Throughput button on the AP Web GUI screen. This behavior looks like a bug that may be fixed in the future. If you are deploying 802.11g APs from any vendor and require connectivity for 802.11b-only clients, we highly recommend that you separately test connectivity for each type of these clients and do not take interoperability between 802.11b and 802.11g protocol implementations for granted.

MAC 802.11 Header shows that this is a *Management* type frame with Subtype *Beacon*. This frame is a L2 broadcast frame with the source MAC address of the AP radio interface 00:0F:23:D1:C9:70 and destination MAC address FF:FF:FF:FF:FF:FF. The Basic Service Set ID is also AP radio interface MAC address 00:0F:23:D1:C9:70. The next field shows that the beacon interval for this AP is set to 100 as a default value. Beacon interval is measured in time units (TUs), where each TU equals 1024 microseconds, so the default period between beacons is approximately 100 milliseconds. Beacon interval is a configurable parameter on the Cisco APs, but changing this value is not recommended without careful consideration. The topic can be found on the AP Web configuration page under Radio Interfaces | Radio X | Settings and is called *Beacon Period* (for no apparent reason) or it can be controlled using the CLI radio interface configuration command *beacon period*. Analysis of the *Capabilities* fields shows that this AP uses Wireless Equivalent Privacy (WEP) encryption of some sort for data communications (with the *Privacy* bit set to 1) and that this is really an AP ready to serve clients (with *Extended Service Set* bit set to 1).

Now let's look at the IEs that follow the MAC Header. The IE #0 contains SSID information for the AP. As anyone who ever used WLAN connectivity knows, that SSID is used as a label to identify a particular WLAN. As you can see, our WLAN has a very dull name, *TestWLAN*. As discussed in the chapter of this book devoted to VLANs, the Cisco APs can support up to 16 different SSIDs mapped to different VLANs, but only one of them can be advertised in the AP beacons. The SSID that is being advertised is an invitation to connect to the WLAN, and various operating systems, such as Windows XP, know how to take advantage of it by displaying the network information in the network configuration window and asking the user if he or she wants to connect. Different vendors call such SSIDs by different names. Cisco now calls it a *Guest Mode SSID,* whereas before the company called it *Broadcast SSID.* AP administrators can elect not to advertise any SSID by setting *Guest Mode SSID* under Security | SSID Manager | Global SSID Properties to *None,* and in this case the *SSID* field in the IE#0 will be blank (but the *Length* field will still be shown correctly).

**WARNING**

**Many people believe that disabling broadcast SSID is a great security measure. We know many companies where IT managers requested to shut WLANs down unless broadcast SSID is disabled immediately. As we will see from the frame captures that follow, WLAN SSIDs are always present in many 802.11 management and control frames, even if they are not directly advertised in the AP beacons. If there are users associated to the network, anyone with shareware wireless-sniffing software on a laptop and a little bit of time on their hands can easily observe these "hidden" SSIDs.**

The IE #2 and the IE #50, if present, list all transmission rates supported by the AP. This beacon was sent by an 802.11g AP with none of the rates disabled (set to *No* in the AP Radio Interfaces | Radio X | Settings Web configuration screen), so we can see that the AP advertises all of them. By analyzing this IE, the client can see the supported rates and can select the AP that supports the fastest rates.

The IE #133 is a Cisco proprietary IE. It is transmitted in the beacons of the Cisco APs if the Aironet Extensions found on the Radio Interfaces | Radio X | Settings Web configuration page are set to *enable* (with CLI use command *dot11 extensions aironet* under radio interface configuration). This parameter is enabled by default. The information transmitted in the IE#133 includes name and IP address of the AP, number of clients associated with the AP, AP power setting, bit error rate information, RF transmitter load, number of hops to the wired infrastructure, RF channel plan, and some other parameters.

As you can see, our protocol analyzer could not properly decode these fields beyond a few bytes of the AP name and the number of clients associated with our AP, but the Cisco client devices know how to read it if they receive it. The

Aironet Client Utility (ACU) that provides GUI interface to Cisco Aironet 350 Wireless Client Adapters will show you the name and IP address of the Cisco AP to which this adapter is associated. *Aironet extensions* provide support for various Cisco proprietary functions, including Message Integrity Check (MIC) and Temporal Key Integrity Protocol (TKIP), which improve WEP security, client power rate limiting, world mode support, and other functions. For our discussion here, it is important to know that by reading appropriate fields in IE#133, Cisco clients can extract additional valuable information about the current state of the AP and use it for roaming decisions. If you have Cisco wireless clients in your WLAN, this parameter should be enabled, even if you do not explicitly use any of the advanced features that rely on this parameter, because it helps Cisco clients improve their roaming capabilities. Non-Cisco clients cannot take advantage of this IE, but it is not supposed to harm them. In the rare cases in which non-Cisco clients are confused by this proprietary IE, you may want to disable it.

Currently work is under way to finalize the IEEE 802.11e standard for quality of service (QoS) for wireless LANs. Following the existing drafts of this standard, Cisco is migrating some of the proprietary information that describes the load of AP to the standard IE #11 (as we'll discuss later in this chapter when we talk about Cisco WVoIP phones).

## Probe Frames

Beacon frames provide a lot of information to wireless clients, but clients do not solely rely on them for association and roaming decisions, in part because the AP may chose not to advertise SSIDs. Instead, most wireless clients actively scan airwaves in search of the APs that can become potential roaming destinations. They do that by periodically broadcasting *probe-request* frames on all RF channels that they support (11 channels in the United States) and waiting for *probe-response* frames from the adjacent APs. Figure 3.3 shows a capture of the *probe-request* frame.

**Figure 3.3** *Probe-Request* **Frame from a Cisco CB21AG Client**

```
Packet Info
   Flags:                     0x00
   Status:                    0x01
   Packet Length:            54
   Timestamp:                16:23:36.143988800 XX/YY/ZZZZ
   Data Rate:                2     1.0 Mbps
   Channel:                  1   2412 MHz
```

**Continued**

## Figure 3.3 *Probe-Request* Frame from a Cisco CB21AG Client

```
  Signal Level:          81%
  Signal dBm:            -38
  Noise Level:           0%
802.11 MAC Header
  Version:               0
  Type:                  %00  Management
  Subtype:               %0100  Probe Request
Frame Control Flags:    %00000000
                          0... .... Non-strict order
                          .0.. .... WEP Not Enabled
                          ..0. .... No More Data
                          ...0 .... Power Management - active mode
                          .... 0... This is not a Re-Transmission
                          .... .0.. Last or Unfragmented Frame
                          .... ..0. Not an Exit from the Distribution System
                          .... ...0 Not to the Distribution System
  Duration:              0  Microseconds
  Destination:           FF:FF:FF:FF:FF:FF  Ethernet Broadcast
  Source:                00:40:96:A0:37:62  Aironet:A0:37:62
  BSSID:                 FF:FF:FF:FF:FF:FF  Ethernet Broadcast
  Seq. Number:           0
  Frag. Number:          0
802.11 Management - Probe Request
SSID
  Element ID:            0  SSID
  Length:                8
  SSID:                  TestWLAN
Supported Rates
  Element ID:            1  Supported Rates
  Length:                8
  Supported Rate:        1.0  (Not BSS Basic Rate)
  Supported Rate:        2.0  (Not BSS Basic Rate)
  Supported Rate:        5.5  (Not BSS Basic Rate)
  Supported Rate:        11.0  (Not BSS Basic Rate)
  Supported Rate:        6.0  (Not BSS Basic Rate)
```

**Continued**

**Figure 3.3** *Probe-Request* **Frame from a Cisco CB21AG Client**

```
  Supported Rate:          12.0   (Not BSS Basic Rate)
  Supported Rate:          24.0   (Not BSS Basic Rate)
  Supported Rate:          36.0   (Not BSS Basic Rate)
Extended Supported Rates
  Element ID:               50  Extended Supported Rates
  Length:                   4
  Supported Rate:          9.0   (Not BSS Basic Rate)
  Supported Rate:          18.0   (Not BSS Basic Rate)
  Supported Rate:          48.0   (Not BSS Basic Rate)
  Supported Rate:          54.0   (Not BSS Basic Rate)
FCS - Frame Check Sequence
  FCS:                      0xEF840C9F
```

As we can see, a *probe-request* frame contains basic information about the wireless client in the familiar format: the data rates it supports and the SSID it is looking for. The main purpose of these frames is to solicit a *probe-response* frame from the AP. An example of such a frame is shown in Figure 3.4.

**Figure 3.4** *Probe-Response* **Frame from the Cisco AP 1231G**

```
Packet Info
  Flags:                    0x00
  Status:                   0x01
  Packet Length:           128
  Timestamp:                16:23:36.145895800 XX/YY/ZZZZ
  Data Rate:               2    1.0 Mbps
  Channel:                 1  2412 MHz
  Signal Level:            60%
  Signal dBm:              -53
  Noise Level:             0%
802.11 MAC Header
  Version:                 0
  Type:                     %00   Management
  Subtype:                  %0101   Probe Response
Frame Control Flags:       %00000000
```

**Continued**

## Figure 3.4 *Probe-Response* Frame from the Cisco AP 1231G

```
                              0... .... Non-strict order
                              .0.. .... WEP Not Enabled
                              ..0. .... No More Data
                              ...0 .... Power Management - active mode
                              .... 0... This is not a Re-Transmission
                              .... .0.. Last or Unfragmented Frame
                              .... ..0. Not an Exit from the Distribution System
                              .... ...0 Not to the Distribution System
  Duration:                  314  Microseconds
  Destination:               00:40:96:A0:37:62  Aironet:A0:37:62
  Source:                    00:0F:23:D1:C9:70
  BSSID:                     00:0F:23:D1:C9:70
  Seq. Number:               1270
  Frag. Number:              0
802.11 Management - Probe Response
  Timestamp:                 964481477  Microseconds
  Beacon Interval:           100
  Capability Info:           %0000010000110001
                              x....... ........ Reserved
                              .x...... ........ Reserved
                              ..0..... ........ DSSS-OFDM is Not Allowed
                              ...x.... ........ Reserved
                              ....0... ........ Robust Security Network Disabled
                              .....1.. ........ G Mode Short Slot Time [9
microseconds]
                              ......x. ........ Reserved
                              .......x ........ Reserved
                              ........ 0....... Channel Agility Not Used
                              ........ .0...... PBCC Not Allowed
                              ........ ..1..... Short Preamble
                              ........ ...1.... Privacy Enabled
                              ........ ....0... CF Poll Not Requested
                              ........ .....0.. CF Not Pollable
                              ........ ......0. Not an IBSS Type Network
                              ........ .......1 ESS Type Network
```

**Continued**

**Figure 3.4** *Probe-Response* Frame from the Cisco AP 1231G

```
SSID
   Element ID:            0  SSID
   Length:                8
   SSID:                  TestWLAN
Supported Rates
   Element ID:            1  Supported Rates
   Length:                8
   Supported Rate:        1.0   (BSS Basic Rate)
   Supported Rate:        2.0   (BSS Basic Rate)
   Supported Rate:        5.5   (BSS Basic Rate)
   Supported Rate:        6.0   (Not BSS Basic Rate)
   Supported Rate:        9.0   (Not BSS Basic Rate)
   Supported Rate:        11.0   (BSS Basic Rate)
   Supported Rate:        12.0   (Not BSS Basic Rate)
   Supported Rate:        18.0   (Not BSS Basic Rate)
Direct Sequence Parameter Set
   Element ID:            3  Direct Sequence Parameter Set
   Length:                1
   Channel:               1
ERP Information
   Element ID:            42  ERP Information
   Length:                1
   ERP Flags:             %00000010
                           x... .... Reserved
                           .x.. .... Reserved
                           ..x. .... Reserved
                           ...x .... Reserved
                           .... x... Reserved
                           .... .0.. Not Barker Preamble Mode
                           .... ..1. Use Protection
                           .... ...0 Non-ERP Not Present
Extended Supported Rates
   Element ID:            50  Extended Supported Rates
   Length:                4
   Supported Rate:        24.0   (Not BSS Basic Rate)
```

Continued

**Figure 3.4** *Probe-Response* **Frame from the Cisco AP 1231G**

```
  Supported Rate:          36.0   (Not BSS Basic Rate)
  Supported Rate:          48.0   (Not BSS Basic Rate)
  Supported Rate:          54.0   (Not BSS Basic Rate)
Cisco Proprietary
  Element ID:              133  Cisco Proprietary
  Length:                  30
  OUI:                     0x00-0x00-0x84
  Value:                   0x120700FF031100
  AP Name:                 ap1.............
  Number of clients:      1
  Value:                   0x000025
WPA
  Element ID:              221  WPA
  Length:                  22
  WPA Value:
  .@........"...AT   00 40 96 04 00 0B 06 A5 00 00 22 A3 00 00 41 54
  ..aC..              00 00 61 43 00 00
FCS - Frame Check Sequence
  FCS:                     0x896A2406
```

The structure of a *probe-response* frame and information it contains is basically the same as in the beacon frame. The AP will not respond to the client if SSID in the *probe-request* frame does not match any SSIDs it supports. If the match is present, the AP will include the matching SSID in the *probe-response*. The valuable information that a wireless client can obtain from the *probe-response* frame is summarized in the following list:

- Beacon interval
- Receiving signal strength
- RF channel
- Supported data rates
- Whether WEP encryption is used or not
- SSID confirmation

Cisco clients can additionally obtain:

- Number of clients associated to the AP
- Power setting of the AP
- Transmission bit error rate
- RF transmitter load
- Number of hops to the wired backbone (to distinguish directly wired APs from APs in the repeater mode)

The wireless client will discard the beacons and *probe-response* frames that do not have matching SSIDs and WEP security settings. Based on the information contained in the beacons and *probe-response* frames with matching SSID and security settings, a client can build a list of the potential association targets (or roaming targets, if the client is already associated) and select the best target. Then it will go through the association and authentication procedures and will finally get connected (or reconnected) to the WLAN. If any of these processes fail, the client will try the next eligible target from the list.

Let's now look at the criteria and processes that Cisco wireless clients use to make roaming decisions and to select the best roaming targets.

# Roaming Decisions and Criteria

During the roaming process, the wireless client has to make two decisions:

- Do I need to roam?
- If yes, which potential target is the best AP?

For the client that undergoes the initial client startup, which is also considered a roaming event, the answer to the first question is always positive. But the clients that are already successfully associated with an AP will have to repeatedly go through both of these decisions. The following events will force a Cisco wireless client to make a decision to roam:

- Client missed eight consecutive beacons from the AP to which this client is currently associated. As we have discussed, the client receives information about the beacon interval in the beacon and *probe-response* frames, and it knows when to expect the next beacon.

- Data retry count exceeded. When the number of attempts to send a frame exceeds the value of the data retry counter, the wireless client will initiate a roam. In ACU version 6.2, this parameter is configurable under the RF Network configuration screen that is depicted in Figure 3.5. The default value for data retry count is 16, and it can be adjusted to any value between 1 and 128. It seems that either not many users find this parameter useful or this criteria is becoming obsolete, because in the current version of the ADU software that is used to configure the latest Aironet CB21AG card, this parameter is no longer accessible. Most probably this parameter should be set differently for 802.11a, b, and g protocols, so the numbers are probably now hardwired into the driver.

**Figure 3.5** Cisco ACU v.6.2 RF Network Configuration Screen

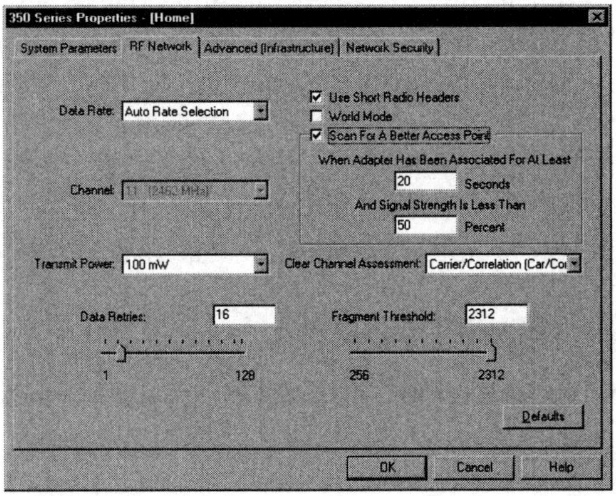


- The retransmit counter connected to the client's data transmission rate reached its predefined threshold. Normally the *Data* frames are transmitted at the highest rate supported by both AP and the client. If both support the same set of 802.11a/b/g protocols and the client's rate is set to *Auto Rate Selection* on the ACU RF Network configuration screen (see Figure 3.5), the possible transmission rates will be those that are set to *Required* or *Yes* on the AP Interface | Radio Interface X | Settings configuration screen. The client always starts communication with the highest common rate, and the retransmit counter is set to 0. Transmission rate will shift to the next-lower common rate if the frame has to be retransmitted three times with the Clear to Send/Ready to

Send (CTS/RTS) mechanism used during the last two retransmissions. If the transmissions at a lower rate were successful and did not involve retransmissions, the communicating parties will attempt to revert to the next higher rate. For every frame that has to be retransmitted at the lower rate, the retransmit counter is increased by 3. For every frame that was transmitted at the highest common rate, the retransmit counter is decreased by 1 until it reaches 0 again. If the retransmit counter reaches a threshold equal to 12, the client will attempt to roam to a different AP unless it has already tried to roam within the previous 30 seconds.

- The client performs periodic scans for a better AP. Previously described roaming decisions were associated with the detection of the transmission problems by the client. But even if there is no transmission problem, the client can just roam to a better AP if it discovers one. This behavior is configurable in the latest versions of the ACU that supports Aironet 350 client adapters (and is not configurable for other Cisco clients). By checking Scan for a Better AP on the ACU RF Configuration screen, you can enable this type of roaming. In ACU 6.2, you can set up additional criteria (see Figure 3.5) to prevent the client from attempting to roam too early after the initial association or too often. The time delay, which equals 20 seconds by default, is applicable only to the initial association; after the delay expires, the client will attempt to roam every second if a better target is found. The second parameter—minimal power threshold before the roaming is allowed—will prevent association "flapping" in case the client has two APs with strong RF signals in close proximity.

## Roaming Target Selection Process

Now that we know the possible reasons that the client decided to roam, let's look at how it chooses the best roaming target. Again, this discussion is applicable only to the Cisco wireless clients (or to Cisco compatible clients that support Aironet Extensions); the process is proprietary, and different vendors may choose to implement it differently. The following description is taken from the Application Note published by Cisco and sounds like a description of the programming algorithm that is implemented in the adapter firmware.

As you remember, as a result of the receiving *probe-response* frames, the client builds a list of potential roaming targets. To compare them to each other, the client needs to start somewhere, so we will introduce a variable called *Current AP* that is defined in one of the following ways:

- First AP in the list of the potential roaming targets, if this is the initial client startup
- AP to which the client is still associated (original AP) if the client is contemplating a roaming decision, provided that it responded to the last *probe-request* frame
- First AP in the list of potential roaming targets if the original AP did not respond to the *probe-request*

The client selects the first AP from the list of the remaining roaming targets and compares its parameters with the first list of criteria:

1. Absolute signal strength must be 20 percent or more if the signal strength of this AP is not less than 20 percent weaker than that of the current AP, or absolute signal strength must be 50 percent or more if the signal strength of this AP is less than 20 percent weaker than that of the Current AP.
2. If this AP is in repeater mode and has more hops to the backbone than the current AP, its signal strength should be at least 20 percent higher than that of the current AP.
3. The transmitter load of this AP is no more than 10 percent higher than that of the current AP.

The AP under consideration should satisfy all the criteria in points 1–3 to be eligible to become a roaming target. If it passes this test, its parameters are compared with the second list of more stringent criteria:

1. Signal strength of this AP is 20 percent higher than that of the current AP.
2. This AP has fewer hops to the backbone than the current AP.
3. This AP has four (or more) less currently associated clients than the current AP.
4. Transmitter load of this AP is 20 percent less than that of the current AP.

If the AP under consideration satisfies any one of these criteria, it will become a new current AP and it will be compared to the next AP from the list of roaming targets. The process ends when the list of roaming targets is exhausted. At this point, the current AP becomes the elected roaming target.

Analyzing this algorithm, we can see that if a really good target is available, it will be found. If no really good targets are available, the client will stay associated to the original AP, if it is still available. If it is not available, the client will associate to the first available AP on the list.

We can also see that the client could roam even if it does not move. The changing conditions of the WLAN, change of load on the adjacent APs, and even appearance or disappearance of the external interference can force the client to make a roaming decision. We can see too that this algorithm will provide some sort of client load balancing between multiple APs.

# Roaming Behavior of Cisco 7920 WVoIP Phones

The Cisco 7920 Wireless VoIP Phone is an 802.11b device that provides cordless phone functionality for the enterprises that deploy VoIP and Call Manager applications. This phone is presented in Figure 3.6. This wireless client device poses special challenges for WLAN designers because it has special RF coverage requirements and unique roaming characteristics.

**Figure 3.6** The Cisco 7920 Wireless Voice over IP Phone

Roaming characteristics of the 7920 phone are a very important factor that will greatly affect the user experience. The phone needs to keep a balance between roaming too often and roaming too late. The phone makes roaming decisions based on the following parameters:

- **Received Signal Strength Indicator (RSSI)** Equivalent of the *Current Signal Strength* parameter of the PC ACU/ADU applications.
- **Quality of Service Basic Service Set (QBSS)** A value describing the AP's current load characteristics.

The WVoIP phone derives QBSS values from the fields of the IE #11 that the AP advertises in its beacons. By default, the Cisco AP does not send this IE. You should enable this functionality by enabling the parameter *QoS Element for Wireless Phones* that can be found on the Services | QoS | Advanced screen. When using CLI, run the command *dot11 phone* in configuration mode. The QBSS IE, based on the 802.11e draft standard, contains three parameters describing current load characteristics of the AP:

1. Number of associated clients (which Cisco AP sends in its proprietary IE #133 anyway)
2. Channel utilization, a portion of the available bandwidth that is currently used to transmit data
3. Rate loss rate, a number of transmitted frames that required retransmission or were discarded as undeliverable

In addition to enabling IE #11 in beacons, this command also enables QoS functionality for Symbol Technologies' Netvision WVoIP phones by activating Symbol proprietary IE #173. Sniffer capture of these fields is presented in Figure 3.7. Please note that wireless sniffing software does not understand and cannot properly decode these proprietary fields.

**Figure 3.7** Part of the AP Beacon Frame with the *QoS Element for Wireless Phones* Enabled

```
SSID
   Element ID:              0  SSID
   Length:                  8
   SSID:                    TestWLAN
Supported Rates
   Element ID:              1  Supported Rates
   Length:                  8
```

**Continued**

**Figure 3.7** Part of the AP Beacon Frame with the *QoS Element for Wireless Phones* Enabled

```
  Supported Rate:          1.0   (BSS Basic Rate)
  Supported Rate:          2.0   (BSS Basic Rate)
  Supported Rate:          5.5   (BSS Basic Rate)
  Supported Rate:          6.0   (Not BSS Basic Rate)
  Supported Rate:          9.0   (Not BSS Basic Rate)
  Supported Rate:          11.0   (BSS Basic Rate)
  Supported Rate:          12.0   (Not BSS Basic Rate)
  Supported Rate:          18.0   (Not BSS Basic Rate)
Direct Sequence Parameter Set
  Element ID:               3   Direct Sequence Parameter Set
  Length:                   1
  Channel:                  1
Traffic Indication Map
  Element ID:               5   Traffic Indication Map
  Length:                   4
  DTIM Count:               0
  DTIM Period:              2
  Traffic Ind.:            0
  Bitmap Offset:            0
  Part Virt Bmap:           0x00
Reserved 11
  Element ID:               11   Reserved 11
  Length:                   4
  Value:                    0x01000001
ERP Information
  Element ID:               42   ERP Information
  Length:                   1
  ERP Flags:                %00000010
                             x... .... Reserved
                             .x.. .... Reserved
                             ..x. .... Reserved
                             ...x .... Reserved
                             .... x... Reserved
                             .... .0.. Not Barker Preamble Mode
```

Continued

**Figure 3.7** Part of the AP Beacon Frame with the *QoS Element for Wireless Phones* Enabled

```
                                   .... ..1. Use Protection
                                   .... ...0 Non-ERP Not Present
Extended Supported Rates
  Element ID:               50  Extended Supported Rates
  Length:                   4
  Supported Rate:           24.0  (Not BSS Basic Rate)
  Supported Rate:           36.0  (Not BSS Basic Rate)
  Supported Rate:           48.0  (Not BSS Basic Rate)
  Supported Rate:           54.0  (Not BSS Basic Rate)
Cisco Proprietary
  Element ID:               133  Cisco Proprietary
  Length:                   30
  OUI:                      0x00-0x00-0x84
  Value:                    0x120700FF031100
  AP Name:                  ap1.............
  Number of clients:        1
  Value:                    0x000025
Symbol Proprietary
  Element ID:               173  Symbol Proprietary
  Length:                   15
  OUI:                      0x00-0xA0-0xF8
  Number of clients:        256
  Load (kbps):              0
  Load (kpps):              0
  Tx power:                 7680
  ntp time:                 0
WPA
  Element ID:               221  WPA
  Length:                   22
  WPA Value:
  .@........"...AT  00 40 96 04 00 11 06 A5 00 00 22 A3 00 00 41 54
  ..aC..             00 00 61 43 00 00
FCS - Frame Check Sequence
  FCS:                      0xBB8F4D6B
```

During initial startup and successful association to the AP, the phone builds and later maintains the table of APs that can become potential roaming destinations. These destinations should have matching SSID and security configuration. The phone actively scans for APs by periodically sending *probe-request* packets on all allowed RF channels and processing *probe-response* packets from APs, as described previously. The phone constantly monitors the state of the RSSI and QBSS of the potential targets and will only consider roaming to the APs that have RSSI greater than 20 and QBSS less than 15.

**Figure 3.8** Sample Site Survey Screen of the Cisco 7920 WVoIP Phone

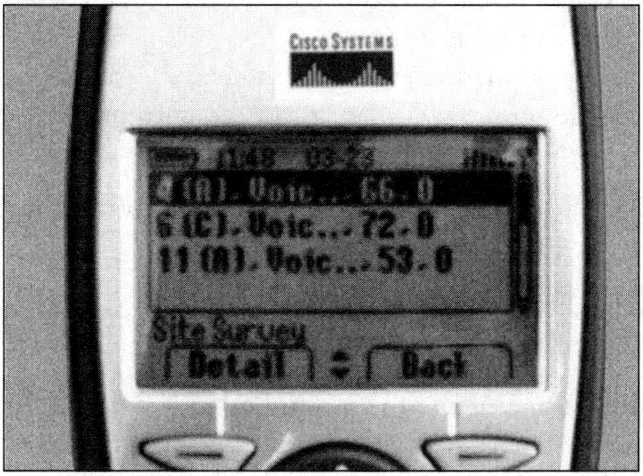


The phone user can see the list of potential roaming targets as well as their current values of RSSI and QBSS by selecting Menu | Network Configuration | Site Survey. An example Site Survey screen is shown in Figure 3.8.

The lines on the screen have the following format:

```
RF Channel, (Status), SSID, RSSI, QBSS
```

Status (c) denotes the AP to which this phone is associated.

The values that the phone shows are averages over a few seconds' interval that can be confirmed by shutting down the target AP. It may take up to 15 seconds to remove the potential target from the table. The phone also has no knowledge whether the target AP has proper connectivity to the wired infrastructure or can successfully communicate to the RADIUS servers.

The following two thresholds are part of the phone's roaming decisions:

- ***RSSI—differential threshold*** The difference in RSSI for two APs that will enable roaming.
- ***QBSS—differential threshold*** The difference in QBSS for two APs that will enable roaming.

*RSSI—differential threshold* and *RSSI—differential threshold* parameters are hard-wired into the 7920 firmware and cannot be seen or adjusted by the network administrator. Their values also depend on the version of firmware that the phone is using. For the current version of firmware 7920.3.3-01-06, both of these thresholds are set to value 15.

Now that we have defined relevant parameters, we can discuss thc 7920 roaming decisions. The phone will roam from the current AP to the potential target if:

1. More than three beacons were lost and there was no reply to the Unicast probe from the current AP
2. The *RSSI—differential threshold* was reached.
3. The *differential threshold* was reached.

Normally if the phone user walks around an area with uniform RF coverage, she will at some point start walking away from the AP where her phone is associated and approach a different RF. The RSSI value of her "home" AP would decrease and the RSSI value of the next AP would increase until the *RSSI—differential threshold* is reached and the phone will roam. If there is more than one roaming candidate, the phone will start with the AP with the next highest RSSI. The APs that advertise QBSS are considered better roaming targets than those that do not. If the client fails to associate to the first candidate, it will try the next one. According to Cisco, the average roaming times for 7920 phones are 100ms with static WEP keys and 200-400ms with LEAP authentication against local database, depending on the server load. During the roaming process, a certain amount of data will be lost and the voice quality may be affected for a short period of time.

### Designing & Planning…

#### Plan to Deploy a Local RADIUS Server

If Lightweight Extensible Authentication Protocol (LEAP) is used for user authentication on WVoIP phones, it is recommended that you use a locally installed RADIUS server and have the user database defined locally on that server. Queries to remote servers, especially located across the WAN, can increase authentication time during wireless roaming and make roaming time more unpredictable.

To minimize roaming times, the phone should always have roaming candidates in its table. This should be achieved by proper design of the RF coverage. The designs in which coverage of one AP abruptly ends and the coverage of another AP abruptly starts should be avoided. This situation may exist in buildings that have metal walls with metal doors or corridors with sharp turns and walls built with reinforced concrete.

### Configuring & Implementing…

#### Provide Smooth Roaming for WVoIP Phones

When designing wireless coverage for WVoIP phones, remember that these phones do not like sudden loss of coverage without having a new potential roaming target. In such cases, it may take the phone a few seconds to reestablish connectivity to the network. Make sure that the phone roamed to a new AP before coverage provided by the old AP is totally lost. If you have an obstacle that sharply breaks access to an AP before a new one is reachable, you may need to increase power level, rearrange existing units, or add a new one.

# Cisco Solutions to Speed the L2 Roaming Process

Starting with the AP IOS Release 12.2(11)JA, Cisco introduced two new proprietary software features that together make up the Fast Secure Roaming (FSR) paradigm. The first one is improved efficiency of channel scanning by the wireless client. The second one is fast client reauthentication using the Cisco Centralized Key Management (CCKM) process. All Cisco wireless clients except wireless bridges (which really better not roam anywhere!) can take advantage of these features, provided they run the proper version of software. For the Cisco Aironet 350 client adapter, the Installation Wizard 1.1 or later will provide support for FSR in conjunction with 1200 and 1100 Series APs. The same features should be available later on the AP 350 platform running IOS. In the future, the wireless clients certified under Cisco Compatible Extension Program v.2 should also be able to take advantage of the FSR.

## Improved Client Channel Scanning

As we discussed in the previous sections, Cisco wireless clients continuously scan all available RF channels in search of potential roaming targets. They do that by sending *probe-request* packets and receiving *probe-response* packets from the adjacent APs. The scanning involves switching radio to a new channel, possibly waiting for the available time slot, sending a probe, and waiting for responses. This process is not very efficient, since it consumes valuable transmit/receive time slots that could otherwise be available for sending and receiving data. The Cisco wireless client scans every channel for approximately 37ms, so it takes, for example, over 400ms to complete one full scan of all 11 channels available in the United States.

To improve roaming, Cisco introduced additional communication processes between its wireless clients and APs and changed the algorithm of client channel scanning with the objective to speed the channel-scanning process and the selection of the best roaming target. These goals were achieved by the following:

1. APs now build the lists of adjacent APs and their channels and communicate this information to the clients. Clients now have an option to scan only channels that may have potential roaming targets.
2. Clients may now elect a roaming target before discovering all adjacent APs that could be potential roaming targets (i.e., to complete a scanning cycle faster).

The following mechanism is used to build the list of adjacent APs:

1. On reassociation, Cisco clients now send additional information to the new AP about the old AP. This information includes the time since it lost its previous association, RF channel, and SSID. The new AP uses this information collected from all newly associated clients to build a list of the adjacent APs. If the client lost previous association more then 10 seconds ago, the information from this client is not included in the list (the old AP may be too far away). The AP can store information about up to 30 neighbors, and it will be aged out after 24 hours.
2. As part of the client reassociation process, the Cisco AP will now send the list of adjacent APs and its channels to the client. Analyzing the captures of the *association request* and *association response* frames between the Cisco wireless client and the AP running the latest software, we can see that these frames now contain Cisco proprietary IE #133 that is probably used for neighbor AP information exchange. The client will now use the adjacent AP list, depending on how busy the client is according to the following algorithm:

   - If the client is idle (did not receive a Unicast packet within the last 500ms), it will not use the information about the adjacent APs but will scan all available channels as usual.
   - If the client is busy (received at least one Unicast packet within the last 500ms), it will first scan only the channels listed in the list of adjacent APs. If no better APs are found (as per the algorithm described in the section, "Roaming Target Selection Process") the client will revert to scanning all available channels. The scanning will stop in 75ms if one (or more) better APs are found. This is called Fast Roam.
   - If the client is busy and is contributing nonzero percentage load to the cell where it is currently associated, it will execute a Very Fast Roam that is identical to the Fast Roam except the scanning will stop as soon as the first better AP is found.
   - The client also builds a local list of adjacent APs based on the results of its previous scans of all channels. If it needs to execute a Fast Roam or Very Fast Roam but it has never received the list of adjacent APs from its parent AP, it will use its local list to execute fast roaming.

This improved channel-scanning functionality requires no special configuration. It should increase effective throughput of the clients and will speed the overall roaming time, especially in cases when connectivity to the parent AP was suddenly degraded or lost.

# Fast Reauthentication Using CCKM

The L2 roaming process involves full reauthentication of the wireless client to the new target AP. If the WLAN uses so-called network authentication—a centralized RADIUS server that verifies credentials of the wireless clients—the authentication process can take between 200ms and 1.2sec. Exact time that may be required for network authentication depends on the specific authentication protocol in use, location of the RADIUS server and user database that holds user credentials, and the current load of the servers involved in this process.

As discussed in the chapter of this book devoted to WLAN security, most network authentication schemas that are currently in use rely on one of the variations of the Extensible Authentication Protocol (EAP) that is part of the IEEE 802.1x standard. The EAP paradigm includes the following entities: Supplicant, Authenticator, and Authentication Server. With EAP implementation of WLAN security, the Supplicant is a software entity that resides on the wireless client, the Authenticator resides on the AP, and the Authentication Server is a RADIUS server located somewhere on the network. When the AP is configured to accept EAP authentication, it will allow the client to associate but will block all data traffic until the authentication process is complete. The AP will then challenge the client to provide authentication credentials and will send them to the Authentication Server. Replies from the server will be forwarded back to the client.

The full network authentication process usually requires multiple messages that travel between the Supplicant and the Authentication Server through the Authenticator AP. As a result of these exchanges, the client with the right credentials will successfully authenticate to the AP and will be provided with the unique session key that will be used for the encryption of the data communication between the client and the AP. Details of this process depend on the specific protocol in use. Cisco currently supports at least five different flavors of these protocols: LEAP, PEAP, EAP-TLS, EAP-TTLS, and EAP-SIM. They differ in security of the communication channel between the Supplicant and the Authentication Server, type of authentication that is provided (one way vs. two way), and in other features. What is important here is that these mechanisms were not designed to provide quick reauthentication that is required by the roaming wireless clients.

To speed the network authentication process, Cisco introduced a new entity called Wireless Domain Services (WDS) that is connected to the local L2 subnet and acts as an intermediary between the Authenticator and the Authentication Server.

Currently the WDS entity is implemented inside the Cisco AP IOS software, but in the future it will be migrated to Cisco routers and switches that use more powerful processors. WDS functionality puts additional load on the AP processor, so Cisco recommends selecting an AP with a small potential number of clients. Cisco also recommends limiting the size of the WDS domain to 30 APs, although this is not a hard-and-fast number. To enable the service, one of the APs on the subnet should be configured as WDS AP. It is possible to configure a second AP as a WDS backup and to configure priorities to control which WDS AP will actually provide the service. The rest of the APs on the subnet function as WDS clients. They do not have hardwired IP addresses of the WDS APs; they discover them using L2 multicast messages over wired infrastructure. If the primary WDS AP fails, the standby WDS AP will become primary, but when the wireless client roams, it will initially have to go through the full authentication process.

The architecture of the WDS authentication is shown on Figure 3.9.

**Figure 3.9** WDS Authentication Architecture

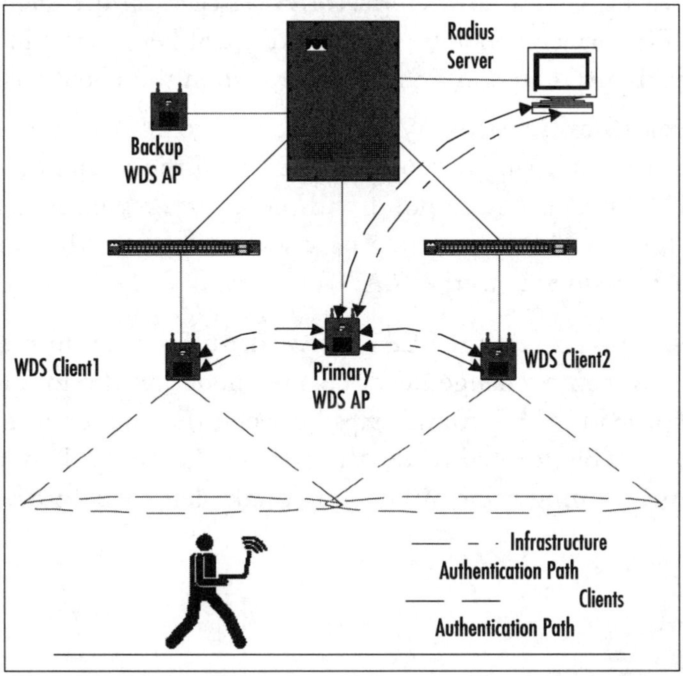

The WDS presence is transparent to the RADIUS server. The WDS functionality is closely aligned with the architecture and key hierarchy that follows the drafts of the 802.11i standard for wireless security. At this point, the fast reauthentication is supported for only Cisco LEAP. All APs on the subnet (including WDS APs) should be configured as LEAP clients for authentication to the RADIUS server through the WDS service.

In simple terms, the fast reauthentication process consists of three stages:

- **Infrastructure authentication** At this stage, all APs acting as Supplicants authenticate to the RADIUS server as LEAP clients through the WDS AP, which acts as an Authenticator. The WDS caches AP encryption keys that are later used to securely distribute additional key material to the APs.
- **Initial client authentication** This process happens when the LEAP wireless client authenticates to the network the very first time. For the client, this process takes as much time as a regular LEAP authentication. After its credentials are verified, the client receives a session key that it uses to encrypt Unicast traffic and a group session key that is used to encrypt broadcast and multicast traffic. The communication process between the client and the RADIUS server again goes through the WDS that acts as an Authenticator. After caching the client's session key, the WDS service generates a few additional keys that will later be used to quickly generate a new session key when the client roams.
- **Fast reauthentication** When the client roams to a new AP that is part of the WDS domain, the new AP will send the authentication request to the WDS, which will respond with necessary keying material that will allow the new AP to generate new session keys for the client without conducting queries to the RADIUS server.

The fast rekeying process for LEAP network authentication requires one round-trip information exchange between the client and the locally installed WDS AP, as opposed to three round trips between the client and the RADIUS server, which is possibly located across the core of the network. According to Cisco, fast, secure roaming using WDS should take less than 150ms.

Designing & Planning…

### If Your WLAN Is Cisco, Go Cisco All the Way

If you have deployed a WLAN based on the Cisco APs, try to standardize on the Cisco hardware for your wireless clients—or at least use hardware certified under the latest version of the Cisco Extension program. As you can see from the information in this chapter, Cisco wireless clients that are deployed on the Cisco WLAN understand Aironet Extensions and thus have superior roaming characteristics. If you are considering purchasing WVoIP phones, we recommend that you at least give Cisco 7920 phones a try. They will also make it much easier for you to configure QoS on the APs to provide priority for VoIP traffic. With non-Cisco WVoIP phones, you will have to do cumbersome manual configuration to achieve the same results.

# Cisco L3 Roaming Solutions

L3 roaming takes place when a client moves between APs attached to two different IP subnets. A graphical illustration of this situation is presented in Figure 3.10.

**Figure 3.10** L3 Roaming Example

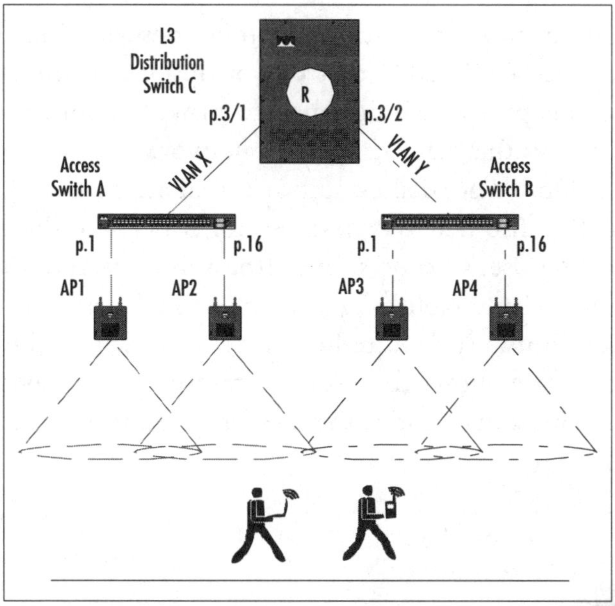

As you can see, when a user moves from left to right, he will first go through the process of L2 roaming between AP1 and AP2, which are connected to the VLAN X, and then he will associate with AP3, which is connected to the VLAN Y. At this point the client will find itself on a different IP subnet and under normal circumstances will lose IP connectivity to the network. Exactly what will happen will depend on the operating system that the client is running and on whether the client is using DHCP to acquire IP addressing information. If the client uses static IP addressing, there is no chance to restore communication to the network until the client's computer is reconfigured with an IP address that's valid for the new subnet. But if the client uses DHCP and the newer OSs such as Windows 2000 or Windows XP, his computer may have a chance to reacquire a new IP address. If the client is a Cisco 7920 WVoIP phone, the phone will lose connectivity to the Call Manager application (will stop receiving SCCP keepalives) and will reapply for a new address via DHCP.

What will happen with applications that the user is running will depend on the type of application. Session-oriented applications such as FTP and Telnet will not survive an IP address change. Web-based and e-mail applications may continue to function normally. Voice calls conducted through WVoIP phones will be lost.

We can see that L3 Roaming is potentially a disruptive process. Cisco offers two solutions to overcome loss of connectivity normally associated with situations n which a client crosses subnet boundaries: Mobile IP and Proxy Mobile IP.

# Mobile IP

Mobile IP (MIP) is an industry-standard protocol described in the Internet Engineering Task Force RFC 2002 that Cisco supports in the IOS software for routing platforms. This protocol is designed to provide connectivity to the network for a client device that has changed its network location but preserved an original IP address. Under normal conditions, a client with such an address will never receive any IP traffic that was destined for it because the network will route this traffic to the user's home subnet (the subnet where the user's IP address really belongs). By establishing a secure tunnel between the routers located on the user's home subnet and on the subnet where the user currently resides, the MIP software allows the user to preserve connectivity to the network. Figure 3.11 represents entities that are involved in the data communication process using MIP.

**Figure 3.11** Data Communication Process Using the Mobile IP Protocol

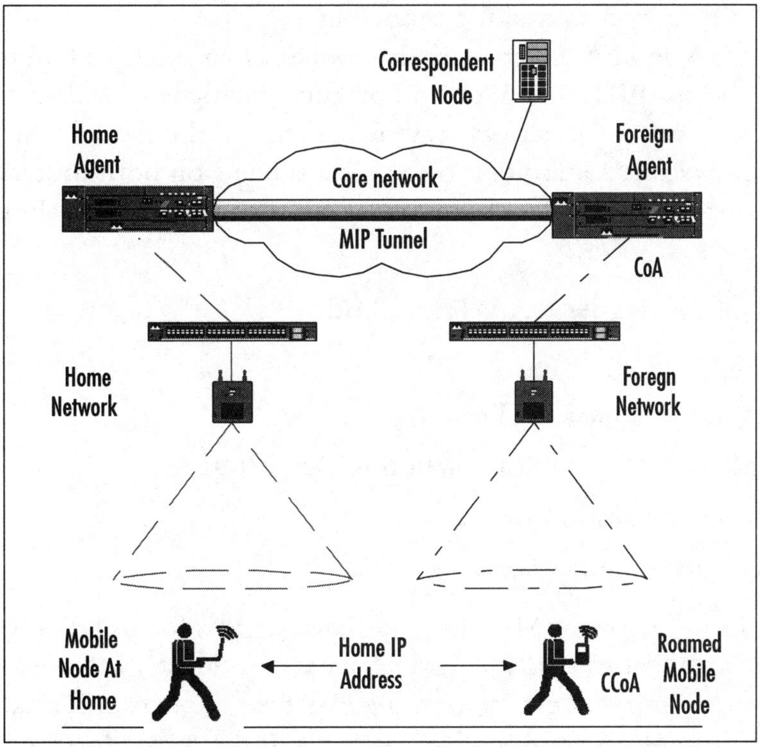


Let's define the participants of this puzzle. Mobile Node (MN) is a computer with installed MIP client software that was originally connected to the Home Network (HN) and still has a Home IP address that corresponds to this subnet. The Home Agent (HA) is a software entity that resides on the router that connects HN to the rest of the network. As a result of the L3 roaming, the MN is now connected to the Foreign Network (FN) that represents a different subnet. Foreign Agent (FA) is another software entity that resides on the router that connects the FN to the rest of the network. The Correspondent Node (CN) is an application server that conducts a session with the MN. The Care-of-Address (CoA) is a temporary address that the FA uses to receive traffic destined for the MN. Co-located Care-of-Address (CCoA) is an alternative to CoA. This is an IP address that MN acquired itself—for example, through DHCP. The CCoA is an IP addresses valid on the FN; CoA may be the FA router interface address or its loopback address.

The MIP communication process consists of three distinct phases: agent discovery, MN registration, and traffic tunneling.

Both the HA and FA advertise their services using Internet Router Discovery Protocol (IRDP), which was originally designed for the routers to advertise themselves as default gateways for clients on the locally attached subnets. IRDP broadcasts are limited to the local subnets on both routers and carry special MIP extensions that contain various types of information about these entities, such as:

- Agent capabilities—HA, FA, or both
- CoA
- Reverse tunneling (RT) support
- Supported tunnel encapsulation (GRE, IPinIP)
- Agent registration lifetime
- Prefix-length extension

During agent discovery, MN does not have to wait for IRDP advertisement but can issue an agent solicitation, and all agents on directly attached subnets should respond. The agent registration lifetime field contains information about the period between consecutive advertisements. If the next advertisement was not received, the MN will send out agent solicitation. MN can understand that it roamed to an FN by analyzing the prefix-length extension that contains the current network address. In this case, it will attempt to acquire a valid IP address that can be used for communications. Two types of such addresses may be used: CoA advertised by the FA is a shared address that can be used by all MNs on a particular FA interface. CCoA is bound to the MN interface and is temporary and unique to this node. The MN with CCoA will establish a tunnel to the HA to carry just its own traffic, but because CoA is located on the router, it can be used to create a shared tunnel to the HA on behalf of multiple nodes on this FN. When MN receives FA advertisements and realizes that it has roamed to a FN, it will begin the registration process.

The MN is preconfigured with an IP address of the HA and a preshared key that is used to encrypt communications with mobility agents. It has information obtained from FA advertisements and now sends a registration request to the HA. The way it does that depends on whether it has obtained the CCoA or not. If CCoA is used, the MN sends a registration request directly to the HA. If the

MN uses CoA, it sends a registration request to the FA, which checks the request for validity, adds the MN to the list of the pending mobile nodes, and forwards it to the HA. If the registration request is invalid, the FA sends a reply to the MN with appropriate error code. On receiving the registration request, the HA verifies that the request is valid, creates the mobility binding association between MN home IP address and its CCoA (or CoA), establishes a tunnel to the CCoA (or CoA if it does not yet exist), and creates a routing entry to send all traffic destined for the MN home IP address to the tunnel. The HA now sends a registration reply, again either directly to the CCoA of the MN or to the CoA of the FA. If the request is invalid, the HA sends back an error message.

The FA receives the message, checks it for validity, adds MN to its visitors list, establishes a reverse tunnel to the HA (if necessary), adds a routing entry to send MN traffic into the tunnel, and relays the reply to the MN. The MN receives the reply (either from the FA or directly from HA), verifies validity of the reply, and is assured that the mobility agents are now aware of its new location. If the registration reply was forwarded directly to the MN CCoA, it will at this point establish the reverse tunnel to the HA (if necessary).

Before its registration lifetime expires, the MN will periodically send out reregistration requests to the HA to update the mobility associations on the HA and FA.

After the registration process completes successfully, the MN is ready to continue network communications transparently for all CNs. Under normal conditions, the MN will send traffic directly to the Correspondent Node (CN), and the return traffic to the MN will be routed to the HN, where the HA will intercept it and send it to the tunnel toward the CoA (CCoA). The IP traffic the MN sends to the CN will always have an invalid source network address, and if the ACLs are configured on the edge routers to block spoofed IP addresses, this traffic will not go through. In this case, the MN will need to establish a reverse tunnel between the CoA (CCoA) and the HA, and the traffic from the MN to the CN will also be first tunneled to the HA and then routed in the regular manner to the CN.

The primary tunneling encapsulation used in MIP is IP in IP, but Generic Routing Encapsulation (GRE) and some others can be used. The MIP's security is provided by establishing security associations between MN, FA, and HA that need to authenticate each other. Centralized authentication using TACACS+ or RADIUS servers is supported. The integrity of the registration messages is protected by the 128-bit shared key.

The MIP deployment requires installation and configuration of the MIP client software on the wireless node, which might not be available for all client platforms. Additional administrative overhead adds cost and complexity to MIP implementations.

## Proxy Mobile IP

To remove the need for the MIP client software and make the L3 roaming transparent to wireless clients, Cisco added new functionality to the AP software called Proxy Mobile IP (PMIP). To implement PMIP, you need to have all standard MIP infrastructure in place, including Cisco routers with the IP Plus feature set, configured HA and FA on the routers, configured security infrastructure to provide security associations between mobile agents, and preferably have tested this configuration using standard MIP clients.

PMIP functionality should be enabled on the APs that will serve roaming clients. Some of the APs should be designated as Authoritative AP (AAPs). Up to three APs can be designated as AAPs, and their addresses should be hardwired into all configurations of all other APs. On the APs, the PMIP should be enabled on the Ethernet interface, Bridge Virtual Interface (BVI), and on selected SSIDs. PMIP does not support VLANs. The security associations settings that include shared keys and range of valid IP addresses can be configured manually or stored on the RADIUS server.

The AP with PMIP enabled needs to know the information regarding the HAs of all potential visiting clients. When the PMIP is first enabled on the APs, they all send information about their local HAs obtained from the IRDP advertisements to the AAP. The AAP builds a subnet map table that lists IP addresses and netmasks of all HAs and distributes this table back to all APs. Whenever a new client with an invalid network portion of IP address associates to the AP, the AP can now compare the client's IP address with the local map and query the AAP if HA information is not found locally. If the first AAP is not available, the AP will query the next configured AAP. The AAPs in turn are responsible for periodic synchronization of the subnet map table among each other.

Once the PMIP is enabled on the AP, it will provide functionality comparable to a regular MIP client. It will start listening to the IRDP advertisements from the HAs and FAs and collecting information about MIP entities available on the network. Once it detects a wireless client that has an IP address with a network portion that does not match the local subnet, it will query the subnet map table to find the HA for the client, acquire a CoA on behalf of the client, and send the MIP registration request to the HA through the FA.

The rest of the PMIP process will follow the standard MIP procedures that will result in establishing a tunnel between the CoA on the AP and the HA for the client. If multiple clients roamed from the same HN to the same AP, they will share the same CoA and the same tunnel. Reverse tunnels are supported as well as the choice between GRE and IPinIP encapsulations.

# WLAN Design Considerations

Currently available L3 roaming solutions have serious drawbacks and limitations. MIP is difficult to configure and support, it requires IP Plus software feature set on the routers that may require hardware upgrade, and it puts more load on the routers. The main drawback is that it requires installation of the MIP stack on client devices, which adds expense and administrative overhead and may not be available for all client platforms. PMIP is supposed to eliminate the need for the client software, but it has its own limitations. This solution does not support VLANs, broadcast, and multicast traffic, and it may have interoperability problems with the DHCP address assignment process.

Additional issues may come up when coverage from the multiple APs that belong to different subnets overlap, as may be the case in a multistory building with different wireless subnets on every floor. It could be very difficult to provide reliable RF coverage horizontally throughout the floor without leaking RF signals to adjacent floors. We now know that wireless clients select their roaming targets based on multiple parameters, and RF signal strength is only one of them. With this design, one can end up with wireless clients constantly roaming between home and foreign subnets and exhibiting suboptimal network connectivity.

It's no surprise that most organizations elected to implement WLANs in a manner that would completely eliminate the need for L3 roaming. The usual solution is to create a single wireless VLAN that spans the whole building and is connected to the L3 distribution switches or to a security gateway of some sort (VPN concentrator or wireless security switch). This solution obviously violates the Cisco campus network design recommendation that dictates the use of separate access switches on every floor representing different VLANs. For security reasons, some organizations have created a separate wired infrastructure to support WLAN by installing a small stackable switch in the telecommunications closets on every floor. (This parallel network can also be used to create an out-of-band management VLAN for the main network infrastructure.) The problem with these solutions is that they do not scale well if the WLAN really has many simultaneous users.

If the wired portion of the WLAN represents a single wireless VLAN, certain steps can be taken to minimize the amount of optional traffic on this VLAN. IGMP Snoop should be enabled on both switches and APs. Another trick is to create an AP management VLAN on the switches, trunk this VLAN to the APs, and put the AP IP address on this VLAN. On the wireless side of the AP, this VLAN should not be mapped to any SSID. This solution will shield the wireless data VLAN from the management traffic.

Another solution is to create multiple wireless data VLANs mapped to different SSIDs and run them parallel to each other on every AP. Different users can be grouped to different VLANs using different user groups on the RADIUS server. This solution will not increase available RF bandwidth (unless you have dual band 802.11a/802.11g APs), but it will partially shield users from each other's traffic. You can also limit the presence of a particular SSID to particular areas of the building.

Specifics of WLAN design will depend on the mobility requirements of the different types of users and the applications they are planning to run. Based on the mobility requirements, we can cluster potential WLAN users into five groups:

- **WLAN users who rarely move** This group includes all applications for which WLAN connectivity was chosen so that there is no need to run wires. Client devices may include desktops with wireless cards or specialized lab or manufacturing equipment that can occasionally be moved around.
- **Hotspot users with laptops** These users came to a specific location to be connected to the Internet. They will not move after they have settled down (and have ordered their cappuccino).
- **Typical office users with laptops** They tend to move around with laptops mostly between their office, their colleague's offices, conference rooms, and the cafeteria. They normally do not run applications while on the move. This type of mobility we can call *portability*.
- **Users of tablet PCs, PDAs, or barcode scanners** Because their computer devices are lighter, these users tend to be more mobile. These users do have a need to run applications while on the move. Depending

on the applications, they will move around a limited area—for example, around a warehouse floor, a manufacturing floor, or a lab.

- **Any users with WVoIP phones** These users are the most demanding. They will want to go everywhere (even to the bathroom) and will want to run their application (talk!) while on the move. These users do require true mobility.

We can conclude from this analysis that only WVoIP phone users may require continuous network connectivity while moving across a large area. All other types of users can afford to lose network connectivity while on the move. So it should be feasible to split a WLAN into multiple L3 domains as long as RF coverage from these domains does not unintentionally overlap.

# Summary

In this chapter we discussed various issues associated with roaming of wireless clients between adjacent APs. We learned about differences between L2 and L3 roaming and current solutions that Cisco offers to provide L3 roaming capabilities. We studied the L2 roaming process in detail and discussed Cisco-specific implementations of this process for Cisco wireless client adapters and Cisco 7920 Wireless VoIP phones. We also looked at the WLAN design implications as they are affected by roaming of wireless nodes.

Many details of the current Cisco implementation of roaming solutions that we discussed in this chapter will probably change as new IEEE 802.11 protocols become ratified and new versions of AP and client software become available. But knowing details of the current Cisco implementation should help you design and troubleshoot your WLAN now and understand the changes that will be introduced later.

# Solutions Fast Track

## Cisco L2 Roaming Solutions

☑ L2 roaming takes place when user moves between APs connected to the same IP subnet.

☑ Wireless clients make the decision to roam based on the current state of wireless connectivity, taking into account multiple factors. Algorithms controlling client behavior in roaming situations are vendor proprietary.

☑ Cisco APs communicate proprietary information to Cisco wireless clients that improves clients' roaming decisions.

☑ Cisco 7920 WVoIP phones require better RF coverage than computer-based wireless clients, and they employ special algorithms to speed roaming decisions.

## Cisco Solutions to Speed L2 Roaming

☑ Cisco is working on solutions to speed L2 roaming. Some new solutions were recently introduced in the AP and client software.

- ☑ The fast client channel-scanning algorithm, based on the information exchange between the APs and wireless clients, helps clients find better roaming targets faster.
- ☑ The WDS entity was introduced to speed client network authentication. WDS uses the CCKM algorithm to locally cache client credentials and quickly deliver them to the client's target AP during the roaming process.
- ☑ In the first implementation, this fast rekeying algorithm supports only LEAP authentication of computer platforms running Cisco client software. The WDS entity currently resides on APs. We can expect that WDS functionality will be migrated to the Cisco switches and routers, and there will be support for more client platforms and more network authentication algorithms.

## Cisco L3 Roaming Solutions

- ☑ L3 roaming takes place when a wireless client crosses IP subnet boundaries during the roaming process. Unlike L2 roaming that is supported natively by the 802.11 protocol, L3 roaming breaks the client's network connectivity and application functionality.
- ☑ Mobile IP (MIP) is a generic standard-based solution to provide network connectivity to clients who have moved to a different network subnet without changing IP address.
- ☑ Proxy Mobile IP (PMIP) is a Cisco proprietary functionality inside AP software that allows wireless clients to take advantage of MIP without having the MIP protocol stack installed on the clients.
- ☑ Both MIP and PMIP solutions have limitations and are cumbersome to deploy and administer.

## WLAN Design Considerations

- ☑ The prevalent WLAN design is to use a single WVLAN per building trunked across all L2 access switches.
- ☑ In the same fashion, additional WVLANs can be created and mapped into different SSIDs when WVoIP phones are used on the network or if

there is a need to segregate wireless users into multiple groups or provide guest Internet access.

☑ A single VLAN solution does not scale well, so keep all unnecessary traffic to a minimum.

☑ A separate native VLAN not mapped to any SSID can be created for AP management (the AP BVI interface will belong to this VLAN).

☑ Most wireless users do not need total mobility across a large area, so a WLAN that consist of multiple large L2 roaming domains will provide required functionality for most applications.

# Frequently Asked Questions

The following Frequently Asked Questions, answered by the authors of this book, are designed to both measure your understanding of the concepts presented in this chapter and to assist you with real-life implementation of these concepts. To have your questions about this chapter answered by the author, browse to **www.syngress.com/solutions** and click on the **"Ask the Author"** form. You will also gain access to thousands of other FAQs at ITFAQnet.com.

**Q:** Can Cisco 7920 phones take advantage of Fast Secure Roaming (FSR) using WDS?

**A:** Currently only Cisco adapter cards with appropriate firmware versions can use this feature. But the roaming algorithm of the 7920 is tuned for a faster roaming process compared to a regular Cisco wireless adapter card. Support for FSR on 7920 is planned for summer 2004.

**Q:** Does Cisco offer client software that provides Mobile IP support?

**A:** No, it recommends that you use the Mobile IP protocol stack from its partner, Birdstep Technologies. See additional information at www.birdstep.com/wireless_infrastructure/mobile_ip.php3.

**Q:** What kind of security solutions do you recommend for Wireless VoIP phones?

**A:** From the faster roaming standpoint, we recommend putting these phones on a separate VLAN mapped to a separate SSID, using static WEP key on this SSID, and configuring ACLs on the router interface for this VLAN to allow only necessary traffic through.

**Q:** Are any solutions currently available to support L3 roaming for Wireless VoIP phones?

**A:** No. Mobile IP requires installation of the Mobile IP client software, which does not currently exist for 9720 phones. The Proxy Mobile IP feature does not support VLANs, which are required to properly implement QoS on AP.

Chapter 4

# IP Multicast in a Wireless LAN

## Solutions in this Chapter:

- The OSI Model Overview
- Multicast WLAN Deployment Recommendations
- Controlling IP Multicast in a WLAN with APs
- Controlling IP Multicast in a Peer-to-Peer WLAN using Bridges

☑ Summary

☑ Solutions Fast Track

☑ Frequently Asked Questions

# Introduction

The primary reason for implementing multicast in any network environment is to reduce bandwidth consumption when delivering shared advanced technology solutions. Multicast allows a single stream of packets to be distributed to a group of participants rather than creating a packet stream per member. Even though Internet Protocol (IP) multicasting has undergone significant changes, it is still a bit tricky to implement, even in wired local area networks (LANs). Getting it to work right in a wireless LAN (WLAN) environment is even more difficult. Some organizations disallow the use of multicast across their WLANs. Multicast requires a network infrastructure that supports protocols and features such as Protocol-Independent Multicast (PIM), Internet Group Management Protocol (IGMP), Cisco Group Management Protocol (CGMP), and IGMP snooping. Multicast is more complex in a wireless environment due to the half-duplex, shared, carrier sense, multiple access/collision avoidance (CSMA/CA) environment of WLANs. This chapter discusses the issues and configuration elements associated with implementing and optimizing IP multicasting in a WLAN.

There are a few things that must be considered for multicast solutions. If your organization needs to perform network-based video conferencing, real-time news feeds, radio broadcasts, or video on demand, you need to enable multicasting. If you are a home user, you may need multicasting for distance learning, Internet gaming, interactive chat sessions, and Internet jukeboxes. Why is multicasting needed in these situations? Multicasting is a method of sending packets to more than one receiver or interface at a time. The hosts receive the same information but the data is sent only once over the network thus reducing bandwidth consumption. (To better understand multicasting, refer to the Open Systems Interconnect [OSI] model shown in Figure 4.1.)

# The OSI Model Overview

As seen in Figure 4.1, the lowest layer of the OSI model is the physical layer, which is responsible for the actual physical conveyance of signals. The OSI stack extends all the way up to the application layer, where features are utilized to aid system performance. The IP suite maps closely to the OSI reference model.

1. The physical layer passes bits in different forms such as light, radio, or electrical impulses through a network. It is the hardware means for sending and receiving data. Some examples are Asynchronous Transfer

Mode (ATM), Ethernet, and serial links. How the bits and bytes get from a source to a destination is the physical layer's only concern. It does not care about the payload, only the mode in which it gets there.

2. The next layer is the data link layer. Frames are created at the data link layer that contain source and destination addresses; frames also furnish the transmission protocol, management, and any errors in the physical layer and flow control. There are two sublayers to the data link layer; the Media Access Control (MAC) layer and the Logical Link Control (LLC) layer.
    - The MAC layer controls how a computer gains access to the data and permissions needed to transmit the data in question.
    - The LLC layer controls error checking, flow control, and frame synchronization.
3. The network layer provides routing capabilities. Routing and forwarding are the primary functions of this layer. Congestion control, error handling, packet sequencing, and internetworking are also handled at the network layer.
4. Above the network layer is the transport layer. The transport layer transfers data between systems and/or hosts in a transparent fashion, and is also responsible for flow control, ensuring there is a complete transfer of the data and error recovery.
5. The session layer manages connections between applications. It establishes the session, manages the session for communication and transfer of data, and terminates the session.
6. The sixth layer of the OSI model is the presentation layer, which translates the packets from the lower layers into an acceptable format for the application layer.
7. The application layer focuses specifically on applications such as e-mail, file transfers, File Transfer Protocol (FTP), and Telnet.

**Figure 4.1** OSI and TCP/IP Models

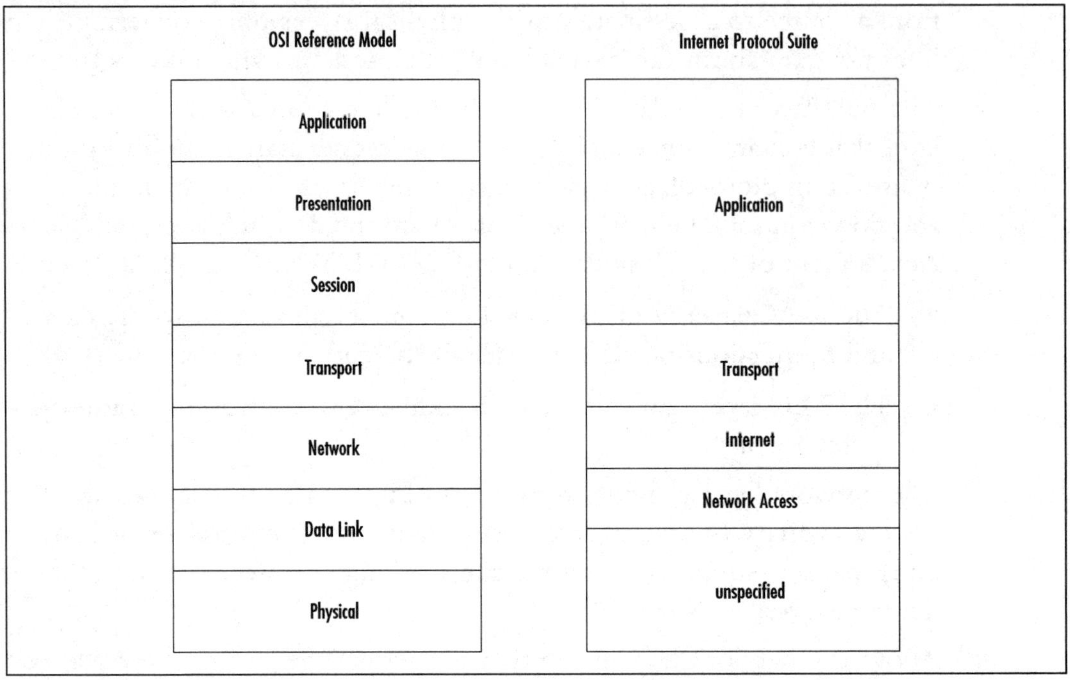


# Data Communication Methods

Once you understand the different layers you need to understand the different modes in which you can transmit data. Typically there are three methods used for communication:

- Unicast
- Broadcast
- Multicast

## The Unicast Method

The unicast method is a one-to-one communication process. Consider Figure 4.2: Host A wants to communicate with Host B. This request is sent via the 192.168.10.*x* subnet. Host B receives the request and responds via the subnet. Host B's response is directed to Host A's interface address. If Host A were on a different subnet, Host B's response would be directed to the router's interface instead.

**Figure 4.2** Unicast

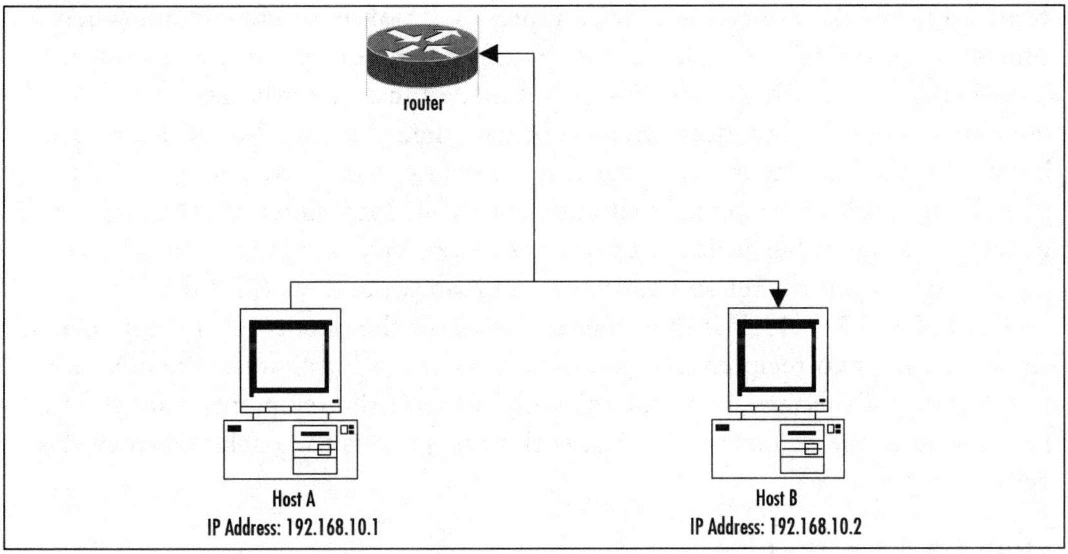


What if this was the case of a webcast lecture and the only reliable method of communication was unicast? Universities use webcasts to deliver lectures from professors or guest speakers remotely, which allows the university to deliver a class to different geographic locations from a single location (see Figure 4.3). Assume each of the hosts at the different university locations represents 30 students. Notice the amount of traffic caused by one lecture sent via unicast.

**Figure 4.3** Unicast to Multiple Recipients

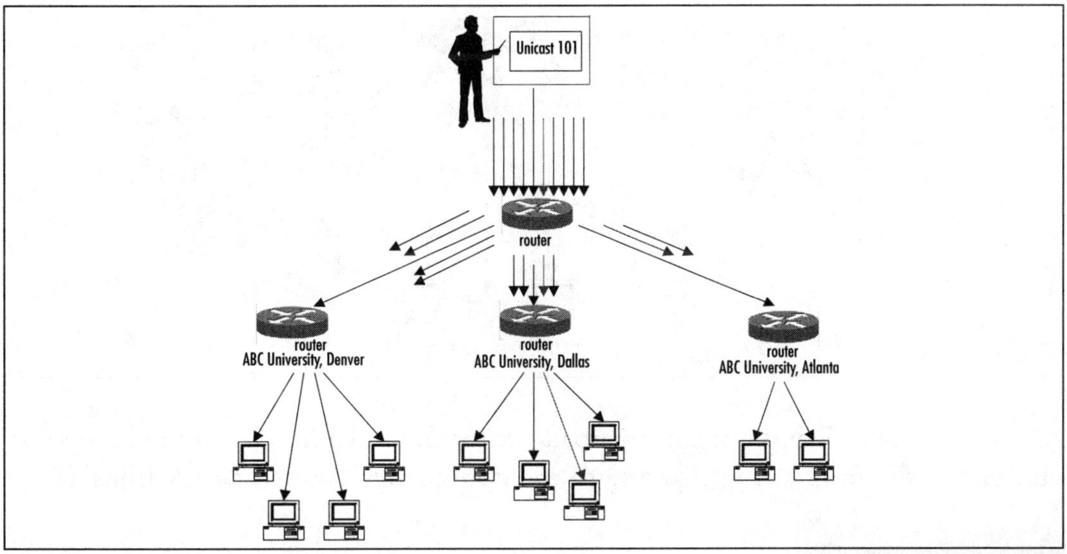

The use of multiple data streams can be extremely taxing on resources and overload networks. For single users, a unicast solution is acceptable but when unicast is transmitted to multiple hosts within a network it can cause problems. The method in which the transmission is made must be optimized.

What about the broadcast method of data transmission? Would this be any better? One major drawback is that it is almost impossible to send broadcast packets to multicast recipients that are not in your local network. Broadcast is easier to configure but it does not save resources. When an application broadcasts, it sends each packet and address to a broadcast address (ffff.ffff.ffff or 255.255.255.255), which is then sent out to all of the clients on the network. If there are only two recipients on a network there is a lot of wasted bandwidth (see Figure 4.4) because, even though only some of the recipients want the transmission, all of the clients on that network must process the packets as they flow across the network.

**Figure 4.4** Broadcasting

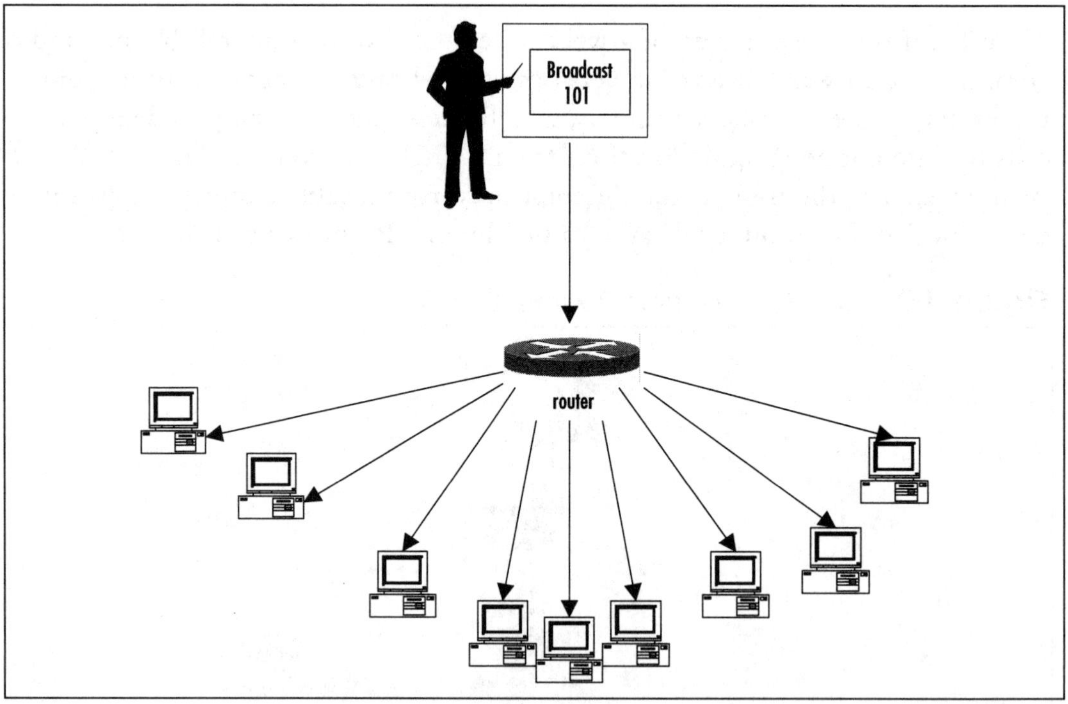


Instead of utilizing unicast or broadcast for the university, it would be better to send traffic from a single source to multiple predetermined destinations. IP multicasting can provide this. RFC 1112 discusses the standard for IP multicasting over the Internet and over local networks (see ftp://ftp.rfc-editor.org/in-notes/rfc1112.txt). It allows for a single source to transmit to multiple destinations without having multiple data streams. This reduces the impact on network resources (see Figure 4.5).

**Figure 4.5** IP Multicasting

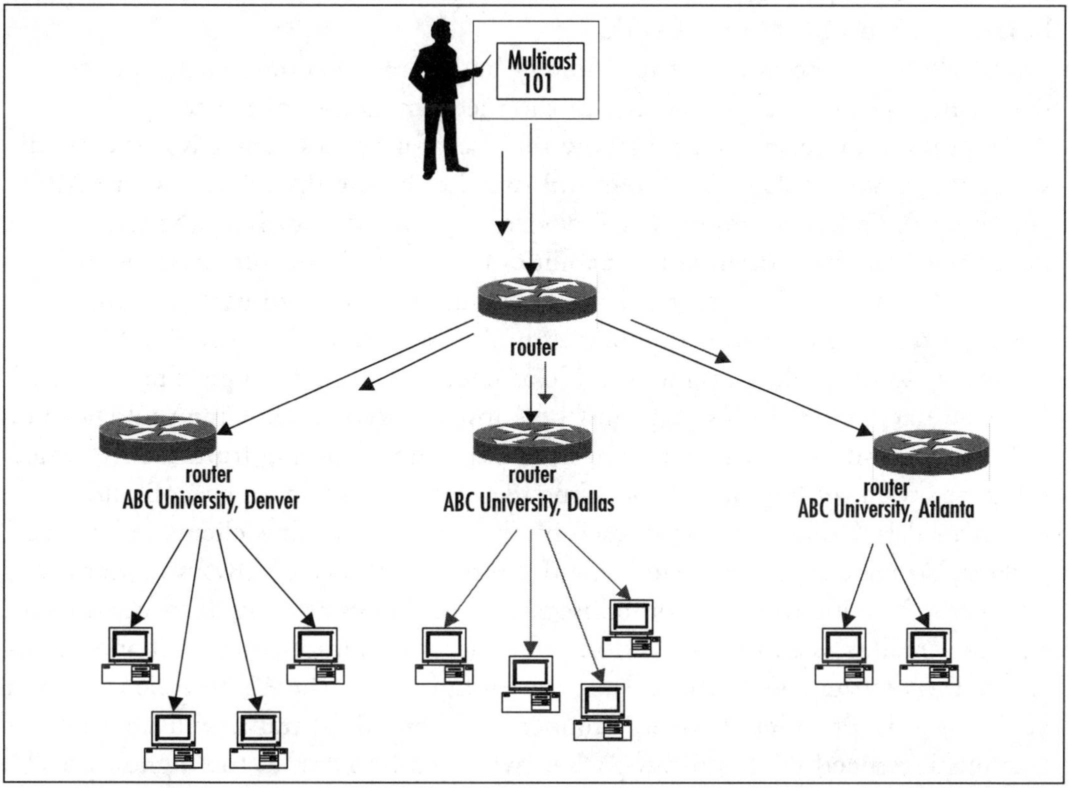


Notice the original single data stream. When the path splits, the Cisco router forwards the packets and sends them to different destinations or multicast groups. The result is a highly efficient method of delivery. Multicasting works in WLANs as well, but there are some issues that must be considered due to the shared nature of WLANs. The following sections discuss some deployment recommendations, WLAN configurations for IP multicast, how to control IP multicast with access points (APs), and how to control IP multicast in a peer-to-peer WLAN using bridges.

# Multicast WLAN Deployment Recommendations

There are many benefits to IP multicasting. Besides optimizing network performance, multicasting supports distributed applications such as video conferencing and distance learning. It reduces the cost to deploy applications by conserving bandwidth through targeted transmissions to intended recipients instead of broadcasting to the entire network.

WLANs can use multicast technology, but there are certain challenges to overcome due to the shared nature of the medium. Some Information Technology (IT) departments disallow the use of multicast across WLANs by filtering that traffic at Layer 3 routers and switches before the packets get to APs and bridges. This is because WLAN devices typically do not have the necessary performance and configuration capabilities to effectively handle multicast filtering. APs are basically just translational bridges that forward packets from Ethernet to 802.11 ports. If support for multicast is desired on the WLAN, it is important to only allow as much traffic as needed via filters to get the job done.

There are several issues that must be considered when allowing multicast in a WLAN environment. Broadcast and multicast frames coming from the Ethernet interface are never fragmented or acknowledged. This includes beacon and broadcast data frames, which are used by the APs and wireless clients for device and session management. Broadcast and multicast packets are always forwarded out the 802.11 interface. In comparison, unicast frames are only forwarded if the destination address exists on the other interface. In addition, broadcast and multicast traffic use the lowest speed data rate configured on the AP. In other words, if you have a data rate set at "basic," multicast and broadcast traffic will use that transmission speed while unicast packets will communicate at the highest possible speed allowed. For example, if you have configured the four 802.11b rates (1, 2, 5.5, and 11 Mbps) as basic, basic, basic, and basic (the default setting for maximum throughput), multicast packets will communicate at 1 Mbps while unicast will occur at 11 Mbps, if possible. Overall throughput will be significantly reduced because it will take the 1 Mbps transmission approximately 11 times longer to occur than the 11 Mbps transmission and, while the 1 Mbps traffic exists, the 11 Mbps traffic has to wait for the other to complete due to CSMA/CA backoff mechanisms.

Because APs forward all multicast packets, wireless clients associated with that AP can see those packets regardless of whether they are part of the multicast group or not. In addition, clients can see multicast packets from all virtual local area networks (VLANs) in WLAN environments, not just the VLAN they are on. It is important to implement Wireless Encryption Protocol (WEP)-, Wireless Protected Access (WPA)-, or Extensible Authentication Protocol (EAP)-based encryption to prevent eavesdropping by hackers. For VLANs, there is a unique encryption key that must be configured during the VLAN setup. This allows clients to decrypt only one VLAN session at a time.

When considering enabling IP multicast in a WLAN environment there are a few recommendations. All Cisco Aironet products support multicast but some have more advanced feature sets than others depending on whether they run VxWorks- or Internet Operating System (IOS)-based operating systems. A big part of deploying a multicast solution for a Cisco wireless network is the number of users you will be servicing and the type, size, and amount of multicast packets that will be traversing the WLAN.

Designing & Planning…

## Planning Recommendations for Multicast

Follow these recommendations to optimize WLAN multicast traffic and utilization:

- Multicast and broadcast packets have a lower reliability than unicast packets because the AP receives no link-layer acknowledgements for them. Therefore, it is possible that even with expert level configuration knowledge, some client devices may still experience problems with multicast in WLAN environments.
- Verify that multicast applications work in a WLAN environment prior to rolling them out companywide.
- Roaming in a WLAN environment will drop multicast sessions. This is because the client also roams from one switch port to another, where the next AP is connected. To prevent this, you must do one of the following when IGMP snooping is enabled on the switch:
    - Enable the **Send IGMP General Query** function on VxWorks-based Cisco APs.

Continued

- Leave the IGMP Snooping Helper enabled (on by default) on IOS-based Cisco APs.

- Filter multicast traffic where necessary. Create packet filters using access lists on routers and switches if multicast is not needed on network segments. To further prevent multicast on APs and bridges, enable VLAN filtering and IGMP snooping on switch ports, if possible (see previous statement). IGMP snooping disables multicast traffic from flooding all switch ports. APs and bridges that have 802.1Q trunks should be filtered so that only active VLANs are allowed.
- Create multicast protocol filters and MAC filters on APs and bridges if limited support is desired on WLAN segments.
- Establish AP and bridge configurations that allow multicast packets to be sent at the highest data rate possible. If significant dropped packets are experienced, a slower data rate will need to be configured to allow a better signal-to-noise ratio (SNR).
- WLAN clients, typically laptop and personal digital assistant (PDA) configurations, may be configured with power-saving capabilities such as Fast or Max power saving protocol (PSP). This will potentially cause multicast jitter and delay, depending on the amount of traffic being received. Buffering occurs during power-save modes and packets are only sent or received once every 200 ms (configurable) when the client card is "awake."
- Do not employ virtual private network (VPN) solutions where multicast is needed. VPN packet encryption and tunneling prevents multicast from working.

## Configuring Multicast and Broadcast Minimum Data Rate Settings in IOS

Some network applications require software clients to maintain a consistent data rate back to the server for connectivity. For other applications, you may need to configure a minimum data rate in order to get the performance you require. Follow these steps to configure the minimum data rate settings in IOS:

1. Begin by entering global configuration mode. Type **configure terminal**.

2. Then enter into interface configuration mode for the radio interface being configured. Type **interface dot11radio 0.**
3. Next, enter the data rate to be either **basic** or **enabled**. You can also enter **throughput** to enhance throughput or **range** to enhance the range. These are optional configurations. Type **speed {[6.0] [9.0] [12.0] [18.0] [24.0] [36.0] [48.0] [54.0] [basic-6.0] [basic-9.0] [basic-12.0] [basic-18.0] [basic-24.0] [basic-36.0] [basic-48.0] [basic-54.0] | range | throughput}**.

**NOTE**

Remember, these are optional settings. Using the **speed** command with the distance sets it in enabled mode. Using the command with basic in front of the distance sets it to basic mode (example: **speed basic-6.0**). Entering **range** or **throughput** automatically optimizes the radio for range or throughput. The bridge sets the lowest data rate to basic and all other rates to enabled when you use the value **range**, and the value **throughput** sets all data rates to basic.

4. Return to EXEC mode and save the configuration. Type **copy running-config startup-config**.

To disable the data rates type **no** in front of the command.

The following example shows how to set up the bridge for basic 24 Mbps service only:

```
wb1400# configure terminal
wb1400(config)# interface dot11radio 0
wb1400(config-if)# speed basic-24.0
wb1400(config-if)# end
```

Data rate 24 is set to basic, and the rest of the data rates are set to enabled.

This example shows how to disable data rate 48.0:

```
wb1400# configure terminal
wb1400(config)# interface dot11radio 0
wb1400(config-if)# no speed basic-6.0 basic-9.0 basic-12.0 basic-18.0 basic-
24.0 basic-36.0 basic-54.0
wb1400(config-if)# end
```

Data rate 48 is disabled, and the rest of the rates are set to basic.

**NOTE**

Remember, broadcast and multicast traffic use the lowest speed data rate configured on the AP. In other words, if you have a data rate set as "basic," multicast and broadcast traffic will use that transmission speed while unicast packets will communicate at the highest speed allowed. For example, if you have configured the four 802.11b rates (1, 2, 5.5, and 11 Mbps) as "basic, basic, basic, and basic" (the default setting for maximum throughput), multicast packets will communicate at 1 Mbps while unicast will occur at 11 Mbps, if possible. Overall throughput will be significantly reduced because it will take the 1 Mbps transmissions approximately 11 times longer to occur than the 11 Mbps transmission, and while the 1 Mbps traffic exists, the 11 Mbps traffic has to wait for the other to complete due to CSMA/CA backoff mechanisms.

# IP Multicast WLAN Configuration

Configuring multicast in a WLAN is not difficult if the multicast servers already exist and the network infrastructure supports it. The question to be answered is whether or not you want multicast in your WLAN environment at all. Once you have made that decision, the following suggestions will allow you to determine how to best implement or prevent multicast. Try to choose the most appropriate device when determining where to filter or allow multicast.

Configuring & Implementing...

### Multicast Filtering for WLAN Implementations

It is important to understand the types of multicast traffic that will be allowed in the WLAN environment and filter accordingly. WLANs have limited shared bandwidth. This makes it difficult to successfully implement blanket coverage of all multicast streams. Multicast traffic can consume significant bandwidth unless low-rate streams are made available as alternatives to higher-rate streams. Filtering allows for the ability to manage multicast traffic based on WLAN capabilities and network engineering requirements.

# Controlling IP Multicast in a WLAN with APs

The best way to implement IP multicast filtering is at the router or switch before that traffic gets to the WLAN segment (see Figure 4.6). The network engineer can use the **ip multicast boundary** command tied to an access list to accomplish this. Only the packets coming from the allowed multicast address will traverse the network segment.

**Figure 4.6** Implementing IP Multicast Filtering

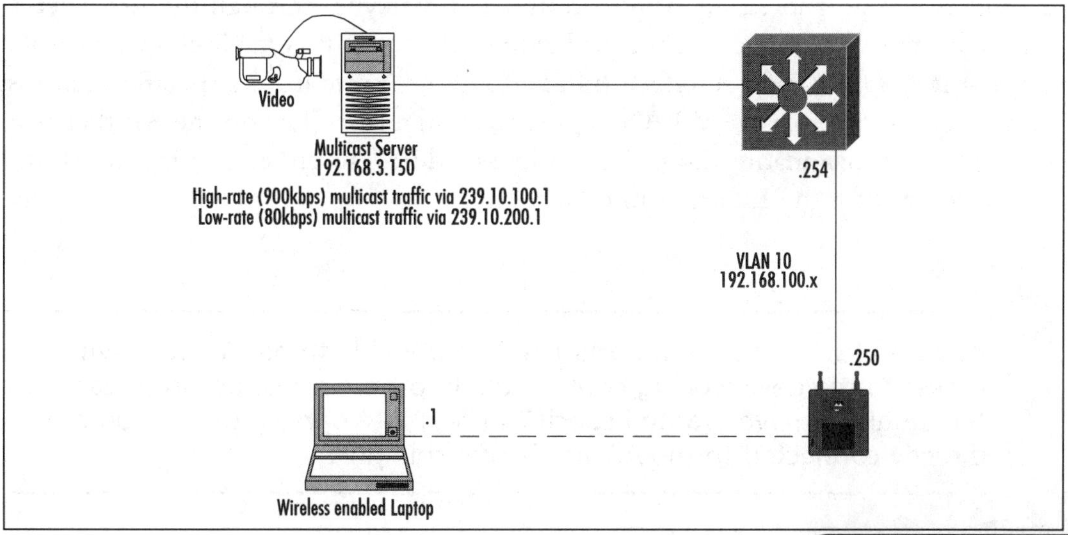

The multicast server sends high-rate multicast traffic via 239.10.100.1 and low-rate traffic via 239.10.200.1. To implement the filter, add the following IOS commands to the Layer 3 switch:

```
Interface vlan10
  Ip address 192.168.100.254 255.255.255.0
  Ip pim sparse-mode

Ip multicast boundary WLAN

Ip access-list standard WLAN
  Permit 239.10.200.1
```

PIM uses unicast routing information to perform multicast forwarding. PIM sparse mode distributes information about active sources, in this case the multicast server. Sources associate with the PIM-SM rendezvous point (RP), in this case the Layer 3 switch. Multicast receivers (the wireless client) register with the RP by sending a join message, thus enabling them to receive the multicast traffic. The multicast boundary designates which ports or VLANs receive multicast packets. In this example, VLAN 10 is allowed to receive multicast packets only from the low-rate server at 239.10.200.1. Packets from the high-rate address are blocked and never traverse the WLAN segment.

# Protocol Filters

Protocol filters, or access lists, can be used to allow or deny specific protocols by IP address, MAC address, or IP port number (Ethertype) through the Ethernet and radio ports of an access point or bridge. Filters can be established using IOS or via the graphical user interface (GUI). To allow traffic from a specific multicast server address to cross the WLAN segment, configure a filter on the AP that allows packets originating from that multicast address to either exit the wireless interface or enter the Ethernet interface.

**NOTE**

Be careful when configuring filters. It is possible to block your own access to the device being configured. In other words, do not create filters until you have granted specific access for yourself, unless you are directly connected to the AP via the console port.

# Controlling IP Multicast in a Peer-to-peer WLAN using Bridges

WLAN bridging helps extend a wired network. Cisco's Aironet BR1400, BR350, and BR1300 series bridges provide this capability. The BR1400 series bridge operates in the 5.8 Gigahertz (GHz) Unlicensed National Information Infrastructure (UNII)-3 band, conforms to the 802.11a standard, and has a maximum data rate of 54 Mbps. The BR1400 bridge is designed for the outdoors and is self-contained in a single unit. It supports point-to-point, point-to-multipoint, and redundant configurations. The BR350 bridge is an 802.11b device and has a maximum data rate of 11 Mbps while the BR1300 is an 802.11g device with a maximum data rate of 54 Mbps. Both the BR350 and BR1300 bridges operate in the 2.4 GHz spectrum.

## Point-to-point Bridging

Point-to-point bridging configuration is fairly straightforward once the bridges are mounted and communicating properly. With two or more bridges, at least one becomes the root bridge. A bridge becomes a root if, during installation, it does not recognize another bridge. If it recognizes an existing bridge, it becomes a non-root bridge and associates to the root bridge. Non-root bridges associate to a root bridge (see Figure 4.7).

**Figure 4.7** Point-to-Point Configuration.

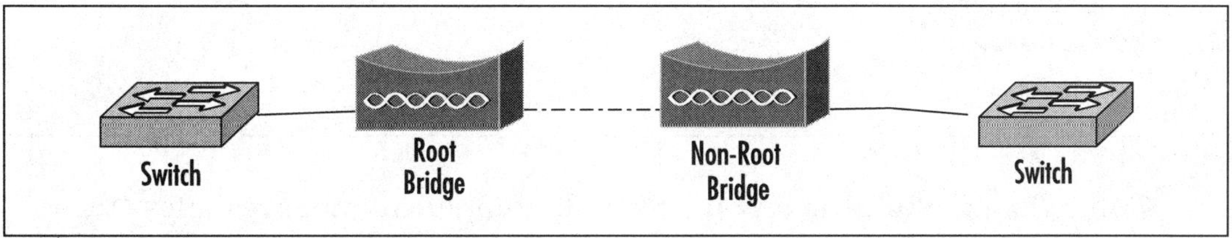


**NOTE**

If you are connecting the bridge to one or more flat networks with more than 256 users, Cisco recommends you use a router to connect the bridge to the network(s).

Once again, to prevent improper multicast traffic from traversing the bridged segment, use the **ip multicast boundary** IOS command on the Layer 3 switch on the multicast server side of the network. If there is a router on the non-multicast server side (i.e. the remote side), the command can also be placed there. This will prevent clients from issuing a join message to the RP for undesired multicast groups.

# Point-to-multipoint Bridging

Point-to-multipoint configurations have two or more non-root bridges associated to the root bridge. Up to 17 non-root bridges can be associated to a single root bridge (see Figure 4.8). The drawback here is that all of the non-root bridges must share the available bandwidth. Too many non-root bridges could significantly degrade network performance across the WLAN bridged segment. .

**Figure 4.8** Point-to-multipoint Bridging

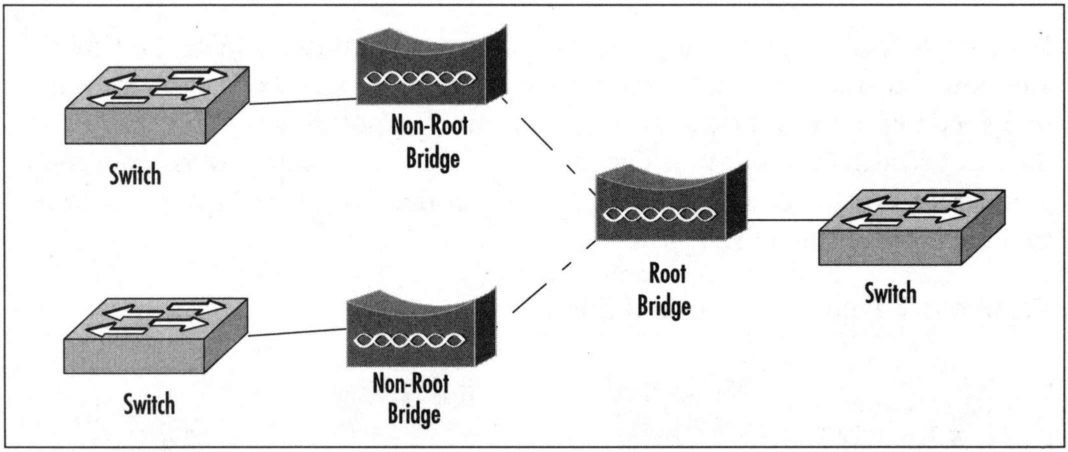


Controlling multicast in a point-to-multipoint environment employs the same concepts that were discussed for point-to-point configurations. Add the **ip multicast boundary** command to routers or Layer 3 switches before the multicast packets get to the WLAN network, or add them to the router at the remote side to prevent clients from joining unwanted multicast groups. A combination of configurations may be needed in a point-to-multipoint environment, depending on bandwidth availability and network engineering requirements.

Once configured, make sure multicast traffic is properly allowed or denied by running IOS show and debug commands such as **show ip mroute**, **show ip mroute active**, and **debug ip igmp <ip address>**.

# Configuring Reliable Multicast for Workgroup Bridges

A workgroup bridge is different from a WLAN bridge. Cisco Aironet 350 Workgroup bridges are used to wirelessly extend the network to wired Ethernet clients. The workgroup bridge associates to an AP the same way a client does, but provides connectivity for a maximum of eight clients, rather than just one. The clients that use this connection share the bandwidth of the wireless connection, a maximum of 11 Mbps for the Aironet WGB350 Workgroup bridge.

Typically there is only one or, at most, a few workgroup bridges associated with an AP. In the rare case that more than 20 workgroup bridges need to associate to an AP, a configuration change must be made. The **Reliable multicast messages from the access point to workgroup bridges** setting attempts to maintain reliable delivery of multicast message to a maximum of about 20 workgroup bridges. If the intent is to exceed 20 workgroup bridges on one AP, this setting will need to be enabled (it is disabled by default). When you increase the number of workgroup bridges past the 20-unit mark, the reliability of delivery must be reduced. When reliability is reduced, the AP cannot confirm if multicast packets reach the intended workgroup bridge. A workgroup bridge at the furthermost edge of the AP's coverage area could lose connectivity. The **Reliable multicast messages from the access point to workgroup bridges** setting is found on VxWorks-based APs and bridges.

**NOTE**

This feature is not supported on the 5-GHz radio.

In AP1200 and AP1100 APs, a similar command exists to increase reliability of multicast to associated workgroup bridges. To configure reliable multicast messages in IOS-based APs and bridges, follow these steps:

1. From EXEC mode, enter global configuration mode and then configure the terminal. Type **configure terminal.**

2. Enter into interface configuration mode for the 2.4-GHz radio interface being configured. Type **interface dot11radio 0**.
3. Enable reliable multicast messages to workgroup bridges. Type **infrastructure-client**.
4. Return to EXEC mode and save the configuration. Type **copy running-config startup-config**.

# Summary

It is important to understand the challenges WLANs bring to network engineering. Most people consider WLANs to be an extension of their wired network where everything works exactly like it does in the "normal" environment. This is not always true. There are many things to think about when implementing WLANs compared to wired networks. Topics of great importance in WLANs (and not to be minimized when implementing wired networks) are security, bandwidth/throughput requirements, protocol support, and application reliability. This chapter considers several issues network engineers will uncover when implementing IP multicast solutions in WLAN environments.

The first section deals with the basics of multicast and how multicasting fits in with the OSI model. The next section discusses many of the primary issues involved in designing, planning, and implementing multicast in WLANs. Many specific recommendations are given that will help the network engineer prevent or resolve multicast issues in this environment. Several IP multicast configuration scenarios are given with diagrams to assist network engineers in rolling out their own scenarios. Additional scenarios cover multicast implementations using APs, bridges, and workgroup bridges.

The key to success in enabling multicast in WLANs is to first understand the requirements of the people who are multicasting the sessions to their clients and then understand the needs of the end users who will receive that information. Once that is done, the network engineer can develop a solution that will best deliver the multicast sessions throughout the network, whether it is over the wired network or the WLAN.

# Solutions Fast Track

## The OSI Model Overview

☑ Unicast, broadcast, and multicast are three methods of data transmission.

☑ Multicasting is based on IP addresses that originate in the 224.0.0.0 to 239.255.255.255 range.

☑ Multicasting allows for a single source to transmit to multiple destinations without having multiple data streams, thus reducing the impact on network resources.

## Multicast WLAN Deployment Recommendations

- ☑ There are many challenges in rolling out multicast applications in a WLAN environment.
- ☑ Allow only as much traffic as needed via filters to get the job done in a WLAN. Apply filters where necessary at the most optimized device.
- ☑ Make sure your multicast applications work in a WLAN environment prior to implementing them companywide.
- ☑ Create WLAN configurations that allow multicast packets to be sent at the highest data rate possible.
- ☑ Turn off power-saving modes if WLAN clients will receive multicast sessions.
- ☑ Do not use VPNs where multicast is needed.

## Controlling IP Multicast in a WLAN with APs

- ☑ Filter accordingly. If no multicast is needed on the WLAN, filter those packets at routers or switches prior to them entering the WLAN segment.
- ☑ If multicast is needed in the WLAN, allow only low-rate sessions across that shared segment.
- ☑ Block high-rate multicast sessions that will saturate the WLAN segment.
- ☑ Be careful when configuring filters; it is possible to block your own access to the device being configured. In other words, do not create filters until you have granted specific access for yourself, unless you are directly connected to the AP via the console port.

## Controlling IP Multicast in a Peer-to-peer WLAN using Bridges

- ☑ Use the **ip multicast boundary** IOS command on the Layer 3 switch or router on the multicast server side of the network.
- ☑ Use the IP multicast boundary IOS command on the remote side to prevent clients from issuing a join message to the RP for undesired multicast groups.

- ☑ A combination of configurations may be needed in a point-to-multipoint environment, depending on bandwidth availability and network engineering requirements.
- ☑ Make sure multicast traffic is properly allowed or denied by running IOS show and debug commands such as **show ip mroute**, **show ip mroute active**, and **debug ip igmp <ip address>**.

# Frequently Asked Questions

The following Frequently Asked Questions, answered by the authors of this book, are designed to both measure your understanding of the concepts presented in this chapter and to assist you with real-life implementation of these concepts. To have your questions about this chapter answered by the author, browse to **www.syngress.com/solutions** and click on the **"Ask the Author"** form. You will also gain access to thousands of other FAQs at ITFAQnet.com.

**Q:** Why should I allow multicast in my WLAN?

**A:** In networking environments, there are several solutions that require the use of multicast, including Cisco IPTV, video conferencing, many e-learning products, real-time news feeds, and radio broadcasts. When implemented properly, multicast can drastically reduce the needed bandwidth for advanced technology solutions. Multicast is not always allowed in WLAN environments, but if it is implemented properly, it can provide enhanced and valuable capabilities for endusers.

**Q:** What are the primary considerations when implementing multicast in a WLAN?

**A:** Allow as little bandwidth consumption as needed in WLAN environments. WLANs are shared mediums that can easily experience throughput degradation if improperly managed. Filter protocols where necessary on the most optimal device. Be flexible with multicast in WLANs. Even the most experienced engineer can have issues with implementing multicast in wireless environments. Be aware of the "gotchas" that are enabled by default.

**Q:** How can bandwidth adversely affect multicast?

**A:** If you have a high-rate datastream traversing a wireless segment, it could easily saturate that segment if allowed. Have your multicast team create low-rate as well as high-rate streams for both the wired and wireless areas of the network.

**Q:** Can I use my assigned IP addresses to perform multicasting?

**A:** No, IP multicast requires addresses within the range of 224.0.0.0 through 239.255.255.255. Some of these addresses are reserved for device and protocol management as well. Understanding how to implement multicast in the wired environment is just as important as understanding how to optimize it for your WLAN environment.

**Q:** How do I prevent unwanted multicast from traversing my WLAN segments?

**A:** Implement protocol, IP address, or MAC address filters on routers or Layer 3 switches before those packets get to your APs and bridges. Implement IP multicast boundaries so clients cannot inadvertently join multicast groups on RPs.

# Chapter 5

# WLAN Guest Network Access

## Solutions in this Chapter:

- Guest WLANs
- Designing a Guest VLAN
- Configuring Guest WLANs
- Access Point and Switch Configuration

☑ Summary

☑ Solutions Fast Track

☑ Frequently Asked Questions

# Introduction

One of the great advantages of a WLAN is the ease with which a guest can sit down and get connected in short order. No need to reboot or install special drivers—if you have a wireless card in your laptop, you can enjoy network services or Internet access as long as you are within range of an access point.

# Guest WLANs

Along with ease of use, WLANs offer a sense of impending danger. Unlike standard Ethernet cabling, you cannot terminate your WLAN at the edge of your office space. Those radio signals will carry until they attenuate (or run out of power). Many news articles have been written about the ubiquity of wireless networks and their level of security or insecurity. And here is the tricky problem: Sometimes the people who need to use the wireless network are not necessarily trusted employees—they are your vendors, visitors, or temporary workers. Giving these people temporary access to your WLAN could be an administrative nightmare because you will have to create 802.1x accounts for them and delete those accounts once your guests are no longer on the network.

As a solution to this problem, we suggest that you consider creating a guest WLAN. This is a virtual WLAN that will allow your guest to access the Internet (or other specific resources) and that will protect your network from casual (or not-so-casual) snooping. Many large companies now provide a guest VLAN for people who need temporary Internet access but do not need access to internal network resources.

Other companies are rolling out guest WLANs as a public service. They have already invested in the wireless infrastructure but are deploying Internet access via a guest WLAN to areas surrounding their facilities. Acts such as these are considered good PR, and although Internet access might be filtered in some way (maybe only Web surfing and e-mail are allowed), it is better than no Internet at all!

Regardless of your reason, if you need or want to set up a guest WLAN, this chapter is for you!

# Designing a Guest VLAN

As with any other network service, rolling out a guest WLAN takes a little preparation and planning. In this section we will give you some ideas to think about when designing and deploying your new guest network. You might want to take some notes or bend the corner of the page so that you can come back here as you make your plans.

# Design

The first step in creating a guest VLAN is determining your area of need. Will your guests be in certain areas of your facility, or is there a need for guest access everywhere? This decision will help you figure out which APs will need to be configured for guest access, since there is no reason to tie up or reserve resources on your APs if there is no need or demand for those VLANs in that area.

The next item to consider is the number of users who will be accessing this guest VLAN at any given point in time. What will their distribution be in relation to each AP? Will they be lumped in certain common areas or in consistently small groups scattered across the WLAN? Coming up with a good number of supportable guests per AP is crucial because you would not want to have the guest traffic overrun traffic from your regular users.

These numbers will also assist you in picking the size of the IP subnet you will be assigning to the guest VLAN. You will want to reserve an IP subnet large enough to cover not only the clients you want to support but the network interfaces you will be adding to this VLAN (router, switch ports, APs, and the like). Presenting a consistent IP subnet to your guests is important because you would not want them to lose their Layer-3 connection if they roam to another AP due to congestion or signal quality.

Another thing to consider is the guest VLAN's SSID. Instead of going with a default SSID like *Guest* or *tsunami*, why not define one that will be unique to your environment? Keep in mind that this guest SSID should not follow the SSID naming conventions you are using elsewhere in your network. You would not want to present a pattern to your guests wherein they could guess the names of your other SSIDs or network devices.

# Topology

It used to be that if you wanted to offer two separate WLANS in a specific area, you would have to roll out two completely separate wireless systems. This presented unique challenges such as channel overlapping, interference between the two systems, and troubleshooting issues between networks, not to mention the cost involved with the duplicate infrastructures (APs, cabling, switches, and so on).

Now we can take what used to be competing WLANs and serve them out of the same AP. This results in major savings, not only in the cost of deployment, but it simplifies ongoing maintenance. To the casual observer, you will look like you have multiple infrastructures. You, however, will know that they are all part of the same system—which makes troubleshooting network anomalies much simpler.

This is great news for service providers, since they can now serve up public Internet to the waiting public as well as provide internal access to employees and other trusted personnel.

## Deployment

One of the handiest things about rolling out a guest VLAN is the ease with which it is accomplished. All that is required is to make sure that your Cisco 1400, 1200, 350, or 340 AP is running either VxWorks firmware release 12.00T or Cisco IOS firmware release 12.2.4-JA or later.

Once you have designed your guest VLAN, you can deploy it to your APs one at a time or en masse, depending on your comfort level. All that is required is some minor configuration on each AP and its corresponding upstream switch.

Keep in mind that the AP will be broadcasting the SSID of the guest VLAN. All other VLANs will have their SSID broadcasts suppressed.

## Guest WLAN Recommendations

From a non-technical perspective, have you thought about placing some placards in the areas covered by your guest WLAN, announcing that it exists, and some pointers on how to get connected? Doing so will boost awareness and provide a mechanism for positive feedback.

On a more technical level, you might also consider creating some access control lists (ACLs) on your upstream switches and routers, preventing traffic on your guest VLAN from being routed anywhere other than to the Internet.

Depending on the amount of broadcast or multicast traffic on your network, you might also consider enabling Internet Group Management Protocol (IGMP) snooping and VLAN filtering on your upstream switch's ports to filter this traffic. Keep in mind that your AP acts as a bridge and will forward all that broadcast and multicast traffic onto the wireless network. Your AP's radio cannot receive at the same time as it is transmitting, so all that extra traffic could be slowing down your overall throughput. Stopping this traffic before it hits the AP could give your WLAN a significant performance boost!

# Configuring Guest WLANs

To configure your guest WLAN from the Command Line Interface (CLI), you first need to open a Telnet or SSH connection to the AP. After logging in and getting into enable mode, you will input the following commands:

```
configure terminal
```

This enters you into global configuration mode. Then:

```
interface dot11radio 0
```

Now you are in interface configuration mode. Subsequent commands will be applied to the interface you have specified (dot11radio 0). For 2.5GHz APs, you will only see a dot11radio 0. If you have a 5GHz AP (802.11a), you will be able to configure a dot11radio 1. It is of interest to note that when you have multiple VLANs on an AP, the dot11radio 0 will have a period and numbers following it. These are subinterfaces, and the numbers will correspond to the VLAN they are tied to.

```
ssid GuestNet
```

This will be the SSID of your guest VLAN. It can be anything up to 32 alphanumeric characters (a–z, A–Z, 1–0). Keep in mind that the SSID is case sensitive and cannot include spaces! Once you have entered this command, you are placed in SSID configuration mode, and any new configuration parameters will be applied to this SSID only. For our example, we will use the SSID of GuestNet:

```
vlan 67
```

This will assign this SSID to VLAN 67, which we have designated as the guest VLAN. Any wireless client who uses the guest SSID to connect to the WLAN will be joined to this VLAN. This VLAN will also need to be configured on your upstream switch. You might want to write this number down, since you will need it in future configuration statements.

```
guest-mode
```

This is the all-important configuration line. Without this, all you will have done is configure yet another SSID and VLAN on your AP. This line tells the AP that this SSID is for guest mode. Now that this has been entered, the AP will include this SSID in its beacons. Additionally, any wireless clients who attempt to associate to the AP without specifying a specific SSID will be permitted to connect to this SSID.

```
exit
```

This brings you back into global configuration mode:

```
interface dot11radio0.67
```

You are now entering the configuration mode for the radio's VLAN subinterface. This vlan-id should be the same as you configured in the preceding step. In our example, we will be configuring subinterface 67.

```
encapsulation dot1q 67
```

In previous configuration statements, we have told the AP that we want to tie a specific SSID to a VLAN. With this statement, we are actually enabling that VLAN (turning 802.1q on).

```
exit
```

Returning to global configuration mode one more time.

```
interface fastEthernet0.67
```

We have enabled 802.1q on the interface for the wireless side. Now we'll enable it on the wired side of the AP. You are now in interface configuration mode for the subinterface for the VLAN on the Ethernet interface.

```
encapsulation dot1q 67
```

We have now turned on 802.1q on the Ethernet interface. Now the AP can send and receive 802.1q tagged traffic for the guest VLAN of 67.

```
end
```

This gets you out of configuration mode. You will be at a standard prompt and have the opportunity to save your configuration by executing the following command:

```
copy running-config startup-config
```

In summary, here are the configuration statements needed to set up your guest VLAN:

```
configure terminal
interface dot11radio 0
ssid GuestNet
```

```
vlan 67
guest-mode
exit
interface dot11radio0.67
encapsulation dot1q 67
exit
interface fastEthernet0.67
encapsulation dot1q 67
end
```

There are other interface and SSID-specific commands that could be entered, but we'll save those for a later chapter.

# Access Point and Switch Configuration

Once you have configured your guest VLAN on your AP, you must make sure that you add this VLAN to the AP's upstream switch. This way, your guests will be able to access the internal network to get out to the Internet.

Once you have logged in to your switch, you can enter the following commands:

```
configure terminal
```

We are entering global configuration mode again with:

```
interface FastEthernet0/2
```

The AP is plugged into port 2 of our switch, so we want to configure this inteface.

```
switchport trunk encapsulation dot1q
```

We want to specify the kind of VLAN encapsulation we will be using. Our choices on the switch are Inter-Switch Link (ISL) and 802.1q, but the AP only speaks 802.1q. (See Chapter 9 for more details on both trunking protocols.)

```
switchport trunk allowed vlan 1,67
```

Here we will tell the switchport which VLANs are allowed to be in the VLAN trunk. VLAN 1 is normally the native VLAN, so it is included along with our guest VLAN of 67.

```
switchport mode trunk
```

Enable trunking on this port:

```
no ip address
```

This is important, since the actual switch interface will not need an IP address. We will configure IP addresses later, but assign them to the VLAN interface on the switch.

```
exit
```

This brings us back into global configuration mode:

```
interface vlan 67
```

Entering interface configuration mode for our guest VLAN:

```
ip address x.x.x.x
```

This gives your VLAN an IP address on the switch. Since this IP address is specific to the VLAN, it does not matter how many physical interfaces are on the switch trunk or bridge the VLAN. This is why we did not give the FastEthernet port that connects to the AP an IP address.

# WLAN Guest VLAN Filtering

Most users of your guest VLAN will appreciate the fact that you have provided them with a way to access the Internet. For some, however, human curiosity might get the better of them and make them attempt to discover some of your private VLANs. Don't rely on "security through obscurity" here—sooner or later, the private VLAN SSID could be leaked. Use a RADIUS server to authenticate your wireless users. This way, even if an end user changes his or her SSID to a private VLAN, that user will be prevented from actually joining that VLAN due to 802.1x authentication errors. (This is discussed in greater detail in Chapter 9.)

# Summary

So there you are! With a little bit of planning and configuration, you are now able to give your guests wireless access to the Internet. Although you might not get a personal word of thanks from them, you can rest assured that you have provided them a welcome service while keeping your internal network secure. Isn't that what this is really about?

# Solutions Fast Track

## Guest WLANs

☑ Allows a user to associate with the AP without specifying a SSID.

☑ Enables non-802.1x clients network access.

☑ Separates noninternal traffic from being able to access internal network resources.

☑ Are specified by using the guest-mode configuration statement from within interface configuration mode.

## Designing a Guest VLAN

☑ Decide what services will be provided (Web only, SMTP, other services).

☑ Determine whether QoS or other network requirements are associated with the services to be provided.

☑ Figure out if any security features need to be implemented other than the VLAN itself.

☑ Create virtual LANs and segment specific types of traffic from one another.

☑ Useful in providing distinct network access to disparate groups of users.

☑ Have to be mapped to a specific interface or trunked.

## Configuring Guest WLANs

☑ Decide on a name for the guest SSID, and choose a VLAN number for it.

☑ Configure the guest SSID with the guest-mode keyword.

- ☑ Turn on 802.1q encapsulation on the radio's subinterface for that VLAN.
- ☑ Turn on 802.1q encapsulation on the Ethernet subinterface for that VLAN.

## Access Point and Switch Configuration

- ☑ Configure 802.1q trunking on the Ethernet port to which the AP connects.
- ☑ Assign IP addresses to the VLANs, not the physical interface ports on the switch.

# Frequently Asked Questions

The following Frequently Asked Questions, answered by the authors of this book, are designed to both measure your understanding of the concepts presented in this chapter and to assist you with real-life implementation of these concepts. To have your questions about this chapter answered by the author, browse to **www.syngress.com/solutions** and click on the **"Ask the Author"** form. You will also gain access to thousands of other FAQs at ITFAQnet.com.

**Q:** How many Guest WLAN's can I have?

**A:** While you could create more than one unsecured WLAN, only the Guest WLAN will have its SSID in the Access Point's beacon messages.

**Q:** I have users who should be connecting to my regular WLAN, but are connecting to my Guest WLAN. Help!

**A:** Check to see if the user's laptop supports 802.1x. If it doesn't, it will join the Guest WLAN by default.

**Q:** My switch only supports ISL trunking. Can I still create a Guest WLAN?

**A:** Not really. None of the Cisco Access Points support ISL trunking. So while you could configure a Guest WLAN on the Access Point, traffic wouldn't go anywhere because the VLAN would not extend past the Access Point.

**Q:** How do I prevent outside people from connecting to my Guest WLAN?

**A:** While the idea of a Guest WLAN is providing Internet access, if you want to limit the area of coverage, you might try lowering the Output Power of the radio.

Chapter 6

# Implementing Cisco Wireless LANs

## Solutions in this Chapter:

- The Cisco Wireless and Wireless-aware Vision
- The Cisco Structured Wireless-aware Network Product Line
- Cisco IOS and WLANs
- Cisco Aironet Access Points (APs)
- Cisco Aironet WLAN Client Adapters
- CiscoWorks Wireless LAN Solution Engine 2.x
- Cisco Wireless Security Suite
- Cisco Secure Access Control Server (ACS) 3.2
- Enhanced Client Network Management Features with Extended Client Support
- Workgroup Bridges
- Wireless Bridges

# Introduction

In November 1999, in an effort to catch up to other vendors' who had already entered the wireless field, Cisco purchased one of the leading wireless technology companies called Aironet Wireless Communications, Inc. Aironet was a leading developer of high-speed wireless products who also played an important role in the Institute of Electrical and Electronic Engineers (IEEE) 802.11 wireless work group. Since then, Cisco has been aggressively expanding their technology in wireless offerings in both hardware and software, providing one of the industry's most competitive wireless solutions.

In June 2003, Cisco again showed its ambition and interest in wireless technology by acquiring a popular wireless company called Linksys. Linksys helped Cisco establish its presence in the small office/home office (SOHO) market by mainly targeting wireless home users. There is currently no plan to integrate Linksys wireless devices with the Aironet product line. Linksys is still operating under its own banner with insight and direction coming from Cisco.

Cisco has a diversified selection of wireless solutions and wireless-aware products that are capable of providing wireless networks to small and enterprise-level environments. This chapter reviews each Cisco product including wireless access points (APs) and wireless bridges and their benefits. It looks at how Cisco's strong wire-networking heritage is being incorporated into its wireless networking offering and its management in order to provide true enterprise-level solutions for both data and voice over wireless. With its new wireless solutions portfolio, Cisco has positioned itself to be a strong future player of wireless technology and to serve as a one-time stop for its customers seeking wired and wireless network solutions.

# The Cisco Wireless and Wireless-aware Vision

Throughout the years, Cisco has played an important role in the network industry, mainly known for its routers and switches. In the past 15 years, Cisco has been diversified its product portfolio by purchasing different technology companies offering products and services such as content networking, voice, security, storage, management and others.

Cisco's vision is to seamlessly integrate wireless networking technology with wired networking by extending technology that was developed for wired networks into wireless networks. Technology such as security, scalability, and manageability that customers have come to expect on wired networks are being extended and provided in wireless networks. This chapter reviews Cisco products and the extended services they provide for wireless networks in more detail.

# The Cisco Structured Wireless-aware Network Product Line

The Cisco structured wireless-aware network product line provides network architects and managers with quality, security, and manageability for small to large enterprise networks. In June 2003, Cisco announced a list of new enhanced wireless-aware network products that help integrate local area networks (LANs) with wireless local area networks (WLANs). Some of the wireless and wireless-aware solutions that Cisco offers include:

- Cisco Aironet Series Wireless APs
- Cisco Aironet Series Wireless Bridges
- Cisco Aironet Series Client Adapters
- Cisco Internet Operating System (IOS) Software
- CiscoWorks Wireless LAN Solution Engine (WLSE)
- Cisco Wireless Security Suite
- Cisco Secure Access Control Server (ACS)
- Cisco Wireless-aware LAN Switches and Routers
- Cisco Wireless Antennas and Accessories

Figure 6.1 shows an example of wireless integration with a wired network. In this figure, Cisco offers all networking products from wireless phones and wireless client adapters to wireless APs and bridges with multiple external antenna options. This figure also shows the routers, switches, and management applications such as CiscoWorks and ACS needed to manage the entire wired and wireless infrastructure from central command.

**Figure 6.1** Structured Wireless Product Line

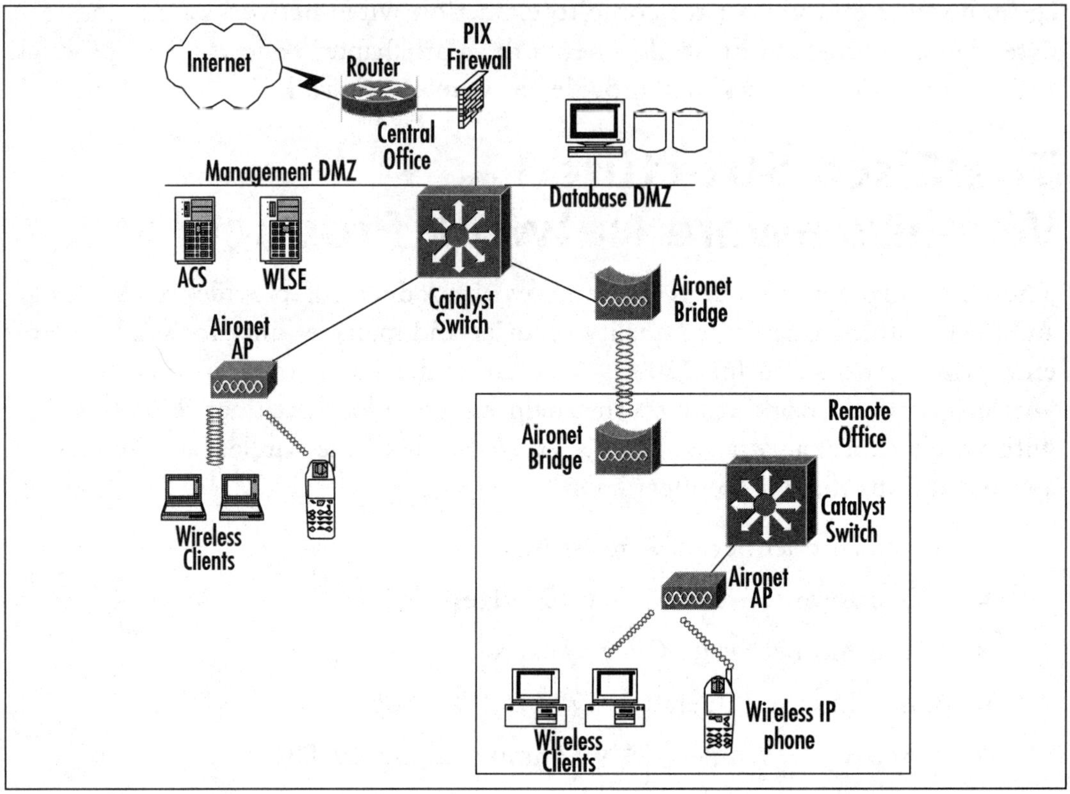


# APs

Cisco wireless APs (shown in Figure 6.1) are used to provide wireless connectivity to clients, mainly serving as a gateway between wireless and wired networks. Cisco Aironet APs support many different functions, such as detail security settings, to ensure confidentiality, integrity, and availability. Cisco APs also support virtual LANs (VLANs) that may be used to provide access for multiple unique wireless clients with different compatibility or policy settings. VLANs and their benefits are discussed in detail in Chapter 9, "Wireless LAN VLANs."

# Aironet Bridges

Cisco Aironet wireless bridges (shown in Figure 6.1) are used to interconnect remote offices with the main site. A slight distinction needs to be made between an AP and a bridge. Generally, APs handle end clients such as laptops with wireless cards or wireless Internet Protocol (IP) phones, while wireless bridges connect networks (wireless or wired) over a greater distance than that handled by an AP. Bridges are discussed later in this chapter in the "Wireless Bridges" and "Workgroup Bridges" sections.

## Client Adapters

Cisco wireless client adapters are used to connect a variety of devices to the AP. Cisco wireless client adapters support 802.11a-, 802.11b-, and 802.11g-compatible network technology. Cisco wireless client adapters are available in CardBus-, Personal Computer Memory Card International Association (PCMCIA)-, and Peripheral Component Interconnect (PCI)-compatible devices to fit any environment. Client adapters are discussed in the "Cisco Aironet WLAN Client Adapters" section later in this chapter.

## Cisco IOS

Cisco IOS is a well-known operating system that runs mainly on Cisco routers and switches. Not long ago, Cisco made the decision to convert Aironet's original operating system (VxWorks) that is run on APs and bridges, into IOS firmware. This change brings many new features and manageability benefits to the Aironet product line and also brings Cisco a step closer to true integration of their wired and wireless network products. Details of IOS upgrades and their integration on Cisco APs and bridges are discussed later in the "Cisco IOS and WLANs" section.

## Wireless LAN Solution Engine

With the continued expansion of wireless products and their deployment, network engineers needed a solution to easily manage their mid-size to large wireless environments. Cisco recognized the need for central management and added Wireless LAN Solution Engine (WLSE) to its popular CiscoWorks management application set. WLSE provides scalable and central management solutions for thousands of deployed Aironet products. WLSE is a welcomed tool among network administrators not only for its central management ability in larger deployments but because it improves security measures such as rogue AP detection. Some administrators install WLSE even if they do not have a legitimate wireless network, solely to use its unauthorized AP detection functionality to protect wired networks. Details of WLSE are covered later in this chapter in the "CiscoWorks Wireless LAN Solution Engine 2.x" section.

## Wireless Security Suite

Past wireless security research and findings caused many companies to hesitate to deploy wireless solutions due to the high security risks. Cisco has continually invested money and time into developing and supporting best security practices in their products.

Cisco wireless clients and APs support all of the industry wireless security standards such as 802.1X/EAP per-user authentication, Temporal Key Integrity Protocol (TKIP) per-packet keying enhancement for Wireless Encryption Protocol (WEP)/RC4 encryption, and much more. Cisco has further added Extensible Authentication Protocol (EAP) support to its popular ACS and has introduced new enhanced rogue AP detection and mitigation measures in its WLSE management engine. Detecting rogue APs is covered in detail in Chapter 8, "WLAN Rogue AP Detection and Mitigation."

## Access Control Server

The Cisco Access Control Server (ACS) is used for central user access management and user Authentication, Authorization, and Accounting (AAA). It supports 802.1X/EAP, Lightweight Extensible Authentication Protocol (LEAP), Protected Extensible Authentication Protocol (PEAP), and other per-user authentication protocols using carrier protocols such as Remote Authentication Dial-In User Service (RADIUS) and Terminal Access Controller Access Control System (TACACS). ACS provides scalability and improves manageability with its local user database and its support for remote user databases such as Lightweight Directory Access Protocol (LDAP), Active Directory, or RSA Security's ACE/Server Solution. ACS and its interaction in wireless environments is covered later in the "Cisco Secure Access Control Server (ACS)" section.

## Cisco Wireless LAN Switches and Routers

Cisco's announcement in 2003 to provide structured wireless-aware network solutions included router and switch integration starting in 2004. Some of the routers included were the Cisco 2600XM and 3700 series. Switches included in this announcement were the Cisco Catalyst 3750, 4500, and 6500 series. This integration will enhance services such as Quality of Service (QOS) and its end-to-end delivery. Further, it will aid in security services such as rogue AP detection and mitigation (the Cisco-compatible switch will be able to work in coordination with WLSE to automatically pinpoint the actual port of unauthorized connection and shut it down).

# Cisco Wireless Antennas and Accessories

Cisco offers a variety of different styles of antennas for both APs and bridges in the 2.4 Gigahertz (GHz) (802.11b/g) and 5 GHz (802.11a) product line. As discussed in Chapter 2, antennas play a crucial role in designing a successful WLAN. Each type of antenna offers a different style of coverage, whether it provides omnidirectional coverage for local users or long-range directional point-to-point and point-to-multipoint coverage between buildings and offices.

Before discussing the available antenna choices, we need to review the terminology used:

- **Decibel (dB)** Unit of measure for ratios describing loss or gain, normally expressed in watts. A decibel is the measurement of power gained or lost between two communication devices. These units are normally given in terms of the logarithm to Base 10 of a ratio.
- **dBi Value** Ratio of the gain of an antenna as compared to an isotropic antenna. The greater the dBi value, the higher the gain. If the gain is higher, the angle of coverage is more acute.
- **Isotropic Antenna** Theoretical construct that an antenna will radiate its signal 360 degrees to cover the area in a perfect sphere. Used as a basic to describe the gain of a real antenna.
- **Line-of-sight** Unobstructed straight line between two transmitting devices. Administrators will most often need a line-of-sight path for long-range directional radio transmissions. Due to the curvature of the earth, the line-of-sight for devices not mounted on towers is limited to 6 miles (9.65 km).
- **Signal Attenuation (Multipath Fading)** Reduction of signal strength based on one of several factors: absorption, diffraction, reflection, or refraction. Absorption is obstruction (such as trees) that soak up the signal so that it is unable to reach the receiver that it is trying to communicate with. Diffraction is when a signal bends around an obstruction that has a reflective quality (such as glass). Reflection is when a signal bounces off a surface (such as a body of water) causing distortion and sometimes canceling the signal.

The sections that follow reference the horizontal and vertical coverage angle of an antenna. (Refer to Figure 6.2 for a diagram of horizontal coverage and Figure 6.3 for a diagram of vertical coverage).

**Figure 6.2** Horizontal Coverage Angle

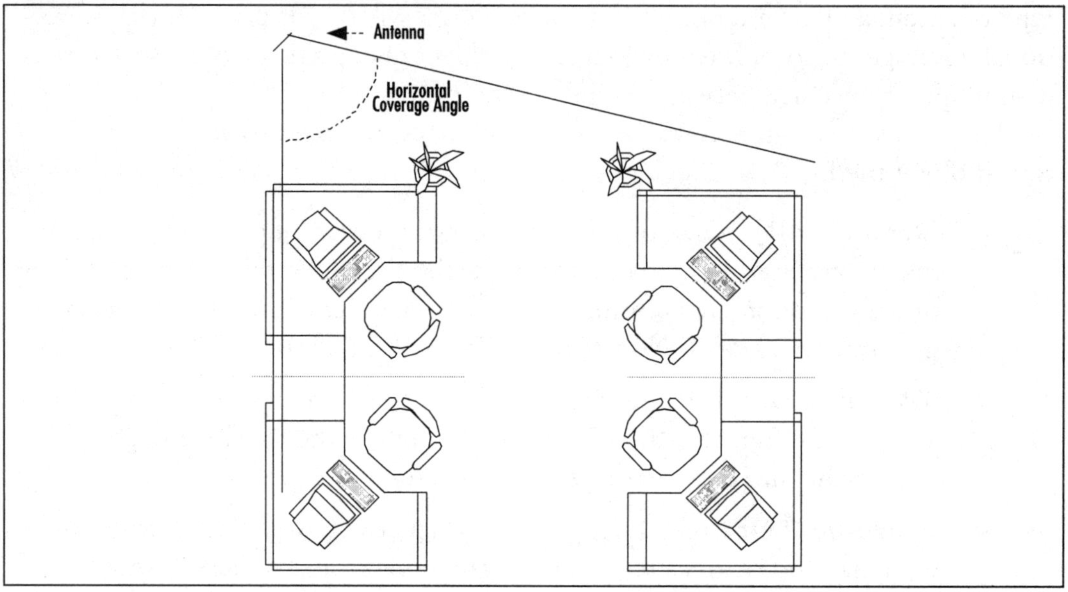


**Figure 6.3** Vertical Coverage Angle

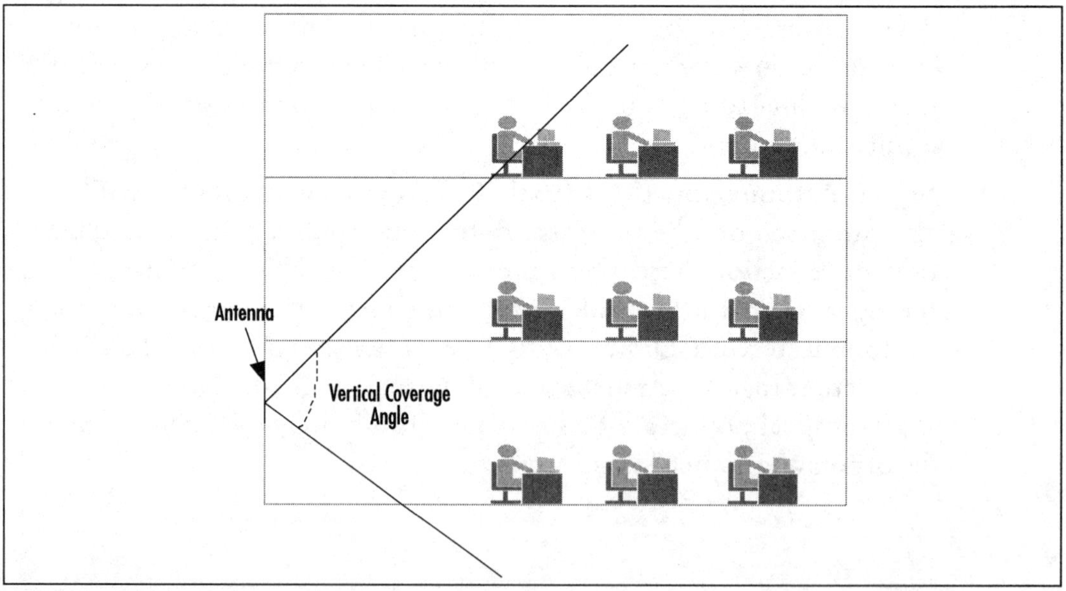

## Ceiling Mount Omnidirectional Antenna 2.4 GHz (AIR-ANT1728)

This indoor AP medium range antenna is designed to provide 360-degree coverage in an office space environment. Specifically, the antenna has 360-degree horizontal coverage and 38-degree vertical coverage. It is cylindrically shaped and is 9 inches long with a 1-inch diameter. It is light enough, (4.6 oz [131 g]) to be hung from a drop ceiling and comes with a mounting bracket specifically for this purpose. It has a 3-foot pigtail of coaxial cable at one end that terminates in an RP-TNC connector. The approximate range provided by the antenna is 497 feet (151 m) at 1 Mbps and 142 feet (44 m) at 11 Mbps.

## Mast Mount Omnidirectional Antenna 2.4 GHz (AIR-ANT2506)

The mast mount is primarily a short-range outdoor antenna. Much like the ceiling mount omnidirectional antenna, the mast mount omnidirectional antenna has 360-degree horizontal coverage and 38-degree vertical coverage. Though it is specifically designed for outdoor short-range point-to-multipoint applications, it can also be used indoors if needed, providing similar performance to the ceiling mount omnidirectional antenna. The mast mount antenna is cylindrical in design, 13 inches long and 1 inch in diameter. The approximate range provided by the antenna for an outdoor bridge configuration is 5,000 feet (1525 m) at 2 Mbps and 1,580 feet (480 m) at 11 Mbps.

## High-Gain Mast Mount Omnidirectional Antenna 2.4 GHz (AIR-ANT24120)

This antenna is much like the mast mount omnidirectional antenna described in the preceding section. It is a medium range outdoor antenna that has a 360-degree horizontal coverage and is designed to be fastened to a mast and used for point-to-point coverage applications. However, the major performance difference comes in the vertical coverage. The high-gain antenna has a vertical coverage of only 7 degrees compared to 38-degree vertical coverage from the normal mast mount antenna. Because it transmits the signal at a smaller angle, more of the energy of the signal is concentrated thus giving the antenna better range. Specifically, the range of this antenna is 4.6 miles (7.4 km) at 2 Mbps and 1.4 miles (2.3 km) at 11 Mbps. Like the previous mast mount antenna, this mast mount antenna is also cylindrical in design, however at 42 inches long it is about four times longer. The diameter of the high-gain mast mount antenna is also slightly larger at 1.5 inches. The cable used as a pigtail on the antenna is a 1-foot RG-8 cable with an RP-TNC connector on the end.

## Pilar Mount Diversity Omnidirectional Antenna 2.4 GHz (AIR-ANT3213)

This antenna is specifically designed to provide omnidirectional service while being unobtrusive. For indoor use, this medium range antenna has two RP-TNC connectors on the end of a 3-foot Siamese coaxial cable. The two inputs allow for the transmission of diverse signals in the event that there is a failure of one of the transmissions. The exterior of the antenna is covered with a tan cloth and when mounted with the mounting brackets that ship with it, it will sit approximately 6 inches from the wall. It has a rectangular shape with dimensions of 10 × 1 inches. The antenna has 360-degree horizontal coverage and 30-degree vertical coverage. The approximate range provided by the antenna is 497 feet (151 m) at 1 Mbps and 167 feet (51 m) at 11 Mbps.

## POS Diversity Dipole Omnidirectional Antenna 2.4 GHz (AIR-ANT3351)

All of the other antennas discussed in this section are specifically designed to work with the APs or bridges, however, this one is designed to work with special Aironet wireless client adapters. Specifically, this antenna works with the Aironet 350 series AIR-LMC adapter that has dual Micro-Miniature Coaxial (MMCX) connectors. (AIR-LMC adapters are discussed later in the "Aironet Client Adapters" section.) These connectors attach to the antenna's 3-foot pigtails (also with MMCX-style connectors) and allow for better signal transmission from the workstation. The antenna provides 360-degree horizontal coverage, 75-degree vertical coverage, and a range of 350 feet (107 m) at 1 Mbps and 100 feet (51 m) at 11 Mbps. Though somewhat irregularly shaped, the diversity dipole antenna has dimensions of 7 inches long, 2 inches wide, and 8 inches high. (See Figure 6.14)

## Diversity Ceiling Mount Omnidirectional Patch Antenna 2.4 GHz (AIR-ANT5959)

This small, 5.3 inches long × 2.8 inches wide × by 1-inch thick rectangular antenna was specifically designed to be unobtrusive in a normal office environment. It comes with a mounting bracket that allows it to be mounted on a drop ceiling, thus allowing for maximum coverage in a cubicle environment. It provides 360-degree horizontal coverage and 80-degree vertical coverage. In addition, it has roughly the same range (350 feet (105 m) at 1 Mbps and 130 feet (45 m) at 11 Mbps) as the standard dipole antenna that comes with some of the APs. It has two diverse transmitting elements accessed via two 3-foot pigtails each with their own RP-TNC connector.

## Directional Wall Mount Patch Antenna 2.4 GHz (AIR-ANT3549, AIR-ANT1729)

Unlike the antennas discussed up to this point, the wall mount patch antenna is not omnidirectional. The horizontal and vertical coverage area for this antenna (AIR-ANT3549) is 60 degrees. The main difference in the installation of an omnidirectional antenna and a directional antenna is that an omnidirectional antenna is installed in the center of the coverage area while a directional antenna is installed at the edge of the coverage area. The wall mount patch antenna is a long-range 700/200 foot (213/61 m) indoor antenna for use with the AP products, which can also be installed as a medium-range outdoor bridge antenna. Specifically, this particular antenna is small, 5 inches square × .5 inches thick, and light colored so that it blends in easily in an office environment. Mounting this antenna requires nothing more than four screws to attach it to a wall. In addition, it has a 3-foot pigtail that terminates in an RP-TNC connector for connecting the antenna to the AP or bridge.

In addition to the patch antenna, another version of this antenna (AIR-ANT1729) is available from Cisco. This version has a larger vertical and horizontal coverage angle (75 degrees horizontal and 60 degrees vertical) for transmission and as such has a lower overall gain. It is appropriate for indoor or outdoor applications, and has a range of 542 feet (165 m) at 1 Mbps and 155 feet (47 m) at 11 Mbps if used with an AP. The final difference between the two is that this antenna is 1 inch narrower in width.

## Diversity Directional Wall Mount Patch Antenna 2.4 GHz (AIR-ANT2012)

Just like the previous patch antenna, this one is also a directional-based antenna. It is designed primarily for indoor use and has a medium range of 547 feet (167 m) at 1 Mbps and 167 feet (51 m) at 11 Mbps. However, this antenna has the advantage of having two radiating elements that each have their own 3-foot pigtail attached to them. This allows you to take advantage of the dual RP-TNC connectors that are supplied on many of the APs. The patch antenna supports 80-degree horizontal coverage and 55-degree vertical coverage. Due to its compact size (4.78 inches × 6.66 inches × 1 inch thick), it can easily blend into the surroundings in an office environment

## Yagi Antenna 2.4 GHz (AIR-ANT1949)

The yagi antenna is specifically designed for outdoor use in a point-to-point application. Cylindrical in shape and relatively compact in size (18 inches long with a 3-inch diameter), the signal from the yagi antenna comes out the end of the cylinder. The yagi antenna is able to obtain its gain through limiting the horizontal and vertical radiation pattern. With a 30-degree horizontal and 25-degree vertical pattern, this antenna can obtain a maximum range of 6.5 miles (10.5 km) at 2 Mbps and 2.0 miles (3.3 km) at 11 Mbps. The yagi antenna also comes with a 3-foot pigtail with an RP-TNC connector at the end.

## Dish Antenna 2.4 GHz (AIR-ANT3338)

The dish antenna is also designed for outdoor use in a point-to-point application. Similar in size (2-foot diameter) and function to the small satellite dishes that are used for television transmission, the dish antenna provides the longest range of any of the Cisco antennas. Specifically, it has a maximum range of 25 miles (40 km) at 2 Mbps and 11.5 miles (18.5 km) at 11 Mbps. As with any antenna, the dish antenna can obtain its longer range by reducing the radiation angle. In this case, both the vertical and horizontal radiation angles are 12.4 degrees. This small area can make aligning two dish antennas over a large distance a difficult task. The dish antenna can be connected to the bridge via an attached 2-foot pigtail with an RP-TNC connector.

## Cisco's 2.4 GHz Antennas Summary

Figure 6.4 displays the 2.4 GHz (802.11b/g) antenna family Cisco offers for bridges and APs that support external RP-TNC connectors. Refer to Table 6.1 for a detailed listing of Aironet 2.4 GHz APs and Table 6.2 for a detailed listing of Aironet 2.4 GHz bridges.

**Figure 6.4** 2.4 GHz Antennas for 802.11g/b Aironet Products

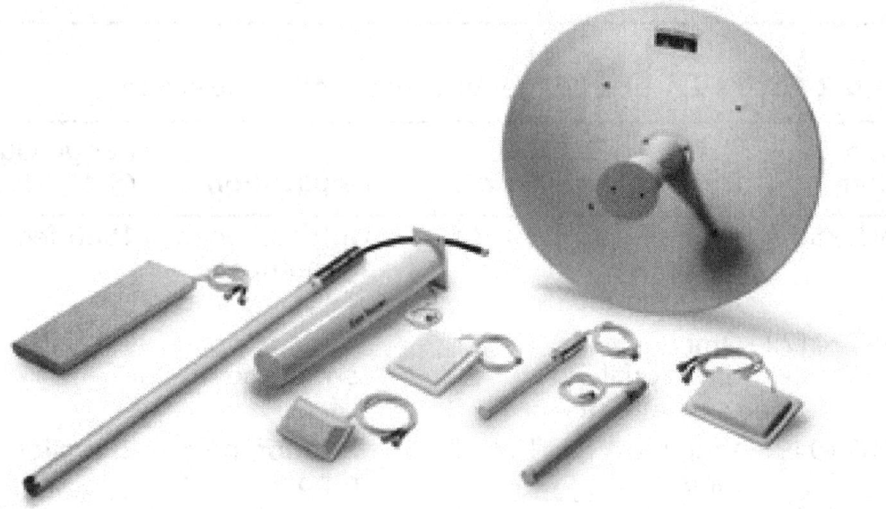



**Table 6.1** Cisco's 2.4 GHz AP Antennas Product Selection

| Product Number | Description | Gain | Application | Range Indoor @ 11 Mbps |
|---|---|---|---|---|
| AIR-ANT5959 | Diversity omni-directional | 2.0 to 2.35 dBi | Indoor, ceiling mount | 130 feet |
| AIR-ANT2012 | Diversity patch | 6.5 dBi | Indoor/outdoor, wall mount | 167 feet |
| AIR-ANT3213 | Diversity omni-directional | 5.2 dBi | Indoor, pillar mount | 142 feet |
| AIR-ANT2410Y-R | Yagi | 10 dBi | Indoor/outdoor, mast or wall mount | 230 feet |
| AIR-ANT1728 | Omni-directional | 5.2 dBi | Indoor, ceiling mount | 142 feet |
| AIR-ANT4941 | Dipole omni-directional | 2.2 dBi | Indoor | 130 feet |

Continued

**Table 6.1** Cisco's 2.4 GHz AP Antennas Product Selection

| Product Number | Description | Gain | Application | Range Indoor @ 11 Mbps |
|---|---|---|---|---|
| AIR-ANT3549 | Patch | 9 dBi | Indoor, wall mount | AP: 200 feet<br>Bridge: 3390 ft |
| AIR-ANT1729 | Patch | 6 dBi | Indoor/outdoor, wall mount | AP: 155 feet<br>Bridge: 1900 feet |

**Table 6.2** Cisco's 2.4 GHz Bridge Antennas Product Selection

| Product Number | Description | Gain | Application | Range Outdoor @ 11 Mbps |
|---|---|---|---|---|
| AIR-ANT2506 | Omni-directional, mast mount | 5.2 dBi | Outdoor, point-to-multipoint | 1580 feet |
| AIR-ANT24120 | Omni-directional, mast mount | 12 dBi | Outdoor, point-to-multipoint | 1.4 miles |
| AIR-ANT1949 | Yagi, mast mount | 13.5 dBi | Outdoor, point-to-point | 2.0 miles |
| AIR-ANT3338 | Dish | 21 dBi | Outdoor, point-to-point | 11.5 miles |

## 5 GHz Antennas

Cisco offers three main antenna choices for its 5 GHz (802.11a) Aironet products. As discussed later in this chapter, the only Aironet product that currently supports 802.11a with external antenna connection is the Aironet 1400 bridge. Although the 1200 Cisco AP also supports 802.11a 5 GHz radio, it comes with its own integrated 5 GHz antenna for its 802.11a radio.

Cisco's three antenna solutions include Omnidirectional, Sector, and Dish. Directional dish provides great point-to-point distance connectivity of up to 23 miles (37km) at 9 Mbps and up to 12 miles at 54 Mbps. Refer to Table 6.3 for more details on Cisco's 5 GHz antenna selection.

**Table 6.3** 5GHz Antenna Solutions for the 802.11a Aironet 1400 Bridge

| 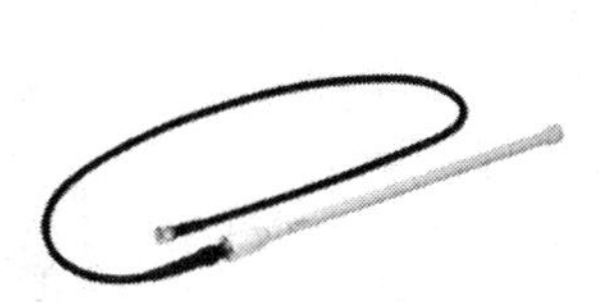 |  |  |
|---|---|---|
| Omnidirectional, mast mount | Sector, mast mount | Dish, mast mount |
| Outdoor, point-to-multipoint | Outdoor, point-to-point and point-to-multipoint | Outdoor, point-to-point |
| 9.0 dBi | 9.5 dBi | 28.0 dBi |
| 2 miles @ 54 Mbps<br>8 miles @ 9 Mbps | 2 miles @ 54 Mbps<br>8 miles @ 9 Mbps | 12 miles @ 54 Mbps<br>23 miles @ 9 Mbps |
| 360-degree horizontal beam width<br>6-degree vertical beam width | 60-degree horizontal beam width<br>60-degree vertical beam width | 5.7-degree horizontal beam width<br>6-degree vertical beam width |
| Weight: 2.0 lb (0.9 kg) | Weight: 1.25 lb (0.6 kg) | Weight: 9.5 lb (4.3 kg) |



## Cisco Wireless IP Phone

Cisco extends its Voice over IP (VoIP) technology into the wireless environment with its 7920 Wireless IP phone (see Figure 6.5). The Cisco 7920 wireless IP phone supports 802.11b in conjunction with Cisco Aironet products such as the 1200 and 1100 APs.

**Figure 6.5** 7920 Wireless IP Phone (LBD02004)

Cisco 7920 fully supports and interconnects with Cisco CallManager and other Cisco voice applications. It is ideal for anyone who must use an IP phone while traveling around a wireless facility. The plan for wireless phones is to be able to hop from a WLAN environment within an in-house office to an outside GSM network in order to get the lowest possible routing cost. The 7920 phone is Cisco's first generation of wireless IP phones. It further supports security features such as 802.1X(LEAP) and 40- or 128-bit Wired Equivalent Privacy to ensure confidentiality.

# Cisco IOS and WLANs

IOS is a well-known operating system that mainly runs on Cisco routers and switches. Not long ago, Cisco made the decision to convert Aironet's original operating system, VxWorks (used in APs and bridges), into an IOS-compatible system. This new IOS operating system brings many new features and manageability benefits into the Aironet wireless product line.

One of the biggest benefits of IOS and its support for wireless Aironet products is manageability. Network administrators who work with other Cisco products such as routers are very familiar with the IOS firmware image and its configuration. Table 6.4 lists Cisco Aironet products that support IOS.

**Table 6.4** Aironet IOS Support List

| Product | IOS | VxVorks |
|---|---|---|
| AP 1200 | Yes | Yes |
| AP 1100 | Yes | No |
| AP 350 | Yes | Yes |
| Bridge 1400 | Yes | No |

# Upgrading from VxWorks to IOS

Those who run compatible non-IOS VxWorks AP operating systems can upgrade to IOS standard software. Cisco supports three main options for upgrading and converting from non-IOS to IOS firmware:

- Using the Cisco Aironet Conversion Tool for Cisco IOS software
- Using a browser and VxWorks
- Using CiscoWorks WLSE for IOS conversion

**WARNING!**

**If you are upgrading a Cisco Aironet 1200 or 350 AP from VxWorks to IOS, you cannot reverse the process. When an AP is upgraded to IOS you cannot go back to the VxWorks operating system.**

## Using the Aironet Conversion Tool for Cisco IOS Software v2.0

The Cisco Aironet Conversion Tool (CAC Tool) is used to upgrade compatible non-IOS (VxWorks) APs into IOS firmware image. It includes the capability to not only convert the operating system firmware but also the actual configuration file settings so that you do not have to reconfigure the AP after upgrade. The 350 series APs non-IOS versions supported by the conversion tool are: 11.21, 11.23T, 12.00T, 12.01T, 12.02T, and 12.03T. The 1200 series APs non-IOS versions supported by the conversion tool are: 11.54T, 11.56, 12.00T, 12.01T, 12.02T, and

12.03T. If you are running older version of VxWorks that is not supported, you must first upgrade your VxWorks firmware and then convert to IOS.

In order for the conversion tool to function properly, the target APs must have Simple Network Management Protocol (SNMP) write-enabled, administration with the admin privilege configured, and the firmware privileges enabled. SNMP is used to obtain configuration settings prior to the firmware upgrade, which are then converted and applied to the new IOS configuration file. However, there are a few exceptions to the configuration settings that must be copied from VxWorks manually prior to the upgrade. These settings will not be converted into the new IOS configuration file using the conversion tool. These configurations include: the WEP keys for both radios and VLAN, the LEAP password for the repeater AP, the password used with the user management configuration, and the AAA server key settings.

When you are ready to upgrade the CAC Tool and the conversion IOS image files can be downloaded from www.cisco.com. Follow all of the instructions prior to upgrading.

## Using Browser and VxWorks

The browser can be used to upgrade a non-IOS AP to IOS firmware without converting the configuration file. Keep in mind that by using this method of conversion the newly IOS-converted AP will lose all of its previous configurations. The steps needed to upgrade from non-IOS to IOS without converting the configuration settings are:

1. Download the **Helper Image** for the 350 or 1200 AP from www.cisco.com.
2. Open the browser connection to your AP and login.
3. Click on **Setup** in the Summary Setup window.
4. Click on **Cisco Services** under Services.
5. Click on **Through Browser** under Fully Upgrade Firmware.
6. Browse to the **helper** image you downloaded from www.cisco.com.
7. Click on **Browser Update Now** to begin the upgrade process.

**WARNING**

When upgrading the AP do not remove power in order to terminate the upgrade process. You risk seriously damaging the AP. The upgrade can take from 3 to 30 minutes; when complete a message appears to indicate that the AP is being rebooted. Reboot should not take longer than 30 seconds.

### Using CiscoWorks WLSE for IOS Conversion

The WLSE supports central management upgrades of IOS and non-IOS Aironet devices. It also supports conversions of non-IOS 1200 and 350 APs to IOS firmware. Just like the CAC Tool covered in the previous section, the WLSE converts configuration files along with the firmware upgrade. The same limitations apply to the configuration file conversion as to the CAC Tool: the LEAP, WEP keys, and AAA server must be manually entered into the conversion template prior to upgrade as they are not automatically converted.

# Cisco Aironet Access Points (APS)

Cisco wireless APs are the heart of the wireless operation, used to provide wireless connectivity to clients and mainly serving as a gateway between wireless and wired networks. Cisco offers two main types of Aironet AP solutions, the 1200 series and the 1100 series. The 350 series Aironet has reached end-of-sale and is no longer being sold by Cisco. Both the 1200 and 1100 series support the new 802.11g wireless standard. The 1200 series is a higher-end solution that supports two installed modular radios and an external antenna connectivity. The 1100 supports one installed radio with an integrated antenna. This section details features of all three of Cisco's series APs.

## Aironet 1200 AP

The Cisco Aironet 1200 AP series is a very flexible and compatible AP that protects future investments with its modular upgradeable radio and antenna solution design. It supports IEEE's industry standard wireless technologies 802.11b, 802.11a, and 802.11g. Up to two radios can be installed inside the 1200 AP providing the flexibility to support all three standards at the same time using only one AP. The 802.11g mini 32-bit PCI radio module can support up to 54 MBs in a 2.4 GHz band. The 802.11g radio can be configured to support both 802.11g and legacy 802.11b standards. The second modular interface can be used to install 802.11a CardBus 32-bit radio, which supports up to 54 Mbps in a 5 GHz band. Figure 6.6 illustrates the flexibility of the 1200 AP with its 802.11g and 802.11a radios installed. Clients and neighboring APs using any of the three 802.11 technologies are all supported providing for great scalability and compatibility.

**Figure 6.6** 1200/a/b/g Scalability

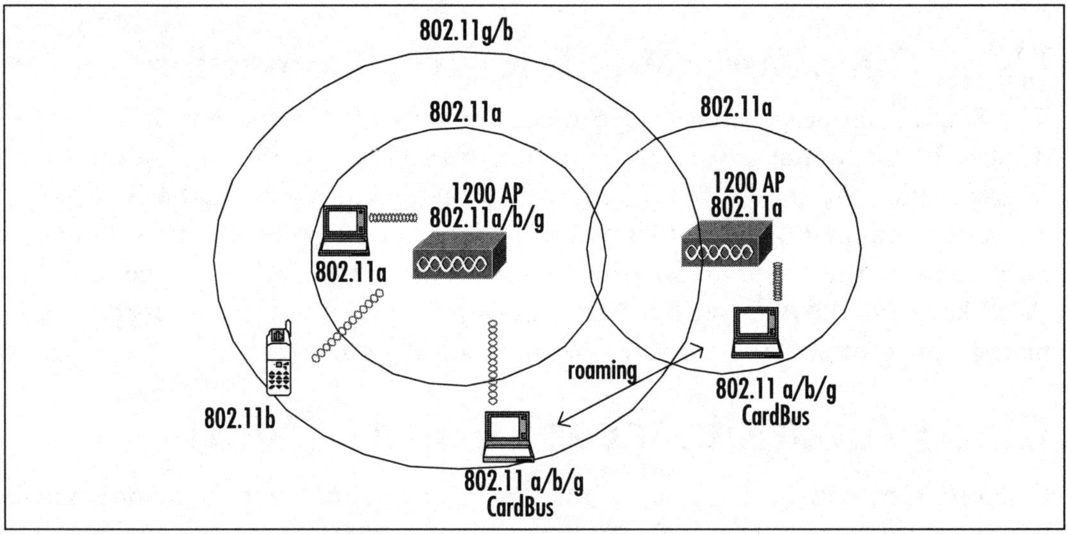


The Aironet 1200 AP can be powered either by direct inline power over its Ethernet cable or by a power supply. The inline power over Ethernet can be either directly connected to a compatible inline-power catalyst switch, a power injector, or an inline power patch panel. Refer to Figures 6.34, 6.35, and 6.36 in the "Aironet Bridge" section in this chapter for a closer look at these three supported power options.

The 1200 AP has an integrated 5 GHz antenna for its 802.11a radio (see Figure 6.7) that can be removed if you are not using an 802.11a radio. It also supports two RP-TNC connectors used for 2.4 GHz 802.11b/g radios (see Figure 6.8) that can be used connect multiple external antenna options.

**Figure 6.7** 1200 5GHz Antenna

**Figure 6.8** 1200 AP With 2x RP-TNC For 2.4 GHz Antennas

The 1200 APs are capable of operating in a –4 to 122 Fahrenheit (–20 to 50 Celsius) temperatures with an 802.11a radio solution and –4 to 131 Fahrenheit (–20 to 55 Celsius) with 802.11b/g radios.

The 1200 AP supports the standard IOS image for easy manageability and configuration. It ensures security with all of the 802.1X/EAP protocol features available and the enhanced WEP protocol features such as TKIP per-packet keying and Message Integrity Check (MIC). Further, the 802.11g radio version is ready to support Advanced Encryption Standard (AES), which is currently being tested and will be available in 2004 with a software upgrade. For more detailed features of the 1200 AP, refer to Table 6.5.

**Table 6.5** 1200 AP Features

| Description | Specifications |
|---|---|
| Supported data rates | 802.11b: 1, 2, 5.5, 11 Mbps<br>802.11g: 1, 2, 5.5, 11, 12, 18, 24, 36, 48, 54 Mbps<br>802.11a: 6, 9, 12, 18, 24, 36, 48, 54 Mbps |
| Supported standard | IEEE 802.11b, 802.11a, 802.11g |
| Indoor range | 802.11g (2.2 dBi dipole antenna) 90 feet @ 54 Mbps; 410 feet @ 1 Mbps<br>802.11a (6 dBi patch antenna) 45ft @ 54 Mbps; 165ft @ 6 Mbps |
| Outdoor range | 802.11g (2.2 dBi dipole antenna) 250 feet @ 54 Mbps; 2,000 feet @ 1 Mbps<br>802.11a (6 dBi patch antenna) 100 feet @ 54 Mbps; 1,000 feet @ 6 Mbps |
| Encryption support | 128-bit WEP with TKIP (802.11g AES ready) |
| Wireless medium | 802.11a Optical Frequency Division Multiplexing (OFDM), 802.11g OFDM, and Direct Sequence Spread Spectrum (DSSS) |
| Media access protocol | Carrier Sense Multiple Access/Collision Avoidance (CSMA/CA) |
| Modulation | 802.11g-DSSS: Differential Binary Phase Shift Keying (DBPSK) @ 1 Mbps; Differential Quadrature Phase Shift Keying (DQPSK) @ 2 Mbps; Complementary Code Keying (CCK) @ 5.5 and 11 Mbps<br>802.11g and 802.11a -OFDM: BPSK @ 6 and 9 Mbps; QPSK @ 12 and 18 Mbps; 16-QAM @ 24 and 36 Mbps; 64-QAM @ 48 and 54 Mbps |
| Remote configuration support | Secure Shell (SSH); Telnet; Hypertext Transfer Protocol (HTTP); File Transfer Protocol (FTP); Trivial File Transfer Protocol (TFTP); Bootstrap Protocol (BOOTP), and SNMP |
| AP can act as dynamic host control protocol (DHCP) client? | Yes |
| Antenna options | 802.11a: Integrated 6 dBi diversity patch<br>802.11g/b: Two RP-TNC connectors |
| Uplink | Auto-sensing 10/100BaseT Ethernet |

**Continued**

**Table 6.5** 1200 AP Features

| Description | Specifications |
|---|---|
| IOS Support | Yes |
| Operating temperature range | 802.11a: –4 to 122 Fahrenheit (–20 to 50 Celsius)<br>802.11b/g: –4 to 131 Fahrenheit (–20 to 55 Celsius) |
| QOS support | Yes |
| VLAN support | Yes up to 16 |
| Security | 802.1X EAP |
| Memory | 16 MB RAM and 8 MB Flash |

Additional accessories such as mounting equipment, locks, and power injectors are available and can be purchased separately from Cisco. For more details and product listings refer to www.cisco.com.

## First-Time Basic Configuration

The Cisco Aironet 1200 AP offers management connectivity over Ethernet or through the console port. If connecting through an Ethernet cable using Telnet or HTTP for the first time, AP's default configuration is set up to automatically obtain IP addresses via DHCP. If you do not have a DHCP server, the AP will pick 10.0.0.1 as its default IP address for the Ethernet port.

**Designing & Planning...**

### Connecting to an AP

It is recommended that when connecting and configuring an AP for the first time, you do not connect it to the production network via its Ethernet port. Use a console port to connect directly to the AP from a terminal server or crossover Ethernet cable for initial configuration settings. Connecting an nonconfigured AP to a production network can cause network issues and can also create a serious security breach that intruders can explore to gain access to your wired connected network.

The default login and password to gain access to the AP is **Cisco** (case sensitive). Note that you do not need to enter the username and password when connecting to the AP via console for the first time, however, you do need the password to gain access to its enable mode where all of the configuration changes to the AP settings are made. Once you are connected via console using the IOS Command Line Interface (CLI), enter the **enable** command and use the default password to gain access. To configure the IP address of the Ethernet port enter the **configure terminal** command followed by the **interface bvi 1** command to gain access to interface configuration mode. Now enter the IP address and netmask you wish to bind with the Ethernet interface by using the **ip address** command.

1. Configure the AP using address 192.168.1.12 in the IOS firmware.

```
ap> enable
Password: Cisco
ap# configure terminal
ap(config)# interface bvi 0
ap(config-if)# ip address 192.168.1.12 255.255.255.0
ap(config-if)# exit
ap(config)#
```

2. Change the default enable password.

```
ap(config)# enable password <new-password-here>
```

3. Assign username **Cisco** with a new password.

```
ap(config)# username Cisco password <new-password-here>
```

4. Save the changes and exit out of enable mode.

```
ap(config)# end
ap# write memory
ap# exit
ap>
```

When using a Web browser instead of the CLI to initially configure your AP, you will be asked to enter both the username and password in order to gain access. Figure 6.9 shows the basic IP address configuration in IOS firmware using HTTP Web interface, and Figure 6.10 shows the administrative control access menu. Refer to these two screen shots to compare the CLI configuration example.

**Figure 6.9** Configuring IP Address using a Web Browser

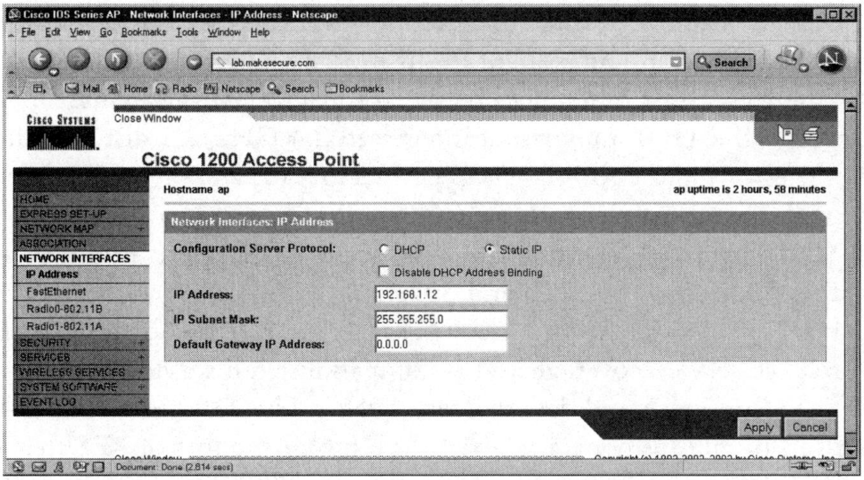


**Figure 6.10** Changing the Administrator and Global Passwords

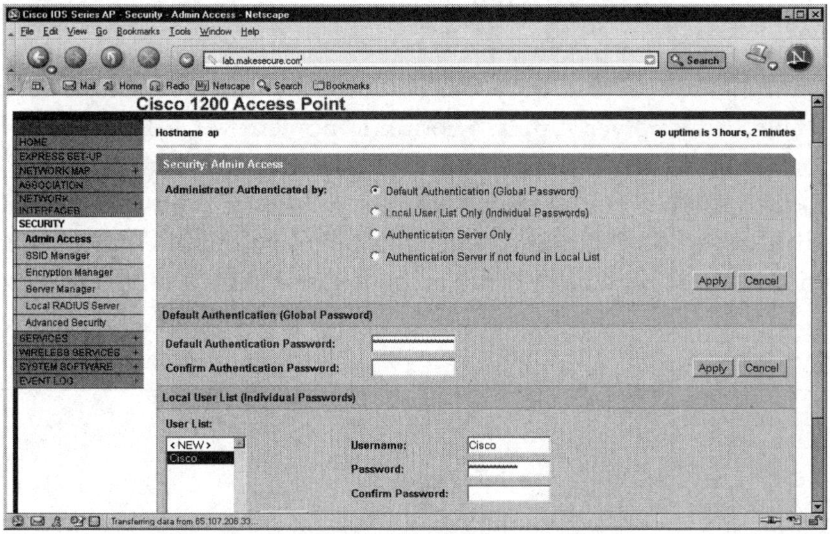


# Aironet 1100 AP

The Cisco Aironet 1100 series (see Figure 6.11) supports a single installed radio that is available in either the 802.11b or 802.11g IEEE standard. If you are running an 802.11b radio in your 1100 series AP you can purchase the 802.11g upgrade kit and easily replace the 802.11b radio inside your unit within a few minutes. The new 802.11g radio is backwards-compatible with 802.11b allowing for easy upgrade and integration in networks where clients are using both 802.11g and 802.11b wireless adapters. Just like the 1200 series AP, Cisco's Aironet 1100 AP proves to be a good future investment for its modular upgradeable design.

There are two major differences between the 1100 AP series and the 1200 AP series. First, the 1100 supports an integrated 2.4 GHz, 2.2 dBi antenna and the 1200 series includes two RP-TNC connectors for a choice of multiple compatible external antennas. Second, the 1100 only supports one installed radio (either 802.11b or 802.11g) while the 1200 series AP supports two installed radios (either 802.11g or 802.11b and 802.11a).

The 1100 series radio (see Figure 6.11) runs on an IOS operating system that provides reliable service coverage and is compatible with services such as QOS to ensure end-to-end voice and data delivery quality. The 1100 series AP installed with an 802.11g radio is compatible with the future support of AES that will be replacing WEP. AES will be available sometime in 2004. The 1100 series further provides security-compatible features such as 802.1X/EAP for per-user authentication, WEP enhancements with TKIP per-packet-keying, and SSH for encrypted administration of the unit.

The Aironet 1100 AP can be powered either by direct inline power over its Ethernet cable or by power supply. The inline power over Ethernet can be either directly connected to a compatible Inline-Power Catalyst Switch, power injector, or inline power patch panel. Refer to Figures 6.34, 6.35, and 6.36 in the "Aironet Bridge" section in this chapter for a closer look at all three supported power options. Refer to Table 6.6 for more detail details on the 1100 series AP.

**Figure 6.11** 1100 AP

**Table 6.6** Aironet 1100 AP Features

| Description | Specifications |
|---|---|
| Supported data rates | 802.11b: 1, 2, 5.5, 11 Mbps<br>802.11g: 1, 2, 5.5, 11, 12, 18, 24, 36, 48, 54 Mbps |
| Supported standard | IEEE 802.11b, 802.11g |
| Indoor range | 802.11g (2.2 dBi antenna) 90 feet @ 54 Mbps; 410 feet @ 1 Mbps |
| Outdoor range | 802.11g (2.2 dBi antenna) 250 feet @ 54 Mbps; 2,000 feet @ 1 Mbps |
| Encryption support | 128-bit WEP with TKIP (802.11g AES ready) |
| Wireless medium | 802.11g OFDM and DSSS |
| Media access protocol | CSMA/CA |
| Modulation | 802.11g-DSSS: DBPSK @ 1 Mbps; DQPSK @ 2 Mbps; CCK @ 5.5 and 11 Mbps<br>802.11g-OFDM: BPSK @ 6 and 9 Mbps; QPSK @ 12 and 18 Mbps; 16-QAM @ 24 and 36 Mbps; 64-QAM @ 48 and 54 Mbps |
| Remote configuration support | SSH; Telnet; HTTP; FTP; TFTP; BOOTP and SNMP |
| AP can acts as DHCP client? | Yes |
| Antenna options | Integrated 2.2 dBi diverse dipole antenna |
| Uplink | Auto-sensing 10/100BaseT Ethernet |
| IOS Support | Yes |
| Operating temperature range | 32 to –104 Fahrenheit (0 to –40 Celsius) |
| QOS support | Yes |
| VLAN support | Yes up to 16 |
| Security | 802.1X EAP |
| Memory | 16 MB RAM and 8 MB Flash |

# Aironet 350 AP

The 350 AP connects individual clients who are wireless-compatible such as laptops, PCs, and wireless IP phones that have wireless client adapters. The AP receives this traffic and formats it to transmit to the wired LAN or another wireless segment. Using multiple APs together, users are able to seamlessly travel between coverage areas without loosing network connectivity.

The 350 series APs come in two versions, the standard and the rugged. The standard AP uses a plastic case with integrated antennas and the rugged AP uses a metal case for outside and harsh conditions with external RP-TNC antenna connectors allowing you to attach different environment external compatible antennas (see Figure 6.12).

**NOTE**

The Cisco Aironet 350 AP has reached end-of-sale. Cisco recommends the 1100 or 1200 AP as its replacement.

**Figure 6.12** 350 AP Plastic (Left), Rugged (Right)

The rugged model has a wider range of operating temperatures (–4 to 131 degrees Fahrenheit [–20 to 55 degrees Celsius]) compared to the plastic case temperatures (32 to 122 degrees Fahrenheit [0 to 50 degrees Celsius]). Both units support inline power over Ethernet cable to extend the flexibility of installment. However, if an inline power injector is used, it loses its plenum rating and gains a smaller operating range of 32 to 104 degrees Fahrenheit (0 to 40 degrees Celsius).

The 350 series APs are equipped with an auto-sensing 10/100BaseT Ethernet uplink port. The AP uses 802.11b technology to communicate with the clients at either 1, 2, 5.5, or 11 Mbps, depending on the signal strength and quality. Speed determines modulation, as shown in Table 6.7. DBPSK and

DQPSK modulation techniques are similar structures. Both are quite different from the CCK technique that is used for the 5.5 and 11 Mbps transmission.

**Table 6.7** 350 Series AP Features

| Description | Specifications |
|---|---|
| Supported data rates | 1, 2, 5.5, 11 Mbps |
| Supported standard | IEEE 802.11b |
| Indoor range | 11 Mbps @ 150 feet; 1 Mbps @ 350 feet |
| Outdoor range | 11 Mbps @ 800 feet; 1 Mbps @ 2,000 feet |
| Encryption support | 128-bit WEP with TKIP |
| Wireless medium | DSSS |
| Media access protocol | CSMA/CA |
| Modulation | DBPSK @ 1 Mbps; DQPSK @ 2 Mbps; CCK @ 5.5 and 11 Mbps |
| Remote configuration support | Telnet; HTTP; FTP; TFTP; and SNMP |
| AP can acts as DHCP client? | Yes |
| Antenna options | Two non-removable 2.2 dBi Dipole with plastic unit. Two RP-TNC connectors for rugged unit (non-antennas supplied with RP-TNC unit) |
| Uplink | Auto-sensing 10/100BaseT Ethernet |
| IOS Support | Yes |
| Operating temperature range | Plastic case AP: 32 to 122 Fahrenheit (0 to 50 Celsius)<br>Rugged case AP: –4 to 131 Fahrenheit (–20 to 55 Celsius)<br>Power injector: 32 to 104 Fahrenheit (0 to 40 Celsius) |
| QOS support | Yes |
| VLAN support | Yes up to 16 |
| Security | 802.1X EAP |

Whichever modulation technique is used for the wireless transmission with the APs, the methodology used for the transmission is DSSS. DSSS uses the entire frequency band (2.4 to 2.497 GHz) for transmission of data, thus allowing for higher throughput than if just one of the frequencies was used. Cisco recently released IOS firmware image support for 350 series APs, which enhanced manageability and integration with wired-aware Cisco networks. Even though the

350 series AP is no longer sold by Cisco, it remains supported under Cisco for several more years.

# Cisco Aironet WLAN Client Adapters

Cisco wireless network cards offer the same functionality as traditional wired NICs, preparing data for transmission from a PC to the network. There are three major series of network cards that Cisco offers. These are:

- Cisco Aironet 350 Series Client Adapters
- Cisco Aironet 5GHz Client Adapter
- Cisco Aironet 802.11a/b/g Client Adapters

## Cisco Aironet 350 Series Client Adapters

The Cisco Aironet 350 series client family adapters (see Figure 6.13) are available in PCMCIA and PCI interfaces. The PCMCIA solution is further divided into PCMCIA with integrated diversity dipole antenna (AIR-PCM35x) and Los Medanos College (LMC) card (AIR-LMC35x) with two MMCX connectors that allow you to connect external antennas. Refer to Figure 6.14 for the external 2.2 dBi antenna (AIR-ANT3351) used to extend the Aironet 350 LMC client adapter range.

**NOTE**

The LMC card is also part of the PCI adapter card used for desktops (see Figure 6.13). It is not advised that you disconnect and use the LMC card from the PCI adapter as a replacement for the factory-built AIR-LMC35x adapter. The LMC card built for the PCI adapter shares a unique boot block version for PCI only and should not be used as PCMCIA.

**Figure 6.13** Aironet 350 Series Family

The PCMCIA card is used mostly in mobile devices such as laptops, and the PCI wireless adapter is used in desktops. The Cisco Aironet 350 clients are a fixed solution that work in conjunction with 802.11b (2.4GHz) APs, bridges, or other wireless client adapters. The 350 series client adapters support the highest range and throughput performance with up to 100 milliWatts (mW) of transmit power. A typical outside range at 11 Mbps is up to 800 feet (244 meters) and up to 2000 feet (610 meters) at 1 Mbps. The indoor range covers up to 130 feet (30 meters) for 11 Mbps and 350 feet (107 meters) for 1 Mbps.

All standard security features such as 802.1X/EAP, 128-bit WEP, and TKIP are supported with Aironet 350 client adapters.

**Figure 6.14** External Client Antenna for LMC Card AIR-ANT3351

## Cisco Aironet 5GHz Client Adapter

The Cisco Aironet 5 GHz client adapter (see Figure 6.15) supports the 802.11a wireless standard and up to 54 Mbps of speed. The adapter is compatible with the Aironet 1200 Series AP that supports 802.11a radio and other 5 GHz compatible clients wanting to communicate.

**Figure 6.15** Cisco Aironet 5GHz Client Adapter (casar_d4)

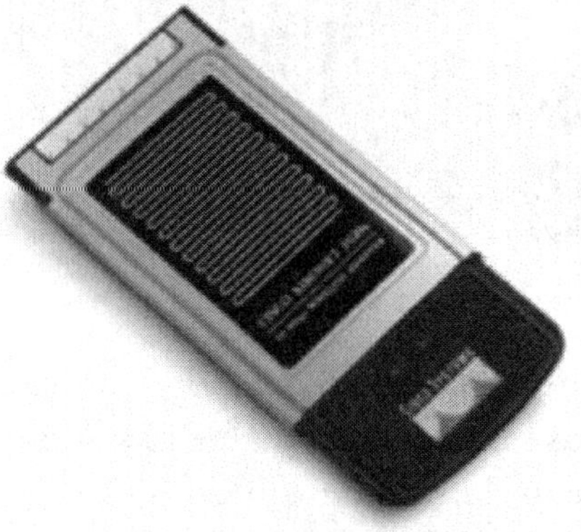



The 5 GHz client adapter has an integrated 5 dBi gain antenna with a typical outdoor range of up to 120 feet (36 meters) at 54 Mbps and 1200 feet (355 meters) at 6 Mbps. It supports all standard security features such as 802.1X/EAP per-user dynamic keying and 128-bit WEP with TKIP enhancements for per-packet encryption. Support of hardware-accelerated WEP encryption allows for maximum throughput with little delay.

## Cisco Aironet 802.11a/b/g Client Adapters

The Cisco Aironet 802.11a/b/g client adapter supports all of the current 802.11 industry standard wireless technology. It can support single mode, 802.11a/g double mode, or all three (802.11/a/b/g) at the same time. With its support for all three radios it is highly scalable and compatible in any wireless environment. Clients are able to reach 54 Mbps in both 2.4 GHz and 5 GHz bands.

Cisco 802.11a/b/g card adapters come in two available interfaces: the 32-bit CardBus used in laptops and the PCI used for desktops (see Figure 6.16) It supports all of the enterprise-level security features such as 802.1X/EAP dynamic per-user authentication and keying. It supports 128-bit WEP encryption and is ready built for AES encryption support.

**NOTE**

AES is a symmetric key block cipher used to encrypt data. The advanced encryption is replacing standards such as Data Encryption Standard (DES) and RC4 in WEP. WEP is currently the standard encryption method used in 802.11 wireless technology. The WEP standard will soon be updated by the more powerful and secure AES. Cisco supports AES in its 802.11g hardware products that will soon be usable with a simple software upgrade.

**Figure 6.16** 802.11a/b/g Client Adapters

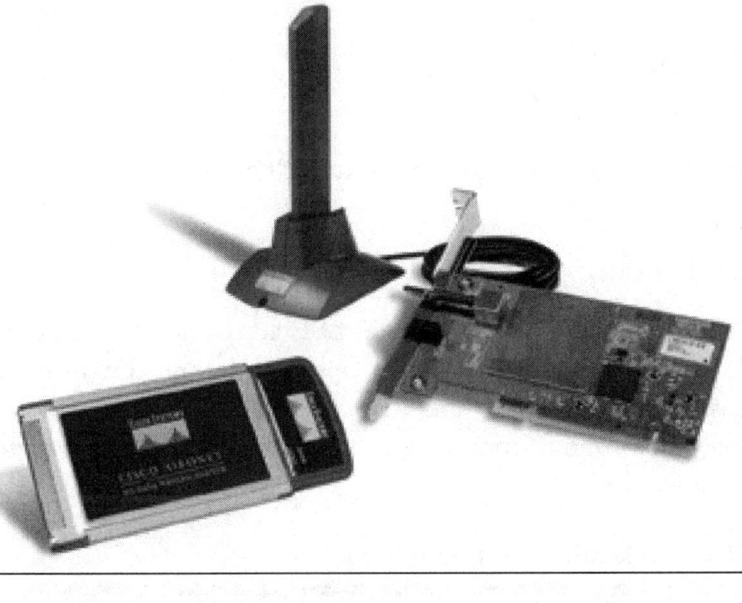



# CiscoWorks WLSE 2.x

Medium to large wireless network deployments require the administration of hundreds to thousands of wireless products such as APs and bridges. Supporting these large WLANs requires hands-on support from multiple network administrators. Cisco recognized this manual unscalable manageability burden in larger networks and so introduced the CiscoWorks WLSE product solution.

WLSE is a hardware and software solution that provides scalable and central management for thousands of deployed wireless Aironet and wired products. The engine simplifies the day-to-day manual operation of an administrator supporting an Aironet wireless environment, with its ability for central management such as pushing configuration templates to hundreds of APs at a time or upgrading the firmware image of every AP.

Cisco WLSE also enhances security by allowing security policies to be applied consistently throughout the wireless network environment from its central management, and continually monitoring and enforcing security settings. It further supports and enhances security with its detection and elimination of unauthorized (rogue) wireless devices introduced into the network. Rogue detection and the use of WLSE to eliminate such a threat is discussed in detail in Chapter 8, "WLAN Rogue AP Detection and Mitigation."

The following are the main core menu functions of the WLSE control dashboard:

- Fault Monitoring
- Device Management
- Device Configuration and Firmware Upgrades
- Reports
- Radio Manager

Figure 6.17 displays a screen shot of WLSE administration graphical user interface (GUI) with overview of all the supported menu functions such as Faults, Devices, Configure, Firmware, Reports, Radio Manager, and Administration. Each menu function is further separated into submenus with additional operational tasks.

**Figure 6.17** WLSE Main Menu Functions

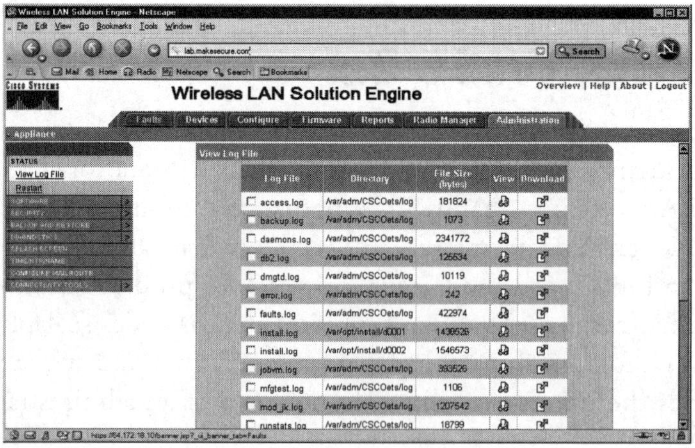

# Fault Monitoring

The Fault Monitoring menu option allows you to manage and monitor your wireless and wired compatible devices. The Faults menu is further separated into four categories: Display Faults, Manage Fault Settings, Manage Network-Wide Settings, and Notification Settings.

**NOTE**

**The use of menus and submenus in the WLSE management GUI can be restricted based on the user access level defined by the WLSE administrator. If you do not see all of the submenus in your browser GUI it is likely that you do not have full administrative access.**

WLSE fault monitoring works based on a defined fault threshold and policy parameters set by the administrator. These parameters are then used to automatically interrogate selected managed wireless or wired devices and compare their received data to the set of threshold and policy settings defined within the WLSE.

For instance, you may define a fault security policy to detect any managed AP without the Public Secure Packet Forwarding (PSPF) configuration turned on. PSPF is a security option used to prevent wireless clients interconnected to the same AP from communicating with each other. The WLSE automatically and periodically checks each selected managed AP by time interval that you specified using SNMP protocol to ensure that PSPF is turned on. If the WLSE fault policy finds an AP without PSPF enabled, it generates a fault alert. Refer to Figure 6.18 for the configuration of a PSPF security check in the Manage Fault Settings menu. Refer to the left side of the screen shot for other possible fault options within the WLSE.

**Figure 6.18** Configuring a Fault Policy and its Settings

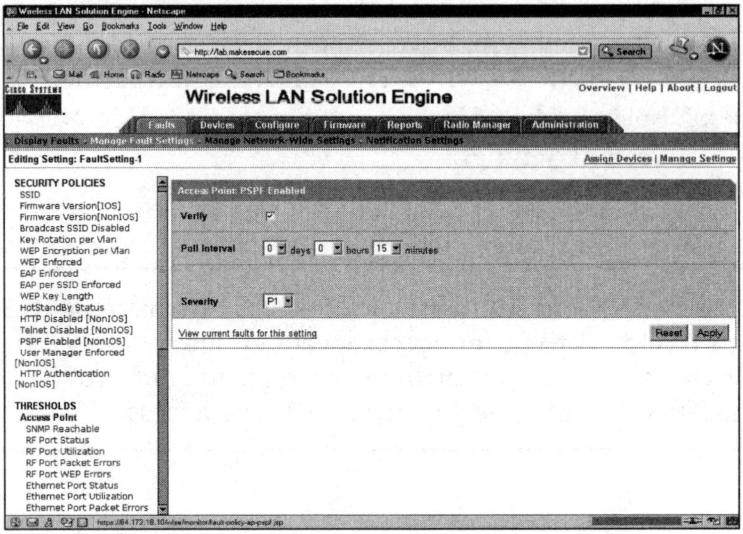

The WLSE fault policy settings include a variety of threshold and policy configurations. Compatible Catalyst switches and Cisco routers that interconnect wireless devices can be monitored and their threshold tested by using the WLSE fault policy. Catalyst switches and routers include threshold options within the WLSE settings such as the central processing unit (CPU), memory, and port utilization. Remote Authentication Dial-In User Server (RADIUS) response times such as authentication can also be monitored from the WLSE to ensure it performs within its defined threshold.

All fault detections that do not pass the defined thresholds or policies are viewed in the Display Faults submenu. The WLSE also has the ability to send out detected fault alerts in the form of a SNMP trap protocol to a remote application such as Netview and to send SYSLOG or e-mail notifications (see Figure 6.19). This is useful if you want to incorporate WLSE faults with other third-party monitoring solutions.

**Figure 6.19** Sending Trap and E-mail Faults to Remote Locations

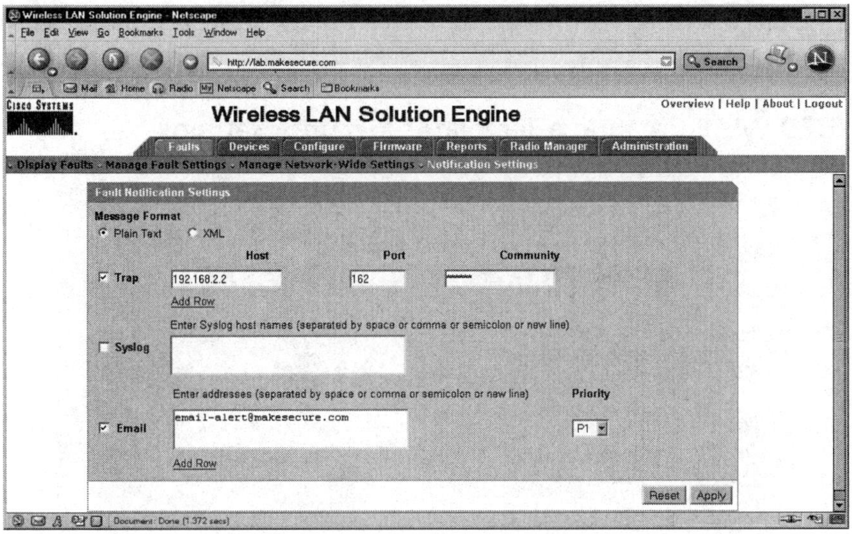


# Device Management

The Devices tab allows you to perform basic device management from the WLSE such as discover new devices and organize devices into manageable groups sets. This menu is where you must add and configure all of your managed devices before you can manage them and use tasks such as monitoring, reporting, configuration, and firmware upgrades within the WLSE.

# Device Configuration and Firmware Upgrades

The Configuration tab in the WLSE mainly allows you to create and archive configuration templates and apply them to a large number of devices at one time. The Firmware tab allows you to upgrade the firmware of IOS and non-IOS APs or convert non-IOS APs to IOS.

## Configuration Tab

The Configuration menu is further subdivided into four categories: Templates, Archives, Jobs, and Auto-Updates. Configuration templates can be either IOS- or non-IOS-based and are used to configure the entire AP or just one specific parameter. Templates include configuration parameters such as Network Settings, Security, Services, Event Log, Wireless Services, and Custom Values that can be applied to Aironet devices. Configuration templates can be pushed to one or multiple managed APs at one time. Refer to Figure 6.20 for a screen shot of the advanced security setting within the template called "Test-Template-AP" and its security settings.

**Figure 6.20** WLSE Creating a Template – Security Settings

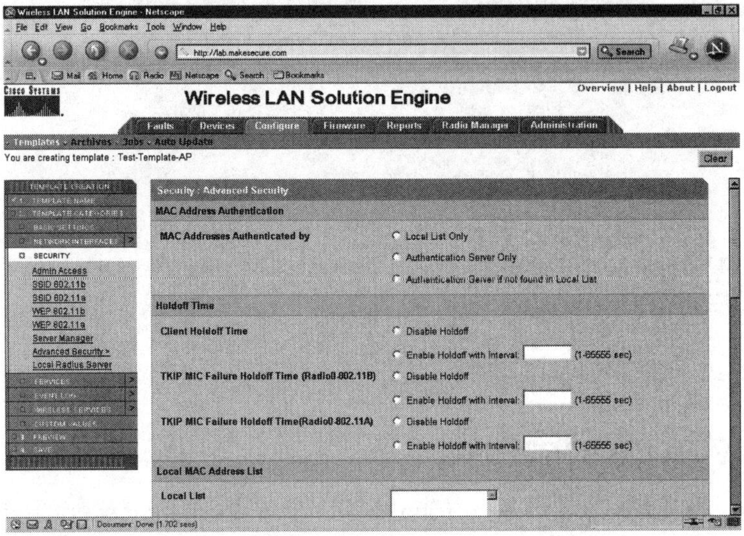


As your wireless network environment becomes larger with multiple distinct group devices and configuration parameters, the Archive feature on the Configuration tab allows you to back up the last four configurations of a device, which can be used in case of fallback from a bad configuration push or to reconfigure the device to its last settings in case of failure.

The Jobs submenu is used to schedule configuration jobs and to archive jobs. The configuration jobs are used to make configuration updates to wireless devices using templates. The WLSE uses Telnet or SSH protocols for IOS-based AP configuration updates. For non-IOS AP configuration updates the WLSE acts as a TFTP server where the AP is triggered via HTTP or SNMP to start downloading the new configuration file from the WLSE directory.

Designing & Planning…

**Open Communication Ports**

As previously noted, the WLSE requires several communication ports such as SSH, Telnet, HTTP, SNMP, and TFTP to successfully communicate with its managed wireless devices. Make sure that these ports are open between your WLSE and its managed devices. When troubleshooting connectivity between your WLSE engine and your managed devices check your firewall or Access Control Lists (ACLs) to ensure that communication is permitted.

Archived jobs are used to back up configuration settings. WLSE uses TFTP to download the configuration. If the downloaded configuration does not include any changes from the previous archive job, the configuration is not saved. Up to four different configuration backups of a single device can be saved in the WLSE.

The last submenu under the Configuration menu is Auto-Updates. Auto-Updates allows you to automatically upload configurations to both APs and bridges. This feature is mainly used when a new AP or bridge is connected to the network with the default factory settings. The DHCP server is used to set up an IP address for the new device and instruct the Aironet device to download the configuration file from the WLSE.

### Firmware Tab

The WLSE can be used to upgrade the Aironet device's firmware version. Updates can be scheduled or activated on demand. The WLSE upgrade firmware options include IOS, non-IOS, and non-IOS to IOS conversion upgrades. The WLSE can be used to upgrade hundred of devices within one job schedule with up to 20 simultaneous multi-threaded upgrades.

## Reports

The Reports menu in the WLSE is used to display report information about the individual or grouped managed devices. The menu consists of six sub-tabs: Device Center, Radio Manager, Wireless Clients, Current, Trends, and Scheduled Email Jobs.

The Device Center allows you to view detailed device information such as network settings, fault status history, configuration history, and firmware history. Figure 6.21 provides a detailed report of one of the 1100 APs displayed from the Device Center submenu.

**Figure 6.21** Reports, Device Center – Detailed Report

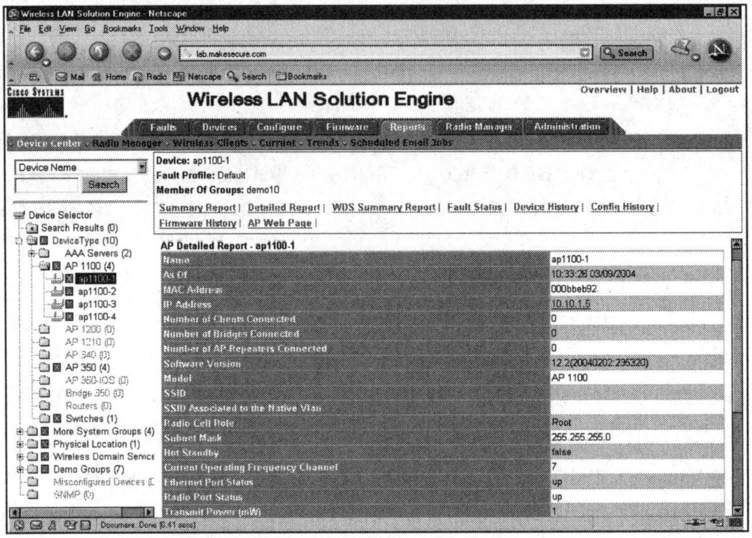


The Radio Manager submenu allows you to view radio reports such as radio parameter settings, paths lost between APs, and channel loading statistics. You can further export these reports in CSV, PDF, or XML format as well as e-mail them directly from the WLSE administrative GUI.

The Wireless Clients submenu allows you to view client information such as the number of clients connected to an AP or how much bandwidth clients are utilizing.

The Current submenu displays current reports from your managed devices. The default frequency is 12 hours in which information reports are collected from each managed device.

The Trends submenu allows you to visualize reports such as radio frequency (RF) or Ethernet transmissions of APs and bridges over a defined period of time. The trend reports can be used to determine peak network utilizations.

The Scheduled Email Jobs submenu allows selected reports to be automatically e-mailed base on a recurring time. This feature can be used to comply with monthly auditing procedures where a security configuration report is required to be sent out for its monthly audit review.

## Radio Manager

The Radio Manager menu is used to pool the managed wireless device's radio information settings and statistics. Information that a radio manager may pool from their managed wireless devices include the RF of the surrounding area in order to detect any interferences or rogue APs. Submenus included under Radio Manager are: Radio Monitoring, AP Radio Scan, Client Walkabout, RM Assistance Configuration, Location Manager, and Manage RM Measurements.

The Rogue AP detection feature is accomplished by the Cisco IOS (1100, 1200) APs and compatible Cisco client adapters taking RF scans in order to detect any broadcast beacons within the coverage area. APs use beacons via RF to announce their presence in a wireless environment. Beacons include information such as the Media Access Control (MAC) address of the advertising AP. When an AP or wireless client detects a beacon from an AP, it sends it to the WLSE Radio Manager where it is compared against a known database of valid and managed APs. If the reported AP beacon and its content does not match any of the valid managed wireless APs in the WLSE database, an alert is generated of possible rogue AP detection.

The Location Manager adds visual effect to the Radio Manager Option menu. Figure 6.22 displays the local manager GUI of a coverage area of four APs after a RF scan of –65 dBm. The WLSE ability for visual effect of radio detections and scans allows for great site survey use. You can import up to 1500 buildings per location with a maximum of 100 floor plans per building and a maximum of 100 APs per floor into the Location Manager in WLSE v2.5.

As mentioned earlier, the WLSE is great for site surveys. During site surveys, APs are placed throughout the facility and activated with the WLSE to perform an RF scan by which they all assume an identical transmit channel and transmit power at the maximum allowed level. The APs then detect each other's presence and coverage area data, which are all reported back to the WLSE for overall visualization and the radio setting calculations for best WLAN performance within the area.

**Figure 6.22** Radio Manager – Location Manager

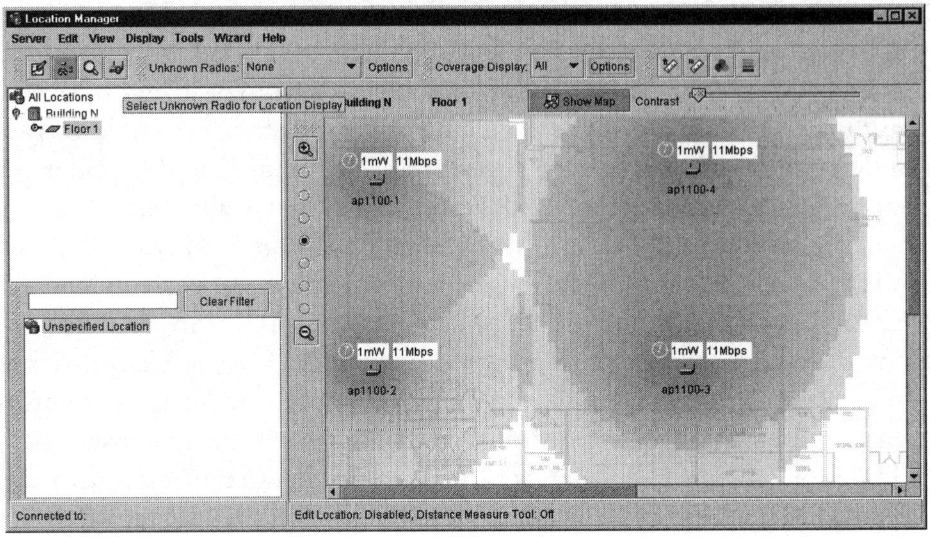


# Cisco Wireless Security Suite

Security features in 802.11 wireless networks can be a major barrier for businesses not wanting to implement wireless technologies. As you will read in Chapter 7, "WLAN Security Considerations," current 802.11 standards have been discovered to be vulnerable to different attacks. Some of the current weaknesses in 802.11 include:

- Static device authentication
- Static WEP encryption method
- Integrity Check Value (ICV) method

In static device authentication, wireless clients are authenticated instead of users. This solution prevents scalability, security, and proper accountability. Wired Equivalent Privacy Protocol, used to provide encryption between wireless clients and APs, has been found vulnerable to attacks where the actual key used could be deciphered in less than 30 minutes on a busy network. WEP uses the RC4 stream cipher (invented by Ron Rivest of RSA Data Security Inc.) for encryption. The RC4 algorithm is a symmetric stream cipher where both parties share the same key to encrypt its data. An Initialization Vector (IV) is a component used with the encryption key to create the ciphered text. Added IV is used to randomize and ensure that the same plaintext data will not generate the same ciphered data.

An IV being sent in cleartext makes WEP vulnerable to intruders that can capture encrypted frames and derive their content. An IV is 24 bits long and provides 16,777,216 possible values. A University of California, Berkeley engineer found that when the same IV is used on two packets that are captured by an intruder, it can be used to derive the contents of the two packets. Packets using the same IV is called a **collision**. For further information, refer to the paper at www.isaac.cs.berkeley.edu/isaac/wep-faq.html.

Fluhrer, Martin, and Shamir (FMS) discovered that including WEP vulnerability in the RC4 key-scheduling algorithm can expose a static WEP key. Due to the RC4 implementation in WEP and its use of a 24-bit IV, different methods can be used in this static pattern to derive the secret WEP key. This FMS attack uses between 100,000 and 1,000,000 encrypted packets using the same static key to derive the WEP key. For further information, refer to the paper at www.cs.umd.edu/~waa/class-pubs/rc4_ksaproc.ps.

A utility called AirSnort was developed to automate the FMS attack. AirSnort operates by passively monitoring encrypted transmissions. When about 5 to 10 million encrypted packets have been captured, AirSnort uses the discovered vulnerability patterns to derive the secret WEP key. AirSnort is a free program that runs on Linux. For further information refer to: http://airsnort.shmoo.com/.

Cisco recognized the worry of wireless security and thus announced the Wireless Security Suite support for all of its wireless product offerings to mitigate such threats as static WEP encryption and authentication process.

## Mitigating Vulnerabilities with the Cisco Security Suite

To mitigate the vulnerabilities covered in the previous section and to allow businesses to implement secure and robust WLANs, Cisco has introduced Cisco Wireless Security Suite. The security suit consists of these core security support options for all Cisco wireless devices:

- EAP Support
- TKIP - Per Packet Key
- TKIP - MIC
- WLSE and ACS Support

Support for EAP protocol in Cisco wireless devices ensures better security and scalability. It allows for per-user authentication and for dynamic per-user

WEP key assignment. Dynamic WEP per-user key assignment can further be configured with the session timeout setting, which causes the WEP key to be regenerated between the AP and the wireless user.

TKIP is used to enhance WEP with per-packet-keying. A temporary key (derived from a hash function using IV) and a base-configured key are being used to encrypt the packets instead of the base-configured WEP key. By implementing TKIP, you do not eliminate the vulnerability of deriving a WEP key. Instead of the intruder deriving the base key that is used to encrypt all of your traffic, the intruder only derives the temporary key of one packet that is different from every packet, thus eliminating some of the WEP vulnerability threat.

Aside from per-packet-keying, TKIP also offers MIC. MIC ensures the integrity of the data and prevents man-in-middle (MITM) attacks from modifying data in transit. MIC adds two new fields inside the encrypted frame: Sequence Number and Integrity Check. Sequence Number is used by the AP to validate packet order and discard any out-of-order packets. Integrity Check is a hash of fields and data of the frame. Any changes made to these fields during transmission will not match the transferred hash and the receiving device will discard the frame.

To complete the Cisco Security Suite, many other security solutions are offered from the standard supported list, such as ACS that provides RADIUS and central user database authentication for EAP protocol and WLSE to audit security policies and detect unauthorized APs within networks.

These new compatible and supported security features for all of Cisco's wireless products ensure secure and reliable WLAN implementations where businesses can feel comfortable implementing new wireless networks.

# Cisco Secure Access Control Server (ACS) 3.2

Cisco Secure ACS 3.2 provides central and scalable user management within wired and wireless environment solutions. It controls who can login to the network device and the privilege levels of users logged in, and provides an accounting of all user activity for audit purposes. It operates as a central RADIUS and Terminal Access Controller Access Control System (TACACS) server. TACACS and RADIUS are network protocols used between the ACS application and its managed device such as the Aironet AP to carry the AAA information. RADIUS is an industry-approved standard protocol and TACACS is a Cisco proprietary.

Cisco ACS 3.2 is available in two configuration options: ACS software 3.2 running on a Windows 2000 Server or ACS software 3.2 running on Cisco's dedicated 1111 platform server. Minimum Windows 2000 requirements to run ACS 3.2 software include: Pentium III 550 MHz, 256 MB RAM and 250 MB disk space. Table 6.8 describes Cisco's dedicated ACS 3.2 1111 platform server specifications.

**Table 6.8** Cisco's Dedicated ACS Server

| Processor | Pentium IV, 2.66GHz |
| --- | --- |
| RAM | 1 GB |
| Hard Drive | 40 GB |
| Ethernet Cards | Two 10/100 Ethernet |
| Disk Drives | 1 Floppy, 1 CD-ROM |
| Size | 1U (1 rack unit) |

Running the ACS 3.2 software application on Cisco's dedicated server versus running it on your own Windows 2000 server has some advantages. One of the advantages includes better security. Cisco's dedicated server engine comes factory pre-hardened and is dedicated to run only ACS applications. It's advantage over open-operation systems such as Windows 2000 server that runs multiple applications by default, is that it does not need to be manually secured and you do not need to maintain software versioning and security patches on a monthly basis.

Cisco's ACS 3.2's core functions in wireless networks include its ability to support several different EAPs such as a RADIUS central server. RADIUS is a communication protocol that carries the EAP protocol. RADIUS is used between the AP and the ACS server.

EAP in WLAN allows you to accomplish tasks such as per-user authentication connecting to the network, mutual authentication between the RADIUS server and the wireless client, and per-user dynamic WEP keying assignments. EAP protocols supported in ACS 3.2 include: LEAP, PEAP, Extensible Authentication Protocol-Transport Layer Security (EAP-TLS), Extensible Authentication Protocol-Message Digest 5 (EAP-MD5), and Extensible Authentication Protocol-Generic Token Cards (EAP-GTC). Details about each of these protocols and their specific uses are covered in Chapter 7, "WLAN Security Consideration."

Figure 6.23 displays an example of the ACS application and its support for RADIUS protocol using LEAP. Each wireless client in Figure 6.23 needs to be authenticated by the RADIUS Server (ACS) in order to gain access. User names and passwords are stored in the ACS application database. ACS 3.2 is also capable of interconnecting with remote databases such as Active Directory or One Time Passwords (OTP) solutions from RSA. The AP in this example will only allow and accept EAP traffic from the wireless clients until they pass their username/password authentication and receive a dynamically generated per-user WEP from the ACS.

**Figure 6.23** ACS and LEAP Authentication

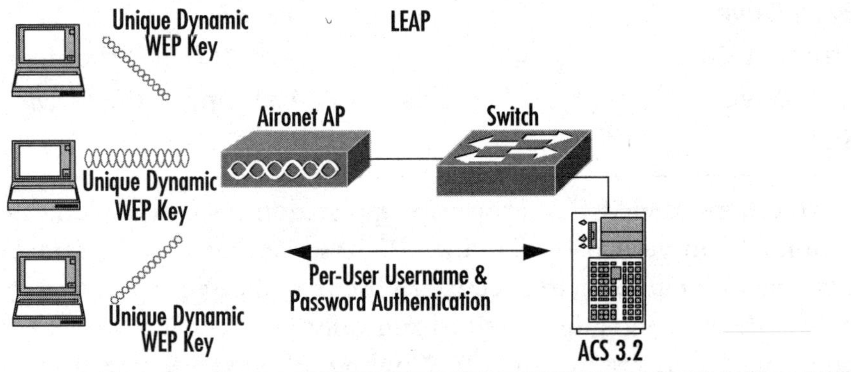


The ACS application is capable of providing central user management and database solutions for all wireless and wired users accessing network resources and devices. For added security, each device such as the Aironet AP needs to be manually added into the ACS application and share common configured security keys in order to communicate.

ACS's support for RADIUS and EAP protocols that allow per-user authentication and per-user generated WEP keys makes your wireless solution much more secure. Without the ACS server and its support, users had to share static common WEP keys and unscalable authentication measures performed by the AP alone.

Cisco ACS 3.2 software is managed using compatible browser and Secure Sockets Layer (SSL) connections. Cisco ACS can support approximately 80,000 users in its own database with its limited required hardware resources. This number can go much higher with an increase in RAM and processing power.

# Enhanced Client Network Management Features with Extended Client Support

Cisco provides several client utilities used with its Aironet client adapters product line. The client network management utilities are bundled with comprehensive and easy-to-use settings for managing, securing, and installing client adapters.

The Cisco Aironet Client Utility (ACU) pictured in Figure 6.24 is used with 350 and 5 GHz series client adapters and provide these core features:

- **Site Survey Tools** The Site Survey and Link Status Meter tools provide graphical information such as signal strength and signal quality between the client and the AP (See Figures 6.25 and 6.26).
- **Troubleshooting** Provides step-by-step visual tests that your adapter takes to associate with the AP such as Radio Test, Association Test, and Authentication Test, and displays its results.
- **Profile Manager** Creates many different unique profiles used in different WLAN environments such as different Secure Set Identifiers (SSIDs), WEP keys, or authentication requirements.
- Supported in many operation systems such as Windows, MAC, and Linux.

**Figure 6.24** ACU Utility

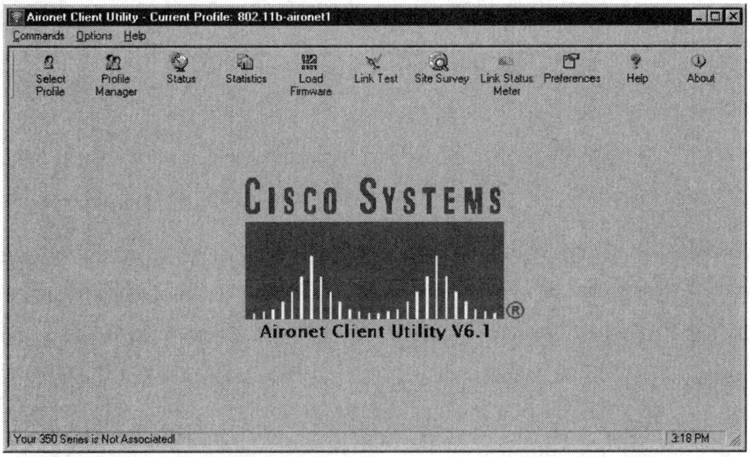

**Figure 6.25** ACU - Site Survey

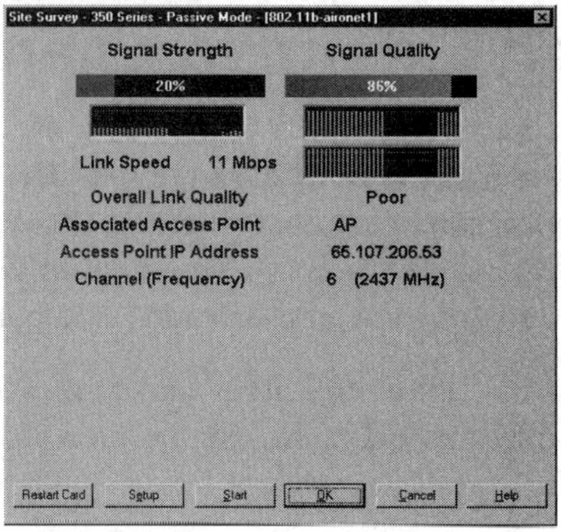


**Figure 6.26** ACU - Link Status Meter

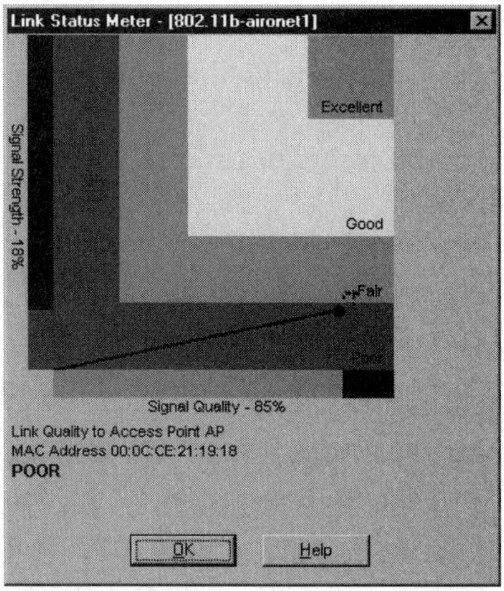


Cisco's new 802.11a/b/g client adapters (reviewed previously in the "Cisco Aironet WLAN Client Adapters" section, support a new set of enhanced client utilities including: Aironet Desktop Utility (ADU), Aironet System Tray Utility (ASTU), and Aironet Client Administration Utility (ACAU). The site survey options covered in the ACU utility has been discontinued in the new client utility set for 802.11a/b/g client adapter cards. All other core features covered under ACU are otherwise supported in the new ADU utility with the addition of AES readiness and enhancement of secure and seamless client roaming capabilities.

# Workgroup Bridges

Cisco workgroup bridges support any Ethernet device such as PCs, printers, and copiers that it then interconnects with a wireless Aironet AP or bridge. The workgroup bridge provides flexibility and ease of use with a ready wireless solution for Ethernet devices that may not have wireless-ready compatibility. The workgroup bridge helps to ensure that no one is left behind, and that all Ethernet-compatible devices can be interconnected with wireless AP.

## Aironet 350 Workgroup Bridge

The Aironet 350 workgroup bridge connects a small number (up to eight) of wired-attached Ethernet stations to an AP or bridge. The 350 workgroup bridge has only one uplink port so that a switch or a hub will need to be used if an administrator wishes to connect more than one Ethernet device. Though the wireless connection usually takes place within a building, it can also be used between remote areas. Figure 6.27 shows a typical workgroup bridge configuration connecting up to eight Ethernet devices with wireless AP. Figure 6.28 further outlines the flexibility and use of a workgroup bridge in a network environment allowing non-wireless-compatible PCs to connect to the wireless AP.

**Figure 6.27** Multiple Hosts (Up to Eight Maximum)

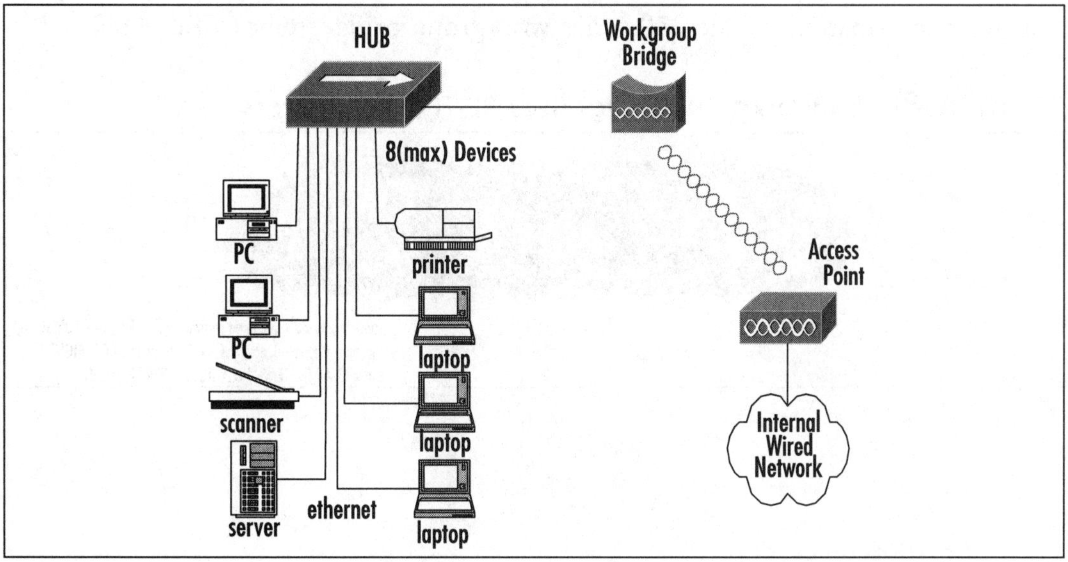

**Figure 6.28** One-Host Support

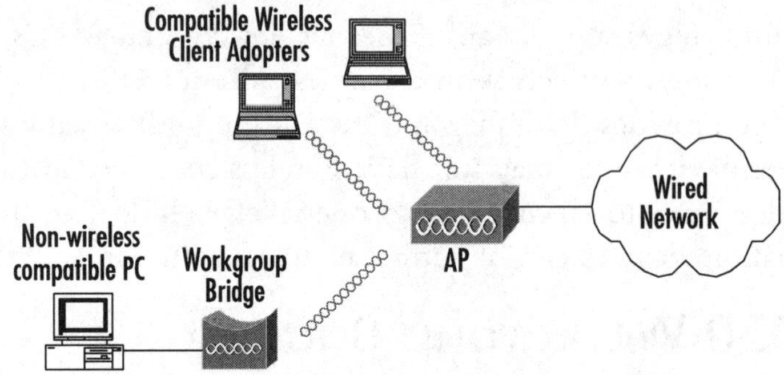


The workgroup bridge is very similar to the other members of the 350 series family. It uses DSSS for modulation and CSMA/CA as the media access protocol, and can communicate at 1, 2, 5.5, or 11 Mbps. The 350 series workgroup bridge does not use inline power. Its only uplink port is a 10BaseT port, which deliberately limits the number of devices it can support. The 350 workgroup bridge is available in a plastic case model only. This bridge is not plenum-rated and is designed for indoor use only. Its operating temperature matches that of the plastic case AP, 32 to 122 degrees Fahrenheit (0 to 50 degrees Celsius). Two antenna options are available: a single non-removable 2.2 dBi dipole antenna (see Figure 6.30), or two RP-TNC connectors with no supplied antennas (see Figure 6.29). For more information on the 350 series workgroup bridge, refer to Table 6.9.

**Figure 6.29** 350 Workgroup Bridge Two RP-TNC Connectors

**Figure 6.30** 350 Workgroup Bridge Single 2.2 dBi Dipole Antenna

**Table 6.9** 350 Series Workgroup Bridge Features

| Description | Specifications |
|---|---|
| Supported data rates | 1, 2, 5.5, 11 Mbps |
| Supported standard | IEEE 802.11b |
| Range | Indoor: 130 feet @ 11 Mbps; 350 feet @ 1 Mbps<br>Outdoor: 800 feet @ 11 Mbps; 2,000 feet @ 1 Mbps |
| Encryption support | WEP 128-bit |
| Maximum number of users supported | 8 |
| Inline power support | No |
| Wireless medium | DSSS |
| Media access protocol | CSMA/CA |
| Modulation | DBPSK @ 1 Mbps; DQPSK @ 2 Mbps; CCK @ 5.5 and 11 Mbps |
| Frequency band | 2.4 to 2.4897 GHz |
| Remote configuration support | Telnet; HTTP; FTP; TFTP; and SNMP |
| Uplink | 10BaseT Ethernet |
| Operating temperature | 32 to 122 Fahrenheit (0 to 50 Celsius) |

# Aironet 340 Workgroup Bridge

The 340 series workgroup bridge is similar to the 350 series workgroup bridge. It can connect a small (up to eight devices) network to a central wired LAN. It has a single 10BaseT RJ-45 connector, so a switch or hub is used to handle more than one device.

**NOTE**

The 340 series workgroup bridge has reached end-of-sale as of July 2003, and will achieve end-of-support in July 2007. Cisco advocates the 350 series workgroup bridge as a replacement.

The remaining configuration details of the 340 series workgroup bridge are outlined in Table 6.10.

**Table 6.10** 340 Series Workgroup Bridge Features

| Description | Specifications |
|---|---|
| Supported data rates | 1, 2, 5.5, 11 Mbps |
| Supported standard | IEEE 802.11b |
| Range | Indoor: 100 feet @ 11 Mbps, 300 feet @ 1 Mbps<br>Outdoor: 400 feet @ 11 Mbps, 1,500 feet @ 1 Mbps |
| Encryption supported | WEP 40-bit or 128-bit |
| Wireless medium | DSSS |
| Media access protocol | CSMA/CA |
| Modulation | DBPSK @ 1 Mbps; DQPSK @ 2 Mbps; CCK @ 5.5 and 11 Mbps |
| Frequency band | 2.4 to 2.497 GHz |
| Remote configuration support | Telnet; HTTP; FTP; SNMP |
| Antenna options | One non-removable 2.2 dBi dipole antenna or 2 RP-TNC connectors |
| Uplink | 10BaseT Ethernet |
| Operating temperature range | 32 to 122 Fahrenheit (0 to 50 Celsius) |

# Wireless Bridges

Cisco Aironet wireless bridges are used to interconnect remote offices with the main central site. [There is a slight distinction that needs to be made again between an AP and a bridge. Generally, an AP handles end clients such as laptops with wireless cards or wireless IP phones, while wireless bridges connect networks (wireless or wired) over a greater distance than that handled by an AP.]

Cisco offers two Aironet bridge solutions: 1400 series bridges and 350 series bridges. The 340 series bridge has reached end-of-life and can no longer be purchased from Cisco. The biggest difference between the 1400 and 350 series bridges is that the 1400 supports 802.11a technology and the 350 series bridge supports 802.11b. The 1400 bridge is the newer generation of Aironet bridges that supports more feature functionality over the 350 series bridge.

## Cisco Aironet 1400 Wireless Bridge

The Cisco 1400 wireless bridge provides a new generation of wireless bridge technology. The 1400 bridge supports data rates of up to 54 Mbps using 802.11a standards with support for installations in harsh outdoor environments. There are two types of Aironet 1400 bridges, one with integrated 22.5 dBi Antenna (AIR-BR1410A-A-K9) and one with an N-Type connector that allows for professional integration of outside external antenna (AIR-BR1410A-A-K9-N) (see Figure 6.31).

**Figure 6.31** 1400 Bridge with Integrated Antenna (Left) K9-N with N-Type Connector (Right)

Cisco offers three optional N-type connector 5.8 GHz antennas that must be purchased separately for the K9-N 1400 Aironet bridge: the Aironet 5.8 GHz 28 dBi dish antenna (AIR-ANT58G28SDA-N); the 5.8 GHz 9.5 dBi sector antenna (AIR-ANT58G10SSA-N), and the 5.8 GHz 9 dBi omni antenna (AIR-ANT58G9VOA-N).

The 1400 bridge uses a power injector (see Figure 6.32) to provide network and power connectivity for the bridge. It is important to realize that there is no RJ45 connector on the 1400 bridge. The bridge uses two dual-coax (F-type) connectors to interconnect with the power injector using coax cable. Coax cable is more suitable for harsh outdoor conditions than standard BaseT Ethernet. Two dual-coax cables, 20 feet and 50 feet, are shipped with the unit. Longer cables can be ordered directly from Cisco.

**Figure 6.32** Power Injector for Aironet 1400 Bridges

A typical installment of a 1400 Aironet bridge is outlined in Figure 6.33. It is important to follow Cisco's detailed installation document provided with the unit when installing the Aironet 1400 bridge in order to have successful implementation and to comply with FCC guidelines.

**Figure 6.33** Typical 1400 Bridge Install Environment

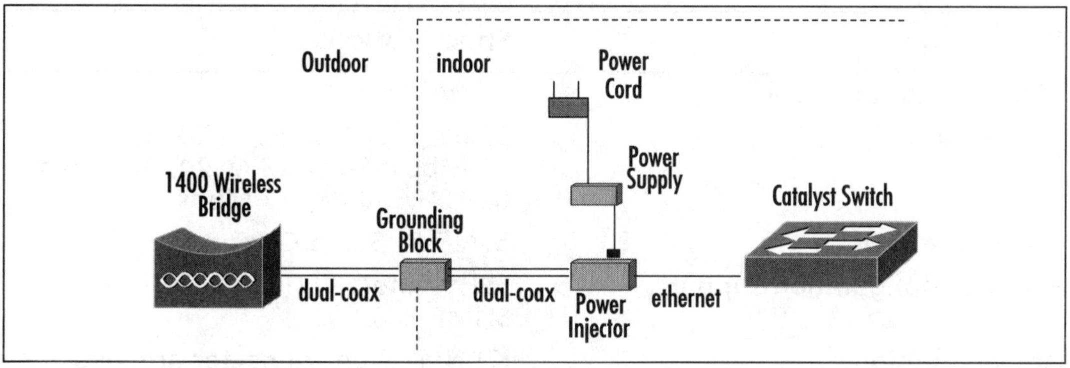


The Cisco 1400 K9 unit with integrated antenna is capable of providing up to 8.5 miles (14 km) at a speed of 54 Mbps and up to 16 miles (26 km) at a speed of 9 Mbps in a point-to-point solution. The 1400 K9-N unit with remote attachable antenna is capable of providing up to 13 miles (21 km) at a speed of 54 Mbps and up to 23 miles (37 km) at a speed of 9 Mbps in a point-to-point solution.

The Cisco 1400 Aironet bridge was built with security features such as 802.1X/EAP mutual authentication, SSH, and AAA capability. It supports 128-bit WEP encryption with TKIP enhancements. The 1400 bridge integrates with the management WLSE engine to provide scalability and manageability. It further supports the IOS operating system, allowing for easy integration into wired networks supporting technologies such as QOS and VLANs. For further details on Cisco 1400 bridges, refer to Table 6.11.

**Table 6.11** Aironet 1400 Bridge Stats

| Description | Specifications |
|---|---|
| Supported data rates | 6, 9, 12, 18, 24, 36, 48, 54 Mbps |
| Supported standard | IEEE 802.11a |
| Range | K9: 8.5 miles @ 54 Mbps; 16 miles @ 9 Mbps<br>K9-N: 13 miles @ 54 Mbps; 23 miles @ 9 Mbps |
| Encryption support | WEP 40-bit and 128-bit |
| Security | 802.1X EAP; TKIP: per-packet + MIC |
| Wireless medium | COFDM |

Continued

**Table 6.11** Aironet 1400 Bridge Stats

| Description | Specifications |
|---|---|
| Media access protocol | CSMA/CA |
| Modulation | BPSK @ 6, 9 Mbps; QPSK @ 12, 18 Mbps; 16-QAM @ 24, 36 Mbps; 64-QAM @ 48, 54 Mbps |
| Frequency band | 5.725 to 5.825 GHz |
| Remote configuration support | SSH; Telnet; HTTP; FTP; TFTP; and SNMP |
| Antenna options | K9-N: N-Type connector antenna support<br>K9: integrated 22.5 dBi antenna |
| Uplink | 100 Mbps over dual coaxial cable<br>Power Injector: 10/100 BaseT Ethernet |
| Operating temperature range | –22 to 131 Fahrenheit (–30 to 55 Celsius)<br>Power Injector: 32 to 122 Fahrenheit (0 to 50 Celsius) |

## Cisco Aironet 350 Wireless Bridge

The 350 series bridge is available only in the rugged, metal case version. These bridges are plenum-rated and can be installed in environmental air space. Their operating temperature range is –4 to 131 degrees Fahrenheit (–20 to 55 degrees Celsius), which allows them to be installed in National Electrical Manufacturers Association (NEMA) enclosures outdoors. The 350 series bridges are powered by inline power through the Ethernet cable; the power supply used in most cases will not be plenum-rated nor have the same operating temperature range. The power supply via Ethernet can be up to 300 feet away, rendering this a non-issue (see Figure 6.34).

**Figure 6.34** Inline Catalyst Switch Support

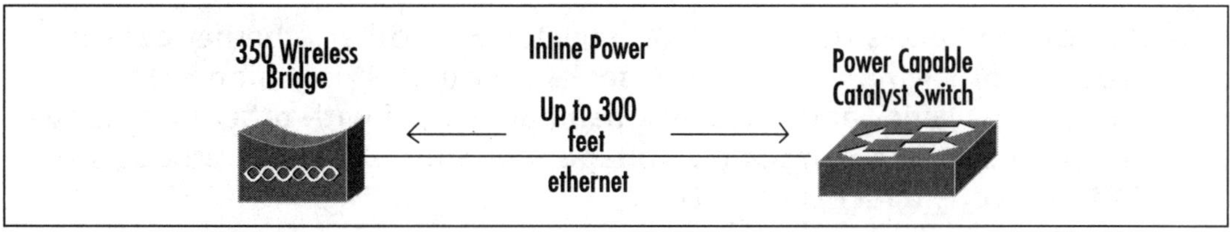


If you do not have a switch without inline power compatibility you can use Cisco's Catalyst Inline Power Patch Panel that interconnects the bridge and the switch providing power for the 350 bridge (see Figure 6.35).

**Figure 6.35** Inline Power Patch Panel Support

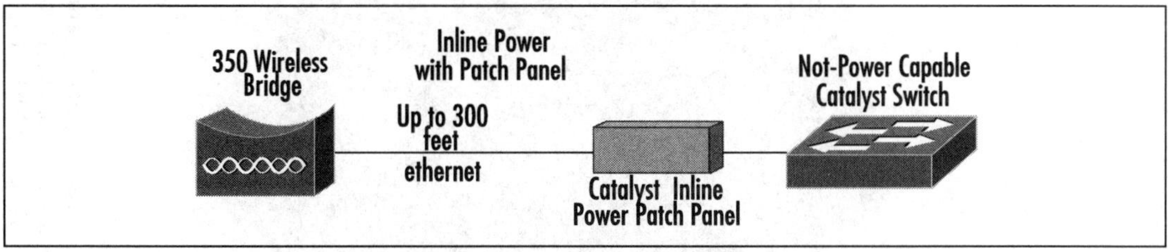


Cisco further provides flexibility of powering the Aironet 350 by using a power injector. The power injector interconnects the switch, bridge, and power source (see Figure 6.36). Figure 6.37 also displays the actual photo of Cisco's power injector for Aironet 350 series products.

**Figure 6.36** Power Injector Support

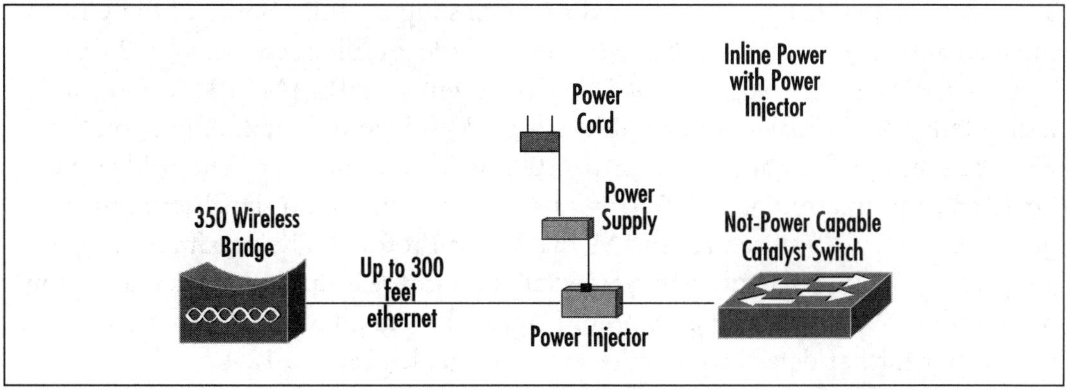

**WARNING**

Do not use Cisco's Aironet power injector with other Ethernet-capable products other than Aironet 350 series products. It is designed for Aironet 350 series and can cause damage if used with other incompatible solutions. The compatible voltage range for 350 series products is 24 to 60 volts direct current (VDC).

**Figure 6.37** Power Injector AIR-PWRINJ for 350 Series

Bridges are typically used in point-to-point configurations, and can support point-to-multipoint topologies as well. Bridges support 802.11b (2.4 GHz) wireless standard technology. No antennas are supplied with bridges, though they have RP-TNC connectors that can be attached to an existing antenna. Some of the Cisco's supplied antennas for the 350 bridge series include: omnidirectional @ 5.2 dBi antenna (AIR-ANT2506); yagi @ 13.5 dBi antenna (AIR-ANT1949), and solid dish antenna @ 21 dBi antenna (AIR-ANT-3338). The only uplink port on the 350 series bridge is an auto-sensing 10/100BaseT Ethernet port. The bridge uses DSSS and can transmit at 1, 2, 5.5, or 11 Mbps depending on signal strength and quality. CSMA/CA is used for the Media Access Protocol. The maximum range (depending on antenna and environmental conditions) is up to 25 miles at 2 Mbps or 18 miles at 11 Mbps. Figure 6.38 displays a photograph of an Aironet 350 bridge. For further details about 350 series bridge, see Table 6.12.

**Figure 6.38** Aironet 350 Bridge

**Table 6.12** 350 Series Aironet Bridge Features

| Description | Specifications |
|---|---|
| Supported data rates | 1, 2, 5.5, 11 Mbps |
| Supported standard | IEEE 802.11b |
| Range | 25 miles at 2 Mbps; 18 miles at 11 Mbps |
| Encryption support | 128-bit |
| Bridge protocol support | Spanning Tree Protocol (STP) |
| Wireless medium | DSSS |
| Media access protocol | CSMA/CA |
| Modulation | DBPSK @ 1 Mbps; DQPSK @ 2 Mbps; CCK @ 5.5 and 11 Mbps |
| Frequency band | 2.4 to 2.497 GHz |
| Remote configuration support | Telnet; HTTP; FTP; TFTP; and SNMP |
| Antenna options | 2 RP-TNC connectors |
| Uplink | Auto-sensing 10/100BaseT Ethernet |
| Operating temperature range | Bridge: –4 to 131 Fahrenheit (–20 to 55 Celsius); Power injector: 32 to 104 Fahrenheit (0 to 40 Celsius) |

# Summary

This chapter provided a comprehensive overview of the Cisco wireless product line. Cisco provides a number of wireless network products, from wireless APs that interconnect wireless clients with the wired network, to bridges that connect networks over great distances. Cisco's wireless product offerings are not limited to APs and bridges. It also includes a rich portfolio of other wireless products such as wireless phones, wireless client adapters, 2.4 GHz and 5 GHz antennas, workgroup bridges, and management applications such as WLSE and ACS.

Cisco strives to provide a standard method of networking by integrating IOS into its compatible Aironet AP and bridge wireless products. A long-supported operation system such as IOS allows for easy manageability and intercompatibility between wired and wireless networks.

Cisco wireless APs and bridges have fairly large coverage areas. These coverage areas can be further expanded with the use of optional external antennas, which can also compensate for obstructions and distances that would otherwise degrade wireless rates. Several antenna options and their appropriate use were covered in this chapter.

Security is on everyone's mind these days. Advances such as EAP and TKIP have done much to improve the security of many wireless networks. Cisco wireless products support these new security mechanisms and enhancements as part of its Wireless Security Suite.

Cisco's vision is to seamlessly integrate wireless networking technology with wired networking by extending technology that has been developed for wired networks into wireless networks. Technology such as security, scalability, and manageability that customers have come to expect in wired networks are being extended and provided in wireless networks.

While a lot of product technology was covered in this chapter, wireless technology is continually expanding at a rapid pace. It is important to keep up on the latest and greatest advances and security implementations to ensure that your wireless networks are secured against the latest threats and provide reliable network connectivity to your clients.

# Solutions Fast Track

## The Cisco Wireless and Wireless-aware Vision

- ☑ Seamlessly integrate wireless technology into wired networks.
- ☑ Provide the same reliable and secure services to wireless environments as what customers have come to expect in wired networks.
- ☑ Become the one-stop solution enterprise for all wireless technology products and services.

## The Cisco Structured Wireless-aware Network Product Line

- ☑ The wireless products Cisco sells and supports include: APs, bridges, antennas, client adapters, wireless phones, and workgroup bridges.
- ☑ CiscoWorks WLSE provides scalable and central manageable solutions for wireless environments.
- ☑ Cisco offers a variety of different styles of antennas for the use of both APs and bridges in the 2.4 GHz (802.11b/g) and 5 GHz (802.11a) product line.
- ☑ The main tradeoff in high gain directional antennas is that it provides greater distance in coverage, but it only in a specific direction.

## Cisco IOS and WLANs

- ☑ The IOS operation system is supported in the Aironet AP 350, 1100, 1200 series and 1400 bridges.
- ☑ Conversions from older Aironet operation system VxWorks can be accomplished using the CAC Tool) or CiscoWorks WLSE, or simply uploading a new image into the AP via a supported browser.
- ☑ IOS is Cisco's core operation system used in devices such as routers and switches. Supporting IOS in Aironet products allows for better manageability for those already familiar with IOS.

## Cisco Aironet Access Points (APs)

- ☑ Cisco offers two major APs: the 1200 series and the 1100 series. Cisco's 350 series AP has reached end-of-sale and is no longer sold by Cisco.
- ☑ The 1200 series AP supports up to two installed modular radios and two RP-TNC 2.4 GHz external antenna connectors with one integrated 5 GHz antenna.
- ☑ Both the 1200 and 1100 AP support electrical power directly over Ethernet cable.
- ☑ The 1200 AP is the only AP that can support all three 802.11a/b/g standards at once. The 1100 series AP supports 802.11b/g.

## Cisco Aironet WLAN Client Adapters

- ☑ Supported client adapters include the 350 series that supports 802.11b, the 5 GHz series that support 802.11a, and the all-in-one 802.11a/b/g client adapter that supports all current wireless standards.
- ☑ Both the 802.11a/b/g client cards and the 350 series client cards come from PCI- or PCMCIA-capable connectors.
- ☑ The only PCMCIA card that currently supports external antenna connections is the 350 series LMC card (AIR-LMC35x) with two MMCX connectors. All other PCMCIA cards have integrated antennas.

## CiscoWorks Wireless LAN Solution Engine (WLSE) 2.x

- ☑ The CiscoWorks WLSE allows you to manage hundreds of Aironet wireless devices from its central command center application.
- ☑ The WLSE enhances rogue AP detection and mitigation by commanding its wireless managed devices to perform regular automatic scans of their coverage area and report any detected APs they find.
- ☑ The WLSE includes the JAVA-driven Location Manager, which allows you to import floorplan maps of a building in correlation to your wireless devices for visual effect and easy location determination.

## Cisco Wireless Security Suite

- ☑ Implementing WLANs alone without any enhanced security measures provided by the Cisco Wireless Security Suite is highly vulnerable to outside attacks.
- ☑ The EAP part of the Security Suite is supported in all Aironet wireless devices. EAP enhances security such as per-user authentication and dynamic WEP key generation.
- ☑ TKIP is used to enhance WEP with per-packet-keying and MIC.

## Cisco Secure Access Control Server (ACS) 3.2

- ☑ Cisco ACS is mainly used as a RADIUS server in wireless environments that are required by EAP solutions.
- ☑ The ACS 3.2 application is supported as a software solution for Windows 2000 server and as a hardware and software solution on a dedicated Cisco server platform.
- ☑ Cisco ACS software can hold up to 80,000 user database records with its minimal hardware configurations. It can also interact with many remote databases such as Active Directory and OTP solutions.

## Enhanced Client Network Management Features with Extended Client Support

- ☑ Cisco ACUs supported in 350 and 5 GHz client adapters include the Site Survey measurement tool.
- ☑ The Troubleshooting option in the client adapter software provides step-by-step visual tests that an adapter takes to associate with an AP. Troubleshooting tests include the Radio Test, the Association Test, and the Authentication Test.
- ☑ New 802.11a/b/g client adapters no longer include site survey tools as part of their client software.

## Workgroup Bridges

- ☑ Cisco offers Aironet 1400 (802.11a) and Aironet 350 (802.11b) series bridges.
- ☑ Aironet 1400 bridges support data rates of up to 54 Mbps using 802.11a standards with support for installations in harsh outdoor environments.
- ☑ There are two types of Aironet 1400 bridges, one with an integrated 22.5 dBi antenna (AIR-BR1410A-A-K9) and one with an N-type connector that allows for the professional integration of outside antennas (AIR-BR1410A-A-K9-N).
- ☑ The Aironet 1400 bridge uses and ships with a power injector that is used to provide network and power connectivity for the bridge. There is no RJ45 connector on the 1400 bridge. The bridge uses two dual-coax (F-type) connectors to interconnect with the power injector using coax cable.

## Wireless Bridges

- ☑ The Aironet 350 workgroup bridge connects a small number (up to eight) of wired-attached Ethernet stations to an AP or bridge.
- ☑ Workgroup bridge provides wireless connectivity to Ethernet-ready devices such as printers, computers, and copy machines that may not be wireless-compatible.
- ☑ Two 350 antenna options are available with the workgroup bridge, a single non-removable 2.2 dBi dipole antenna, or two RP-TNC connectors with external antenna support.

# Frequently Asked Questions

The following Frequently Asked Questions, answered by the authors of this book, are designed to both measure your understanding of the concepts presented in this chapter and to assist you with real-life implementation of these concepts. To have your questions about this chapter answered by the author, browse to **www.syngress.com/solutions** and click on the **"Ask the Author"** form. You will also gain access to thousands of other FAQs at ITFAQnet.com.

**Q:** Can I use my Ethernet cable to power an Aironet 1400 bridge?

**A:** An Ethernet RJ45 connection is not supported in Aironet 1400 bridges. Aironet 1400 series bridges ship with a power injector that is used to provide network and power connectivity for the bridge. The bridge has two dual-coax (F-type) connectors to interconnect with the power injector using coax cable.

**Q:** I run ACS software version 2.6. Will I be able to support Cisco's LEAP authentication protocol?

**A:** Cisco introduced support for some of the EAP family protocols such as LEAP starting in ACS 2.6. You will be able to support LEAP with version 2.6, but it is highly recommended you upgrade to the latest 3.2 version that includes enhanced features and support for EAP family protocols.

**Q:** I am running WLSE version 2.0 and want to support rogue AP detection, site surveys, and the Location Manager feature. What do I need to do?

**A:** All of these features were introduced in WLSE version 2.5. You will need to upgrade to 2.5 in order to support these valuable enhancements.

**Q:** Wireless users are complaining about a very slow LEAP authentication process during peak time hours and about the overall performance. Is there anything I can do?

**A:** Set up fault-monitoring under WLSE to perform automatic threshold health checks such as RADIUS authentication responses and AP/switch CPU and port utilization, and have WLSE alert you when any of the thresholds has been reached. If WLSE detects a slow authentication RADIUS response time, login to your server and check utilization and memory usage. If WLSE detects that your AP interface has high traffic utilization, you may need to implement QOS for RADIUS authentication or add additional APs supporting the increased demand of wireless users. Further, it is advised that you also check the Reports for Wireless Clients under the WLSE menu to make sure that one or two of your clients are not taking up all of your bandwidth resources by downloading large music files.

# Chapter 7

# WLAN Security Considerations

## Solutions in this Chapter:

- Network Security in General
- How Wireless Technology Changes Network Security
- Preventing War Driving and Unauthorized Use of Legitimate Access Points
- Exploring Rogue Access Points
- Designing for Security
- WLAN LAN Extension 802.1x/EAP
- WLAN LAN Extension IPSec
- WLAN Static WEP Keys
- WLAN Security with VLANs

☑ Summary

☑ Solutions Fast Track

☑ Frequently Asked Questions

# Introduction

Security is one of the biggest concerns in any network infrastructure. A public security breach of a business network can scare away current and potential clients causing a company to lose millions in revenue. Today, network architectures built without security components are unacceptable.

Network technology has evolved at a rapid pace, moving the development of security along with it. Constant pressure to assemble the latest security and technology has left many network administrators in a rush to deploy without proper testing and security review, in some cases only to find out that the latest technology has become old by the time of its implementation.

Wireless local area networks (WLANs) are changing the way business is conducted. The use of remote devices such as phones, laptops, and personal digital assistants (PDAs) along with the demand for continued network connection without having the need to plug in, are driving the wireless revolution and adoption of WLAN.

This chapter reviews general network security and best practices that you can undertake to understand and protect your wireless network. It reviews the types of security threats and unauthorized breaches, both on wired and wireless, that you are trying to protect your networks from. It is important to know why you are implementing security features and from whom you are trying to protect you network against prior to implementing any network security architecture.

This chapter also reviews and discusses *war driving* techniques (driving around detecting wireless access points) and the role that they play in reconnaissance that can lead to unauthorized access, as well as how to mitigate these threats to protect your networks against war driving and unauthorized access.

Cisco has developed many new security features to fulfill the need for robust, reliable, and secure wireless networking. Many of you are already familiar with current 802.11 and its Wired Equivalent Privacy (WEP) Protocol vulnerabilities. This chapter reviews the common vulnerabilities in the WEP protocol to help you understand the differences in new enhanced security protocols and practices. You will learn about 802.1x and its security enhancements over 802.11, including its ability to provide scalable centralized security management and its support for per-user authentication. You will review dynamic WEP encryption keys and their implementation to protect private communication between mobile users and access points. You will also learn about virtual local area networks (VLANs) in WLANs and their ability to separate users into different network segments based on security levels and needs.

This chapter also explores the use of virtual private networks (VPNs) in the WLAN. In particular, it looks at implementing Internet Key Exchange (IKE) protocol and IP Security (IPSec) protocol over wireless networks. Using IPSec over WLAN allows you to further secure and protect data on your WLAN.

# Network Security in General

Confidentiality, integrity, and availability, sometimes called "the big three," are the basic tools for the protection and preservation of network security. All three must be ensured for a successfully secured network. Ensuring confidentiality prevents the unauthorized disclosure of communication between two or more parties. Confidentiality is usually accomplished by encryption. The most popular implementations of encryption in today's wireless networks make use of WEP and its Rivest Cipher Four (RC4) algorithm to ensure confidentiality between two or more devices.

Ensuring integrity prevents data from being manipulated during transmission. It is important that data entered by the user has not been inappropriately changed during the transaction without the consent of the user or application. Integrity of data also includes the authentication of the user source. Integrity ensures that the data has been received from a trusted and authenticated origin, and that the sending party confirms and proves who they say they are. Let's look at an example of why integrity is important. Say you pay your bills online with some type of online banking service. You are about to click on the Pay Bill option to wire $48.40 from your bank account to your phone company for this month's telephone service charges. In the middle of the transaction, an attacker (or a man-in-middle [MITM]) intercepts your data and changes the $48.40 to $48400.00, and instead of allocating the monies to the phone company, the MITM inserts a new destination for the funds, in this case their offshore bank account. It is important to prevent someone from listening in on a private communication, not only for confidentiality but also to insure integrity to prevent an attacker from changing data in the middle of a transaction.

Ensuring availability makes certain that information and resources are always available to users. If the network or end system device is not available, it defeats the purpose of having a network. Availability was one of the main concerns and origin of the Internet itself. In 1957, the Department of Defense (DOD) created the Advanced Research Projects Agency (ARPA), which was the beginning of what is known today as the Internet. The ARPA network was primarily built due to the concern for availability. Fear of losing communication between military command centers due to the disruption caused by nuclear attacks helped start the new revolution of network computing. Today, wireless networks are experiencing many availability issues, mainly due to conflicting interference from other mobile devices. Wireless networks such as 802.11b, 802.11a, and 802.11g use public frequencies allocated by the Federal Communications Commission (FCC). Because these frequencies are public and anyone can use them without a license, many manufacturers build and use them in their products. Due to this, WLAN devices find themselves competing for available bandwidth between an army of wireless capable devices.

# Why We Need Security

The Internet and WLANs are rapidly changing the economy, the way we do business, and the way we live. Rapid growth and endless new discoveries constantly improve wireless technology. Business and government leaders recognize the strategic role that the Internet and WLANs play and realize how important it is to be competitive in this new electronic century.

For consumers and businesses to accept wireless networks, they need guaranteed secure methods of communicating, sharing data, and performing electronic commerce. Unfortunately, due to the use of open standards and unfinished security methods in 802.11, some WLAN design standards miss key security components such as dynamic per-user authentication and encryption. Although developers considered the need to provide controlled remote access, privacy for communications, and prevention of attacks to secure WLAN, it was not enough. Many existing wireless networks are not taking advantage of new improved security measures for 802.11 such as 802.1x, and are vulnerable to many different attacks. As with any security implementation, businesses face the daunting security issue of how to implement and constantly update new and improved defensive practices to reduce business vulnerability.

One security challenge that organizations and businesses face is sorting through a wide range of solutions and choosing the right technology for their needs. Every network is unique; therefore, choosing among many different solutions to best meet unique needs is challenging. Luckily, Cisco offers many affordable wireless solutions for home use and enterprise solutions that make it easy for consumers to do so. Cisco's Aironet 1100 access point is replacement for the 350 series, which offers 802.11b or 802.11g connectivity radio and is mostly used at remote branches or home offices. Cisco's Aironet 1200 access point offers an enterprise-level scalability and performance with the support of two installed radios. Radio support for Aironet 1200 includes 802.11a/b/g. For additional detail on Cisco's wireless product listings and their specs visit www.cisco.com/en/US/products/hw/wireless/index.html. After you have selected the right mix of security components for your network environment, the solution must be integrated throughout your enterprise network to be in sync with a single consistent security policy.

As you are no longer physically bound to wire networks in order to access sensitive information on a secure network, it is important to extend the security measures that exist on your wired networks to your wireless networks. You should require and build even greater security measures for wireless connections to ensure data confidentiality and integrity traveling through the air.

## Security Threats

There is a vast range of security threats that networks are vulnerable from. To help understand these different types of network threats, they are categorized into different sections. It is important to understand these threats and their effect on networks to properly secure against them. These threats include:

- Reconnaissance
- Unauthorized access
- Denial of Service (DoS) and Distributed Denial of Service (DDoS)
- Data sniffing
- Data manipulation

## Understanding Reconnaissance Attacks

Reconnaissance is discovering information of a selected target. This information gathering includes mapping network and system devices, scanning for open ports on a system, and finding Internet Protocol (IP) addresses of valuable server services such as Domain Name Systems (DNS), mail servers, and database servers. Reconnaissance includes war driving and is round one of an attack. You must first detect a network before you can exploit it to gain unauthorized access. Techniques such as *social engineering* and *physical detection* also play a part in reconnaissance. Social engineering is the method of calling a user on the telephone and pretending to be someone from within the organization to try to gain information such as phone numbers to dialup access points, WEP keys to wireless access points, logins and passwords to database servers, and anything else that can help an attacker gain access to a corporate network. You would be surprised how many users give out sensitive information when an attacker calls and pretends to be a network administrator working on a problem. The best way to protect your company from social engineering is to constantly educate and audit company personnel.

Let's take a look at some general network techniques for performing reconnaissance. Once an attacker has identified the network they are planning to attack they need to map out system devices that will help them gain access to restricted information. These exploited devices will then help explore a trust relationship to other resources until the attacker reaches their goal. When implementing and configuring security measures on your network, a good practice is to avoid a trust relationship between your outside devices and your inside corporate devices. If a relationship must exist between the two, such as monitoring an outside external perimeter router from a management station that is inside your corporate network, you should limit your communication to specific access ports and their allowed direction flow. Try not to rely on an IP-only trust relationship, meaning a relationship where you trust a device only by its IP. IP addresses are not hard to spoof, especially when you are dealing with User Datagram Protocol (UDP)/IP communication. *Spoofing* is a technique where the attacker tries to impersonate a valid authorized user's IP in order to gain access. Unlike TCP, UDP does not require a handshake to establish communication between two hosts before data is sent. Extra measures should be taken, if possible, to authenticate resources on more than just the IP, such as a Media Access Control (MAC) address or some sort of time key or a biometric parameter.

*Ping sweeps* are used to discover the IP addresses of hosts within a network or subnet. Sweeps are used to perform network mapping of potential targets that can be zeroed in on for more in-depth reconnaissance. The ping command uses an Internet Control Message Protocol (ICMP) Echo Request against a specified network range. If a device exists it replies with an ICMP Echo Reply. NMAP is a great free tool for performing some reconnaissance and is available for free from www.nmap.org.

Once an attacker has a map of your IP addresses they can zero in on devices by port scanning a specific device. A port is an application service that listens on a device for incoming requests over the network. Different applications and services have different port numbers that are regulated by the Internet Assigned Numbers Authority (IANA). If a port scanner reveals that a device is running on port 80, it indicates that the server is most likely running a Web/Hypertext Transfer Protocol (HTTP) application. Further, if it reveals that a device is listening on Transmission Control Protocol (TCP) port 53, it indicates that the server is likely a DNS. Figure 7.1 shows a port scan of a server using NMAP. For a complete listing of the port numbers assigned by IANA, refer to www.iana.org/assignments/port-numbers.

**Figure 7.1** Port Scanning with a NMAP Scanner

```
lab.makesecure.com
#
# nmap -F 192.168.150.1

Starting nmap 3.20 ( www.insecure.org/nmap/ ) at 2003-10-13 20:58 EDT
Interesting ports on 192.168.150.1:
(The 1134 ports scanned but not shown below are in state: closed)
Port       State       Service
7/tcp      open        echo
9/tcp      open        discard
13/tcp     open        daytime
19/tcp     open        chargen
21/tcp     open        ftp
23/tcp     open        telnet
25/tcp     open        smtp
37/tcp     open        time
79/tcp     open        finger
111/tcp    open        sunrpc
512/tcp    open        exec
513/tcp    open        login
514/tcp    open        shell
515/tcp    open        printer
540/tcp    open        uucp
587/tcp    open        submission
4045/tcp   open        lockd
```

Now that the attacker has the IP addresses of both the targeted server and the service ports it is running, they can zero in on the specifics of running the application. The first step of application reconnaissance is to find out the specific type and version. This is often done by connecting into the listening port via Telnet. An NMAP scanner includes a new port-scanning feature that allows for the automatic version reconnaissance of different application ports. By knowing the application version and specifics, the attacker can search for product vulnerabilities and bugs previously announced by the vendor, which they can then use to attack the system or network.

There are many ways of performing reconnaissance. Many of the techniques discussed in this section are performed automatically by Internet boots that the attacker controls. These techniques are programmed robots usually running on already hacked devices that the attacker uses to perform continual scans of the Internet, finding old, unpatched applications and devices they can take advantage of to gain access into corporate networks and businesses.

Wireless networks can be detected in reconnaissance by war driving or sniffing radio frequencies for possible unencrypted communication. The default configuration of wireless access points may beam out special beacon packets used to alert legitimate wireless users to its presence. These beacons can be used and detected by an intruder during reconnaissance for wireless networks. War driving and its techniques are covered in greater detail later in this chapter.

## Unauthorized Access

A network intruder can gain unauthorized remote access into a network through a variety of means. The common target for an attacker is to gain root (UNIX), Administrator (Windows), Level 15 (Cisco IOS), or WLAN connectivity access where they have the power to control the device or to access other networked computers.

In a WLAN, an attacker can gain unauthorized access by war driving. War driving is much like a *war dialer*. Using a war dialer, an attacker scans the phone numbers of a company looking for dialup access points. Automated applications and cheap phone plans make it possible to scan entire area codes in a few days, recording any successful data connection and its initial login prompts which are later exploited as points of unauthorized access entry. A similar approach is taken with war driving, where the attacker drives around looking for wireless signals that beam out unintentionally from a company's headquarters into the streets and parking lots. An attacker can then use techniques to capture these wireless signals and possibly gain unauthorized access into the company's corporate networks.

Weak security authentication practices result in unauthorized access into private networks. A poor password policy can allow a network intruder to gain remote access. It is important to point out that knowing the username is half the battle. Many do not realize that usernames can be as important as passwords. If a username is known, the attacker can try to *brute force* himself in. Brute force is a technique in which the attacker tries every possible word combination in a multi-language dictionary to try to guess a password.

## Designing & Planning…

### Password Policy

To protect your network from unauthorized access, a strong password security policy is a must. A password policy should include the combination and length of password that is to be allowed for use and its expiration period before it must be changed. A good password length is at least 8 characters long with a combination of letters, digits, and special characters. A password should be changed every 60 days to a new unique password. If possible, try to use the One-Time Password (OTP) implementation. OTP allows the user to enter his password plus a unique password that is re-generated every 60 seconds in order to gain access.

Using static WEP keys or a Secure Set Identifier (SSID) alone in a wireless network as the security technique can result in unauthorized accesses. SSID should not be used as a security measure but rather as a measure to differentiate between two or more access points. SSID travels in clear readable format from the wireless client to the access point, allowing for any intruder within the signal range sniffing the frequency a clear view. Intruders can take advantage of poorly configured access point relying only on SSID and WEP keys for security. War driving tools look for a default-configured SSID they can take advantage of.

## Designing & Planning…

### Vendor-default SSID

Vendor-default SSID settings are widely known and should not be relied on as security measures. Even though they are not used to enhance wireless security, you should change these defaults so that they do not enable an attacker to recognize the vendor-specific device. These default SSIDs are:

Continued

- **3COM** comcomcom
- **Bay** Default SSID
- **Cisco** tsunami
- **Compaq** Compaq
- **Dlink** WLAN
- **Intel** 101, 195, xlan, intel
- **Linksys** linksys
- **Netgear** Wireless
- **SMC** WLAN
- **SOHOWare** Same as MAC address
- **Symbol** 101
- **Telectronics** any
- **Zcomax** any, mello, Test
- **Zyxel** 1234

## Understanding DoS and DDoS Attacks

A DoS attack is an attempt to corrupt or disable networks and their services thereby denying the use of network services to legitimate users. An attacker might use a DoS attack to test a system's vulnerability as a prelude to further attacks. A DoS attack can be used as a technique to cover tracks, during or after unauthorized access, by introducing extra traffic into the network. It is not easy to sort out two out of a million almost identical log entries to determine that unauthorized access has been established. A DoS attack can be used to attempt to disrupt the connection between a particular user and a particular service. This is a common technique that attacker's use to impersonate an already authenticated user to gain access to a particular service application or network.

A DoS attack on a WLAN network must be within the signal reach of an access point. It is also possible to attack the access point from its wired side, filling it with useless data, therefore allocating all of its resources. In principle, a WLAN DoS attack could simply involve a transmitter blasting a specific area with 2.4 or 5.2 gigahertz (GHz) bands. Because WLAN's use unlicensed spectrum and are based on inexpensive hardware, these attacks are much more possible. In Figure

7.2, an attacker is transmitting data, causing valid radio communication to bounce off while congesting the access point.

A DDoS attack is an attack using more than one resource to initiate a DoS attack. An attacker can use previously hacked devices and services to send out attacks from multiple resources, increasing its power of success.

**Figure 7.2** Wireless DoS Attack

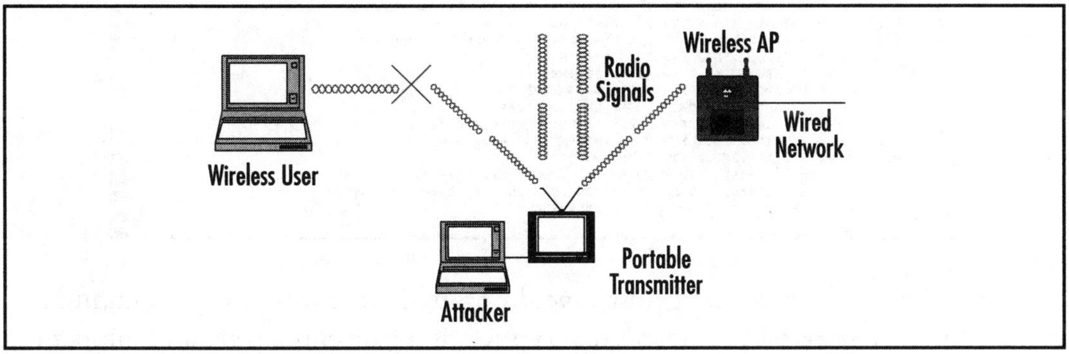


# Data Sniffing

*Data sniffing* is a method of eavesdropping on network communication traffic with a device or utility that analyzes network traffic into readable format. The purpose of data sniffing is to observe and capture communication data between two parties for later analysis.

Wireless networks function in a similar method as repeaters and hubs function on a wired network. Every communication across a wireless network is available and can be viewed by anyone listening within the signal strength. A common way to sniff network communications is to capture TCP/IP packets traveling through the air or on wired networks. Captured data is then used with a protocol analyzer or similar tools that display and analyze captured data into readable format. Unencrypted data traveling across wireless or wired networks can be used to gather intelligence such as SSIDs, login names, passwords, and other useful information an attacker can use to access networks.

Figure 7.3 shows a sniffer capture of a user login to a File Transfer Protocol (FTP) service. FTP is an unsecured method of transferring files and authenticating. As can be seen, the login, password, and actual data that is being transferred are sent in cleartext. The username and password in the sniffer capture are both "admin." A sniffer can pick up this information and use it to gain unauthorized access.

**Figure 7.3** FTP Username and Password Captured with Sniffer

```
lab.makesecure.com
OLD-BROADCAST -> BROADCAST    DHCP/BOOTP DHCPDISCOVER
OLD-BROADCAST -> BROADCAST    DHCP/BOOTP DHCPDISCOVER
OLD-BROADCAST -> BROADCAST    DHCP/BOOTP DHCPDISCOVER
192.168.150.1 -> 192.168.150.2 FTP C port=32781
192.168.150.2 -> 192.168.150.1 FTP R port=32781
192.168.150.1 -> 192.168.150.2 FTP C port=32781
192.168.150.2 -> 192.168.150.1 FTP R port=32781 220 solaris2 FTP ser
192.168.150.1 -> 192.168.150.2 FTP C port=32781
           ? -> (multicast)  ETHER Type=0000 (LLC/802.3), size = 52 bytes
           ? -> *            ETHER Type=9000 (Loopback), size = 60 bytes
OLD-BROADCAST -> BROADCAST    DHCP/BOOTP DHCPDISCOVER
           ? -> (multicast)  ETHER Type=0000 (LLC/802.3), size = 52 bytes
192.168.150.1 -> 192.168.150.2 FTP C port=32781 USER admin\r\n
192.168.150.2 -> 192.168.150.1 FTP R port=32781
192.168.150.2 -> 192.168.150.1 FTP R port=32781 331 Password require
192.168.150.1 -> 192.168.150.2 FTP C port=32781
OLD-BROADCAST -> BROADCAST    DHCP/BOOTP BOOTREQUEST
OLD-BROADCAST -> BROADCAST    DHCP/BOOTP BOOTREQUEST
OLD-BROADCAST -> BROADCAST    DHCP/BOOTP DHCPDISCOVER
192.168.150.1 -> 192.168.150.2 FTP C port=32781 PASS admin\r\n
192.168.150.2 -> 192.168.150.1 FTP R port=32781 230 User admin logge
192.168.150.1 -> 192.168.150.2 FTP C port=32781
           ? -> (multicast)  ETHER Type=2000 (Unknown), size = 375 bytes
           ? -> (multicast)  ETHER Type=0000 (LLC/802.3), size = 52 bytes
OLD-BROADCAST -> BROADCAST    DHCP/BOOTP DHCPDISCOVER
^C#
```

The best way to protect against eavesdropping is to encrypt your communication. In current 802.11 standards, encryption is accomplished by a combination of the RC4 algorithm and WEP keys. The Advanced Encryption Standard (AES) (a more secure encryption algorithm than RC4) will be supported in Cisco's devices sometime in 2004. IPSec can also be used to further protect your communication. IPSec and its use in WLAN are covered later in the chapter. This chapter also looks at new 802.1x practices and new additions to WEP such as Temporal Key Integrity Protocol (TKIP), that help enhance the use of RC4 encryption implementation.

### Data Manipulation

Data manipulation is when an attacker captures data, changes the content of the data, and resends it. The attacker becomes the MITM. To prevent MITM attacks, strong encryption and authentication is used to mitigate such attacks.

## Continuing the Security Cycle

The security cycle is the ongoing process of maintaining the security of your network. Networks are dynamic with rapid enhancements and new vulnerabilities that emerge every day. The network security administrator must keep up with all new threats and continuously approve their network security.

The security cycle breaks security into four phases, as shown in Figure 7.4:

- Securing
- Monitoring
- Testing
- Improving Security

**Figure 7.4** Security Cycle

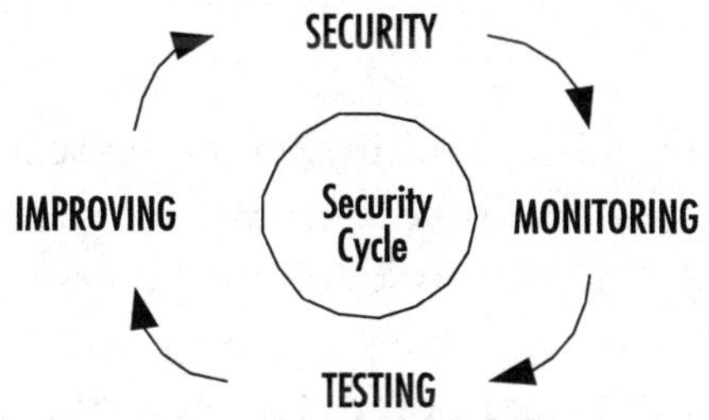


The security cycle is part of the process of keeping up to date with the latest threats and continuously strengthening the overall security of your network.

# Securing

The first step you need to take to activate the cycle is to implement security measures based on your security policy. Your security policy provides the framework of your requirements. This step includes configuring strict authentication and confidentiality. In WLAN networks you should use dynamic WEP keys and per-user authentication with 802.1x to accomplish stronger authentication and confidentiality.

Limit administrative access into your WLAN and network devices. Implement Secure Shell (SSH) rather than Telnet, when you administer and connect to access points. SSH is an alternative to Telnet that is considered standard secure protocol for remote login. SSH provides a secure encrypted connection as opposed to Telnet where login, password, and data is sent in cleartext.

Reduce unneeded services. Services such as FINGER, CDP, NTP, Bootstrap Protocol (BOOTP), and Simple Network Management Protocol (SNMP) should be disabled if you do not intend to use them on your wireless and network devices. If you plan to use services such as SNMP, configure access lists on your access point to restrict everyone but your management servers from using it.

Provide confidentiality with encryption. Confidentiality and integrity are a must for sensitive traffic that needs to flow over unsecured links such as the air. One such approach is to use RC4 encryption in WEP with 802.1x implementation or to use 3-DES with IPSec over VPN.

Secure network boundaries with firewalls to protect and limit the use of company networks and resources. It is a good practice to implement firewalls behind access points to restrict WLAN usage. Firewalls will not only help restrict a wireless users' access into a corporate wired network, but it can also provide a type of accounting where restricted access attempts can be logged.

## Monitoring

After establishing a security policy and implemented security on your network, the next step is to monitor users on your network to catch violations and unauthorized access attempts. Many network administrators ignore this step, mostly because they are to busy or because they can only watch so many log messages during one the day before they are bored. The new 802.1x allows for the use of Remote Authentication Dial-In User Service (RADIUS) that supports accounting for users' logins and time of login. You can also monitor and send system messages from your access points using system logic (SYSLOG) protocol from your centralized monitoring station. These SYSLOG messages include system errors as well as rejected access attempts into an access point.

An Intrusion Detection System (IDS) can be implemented on your network to monitor for potential attacks, backdoors, and viruses coming in from users on the WLAN. IDS watches all network traffic, testing each packet or set of packets against a database of signatures that define different known vulnerabilities. If a match is made against a packet and a known attack signature, the IDS sends out an alarm including detailed information and stores it in a database for future investigation.

## Testing

New security vulnerabilities are discovered daily, and, therefore, you must make sure your network is up to date with the latest security patches and security implementations. One of the ways to test your network is to use security scanners. A network vulnerability scanner scans your network for any security vulnerabilities. Some more popular common free network scanners include the following:

- **Nessus** www.nessus.org
- **NMAP** www.nmap.org

You should constantly test for possible wireless rogue access points connected into wired network that can expose your corporate network. Rogue access points and their threat mitigation are covered later in the "Exploring Rogue Access Points" section.

# Improving Security

The last step in the security cycle is to take all of the information derived from monitoring and testing and improve your network's security. New vulnerabilities and risks are discovered daily. You need to keep up to date and continue this cycle to have a secure network. To stay abreast of the latest security news you should at a minimum perform the following:

- Monitor vendors' news and security patches. Sign up for Cisco security advisory releases and newly discovered bug fixes. To view current bugs and patches for wireless devices, visit www.cisco.com Web site and select **Tools & Utilities** from the technical support page then click on **Software Bug Toolkit** from the troubleshooting tools menu. You will need to be logged in Cisco's Web site in order to use the bug toolkit. Figure 7.5 shows a query in the toolkit for severity 1 and 2 bugs for Cisco 1200 Aironet.
- Sign up to a public security mailing list such as Bugtraq to monitor new reported vulnerabilities and security risks. To view archives or join the BugTraq mailing list visit www.securityfocus.com.
- Review your configuration files periodicaly.
- Verify your security configuration and implementations regularly.

**Figure 7.5** Cisco's Bug Toolkit

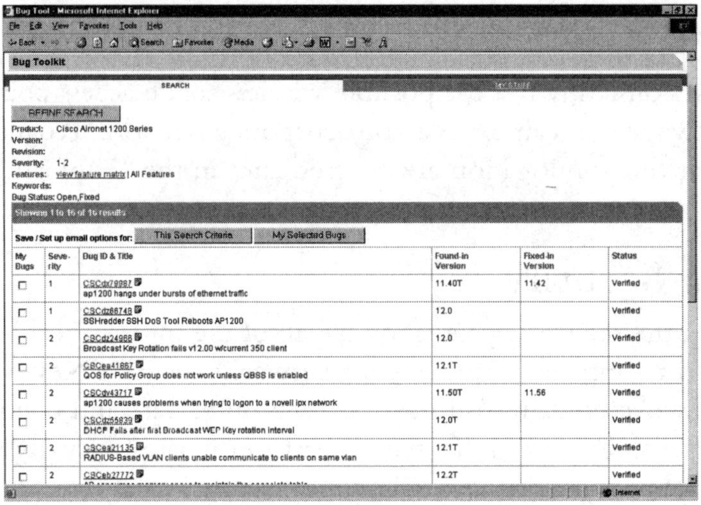


# How Wireless Technology Changes Network Security

Since the standardization of the IEEE 802.11b in 1999, wireless has become more prevalent. Today, wireless technology and networks are widely deployed in places such as corporate offices, airports, cafés, homes, universities, and classrooms. The new 802.11 wireless deployment presents new challenges for security experts and network administrators. Unlike in wired networks, wireless networks and its data travel through the air using Radio Frequency (RF). This presents a new security issue that involves augmenting the new 802.11 standards.

## Overview of 802.11 Standards

Before exploring wireless and its effect on network security, let's review the standards of the 802.11 networks. 802.11 is a member of the IEEE 802 suite, which is a series of specifications for Local Area Networks (LANs). IEEE 802 specifications are focused on the two lower Open System Interconnection (OSI) network layers that include the Data Link layer and Physical Link layer. The MAC on the Data Link layer handles access to shared media and determines how to send the data. The details of transmission and reception is done by the Physical layer and includes the actual wire in wired networks, or the RF in wireless. Refer to Figure 7.6 for 802.11 and its relationship in the 802 standards.

**Figure 7.6** 802.11 and its Relation to the OSI Model

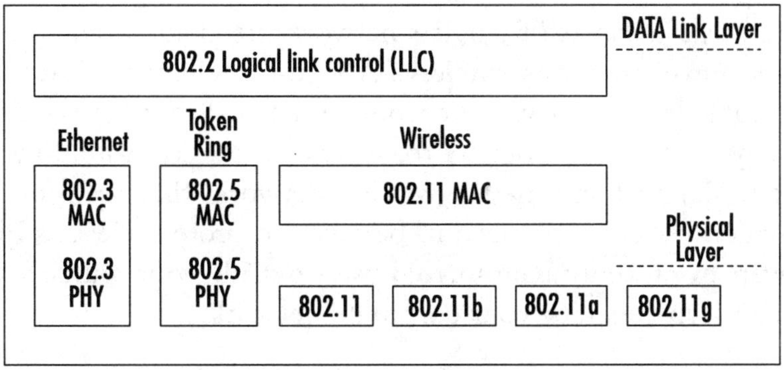


802.11 is another layer that makes use of the 802 and 802.2/LLC encapsulation, just as wired Ethernet or Token Ring technology does. The base of 802.11 includes the 802.11 MAC layer. The physical layers include 801.11 Frequency Hopping Spread Spectrum (FHSS) and Direct Sequence Spread Spectrum (DSSS). In 1999, additional physical layers were added that include the 802.11b which specifies the High Rate Direct Sequence (HR/DSS) and 802.11a Orthogonal Frequency Division Multiplexing (OFDM) technology.

# Shared Network Model

Wireless networks are shared model networks. Unlike in wired networks where media can be separated into segments and isolated into separate domains, wireless signals are available and shared with anyone in the path of the RF signal used to communicate between the wireless user and the access point. It is similar to a hub or repeater operation in wired networks, where if one station connected to a hub sends a packet, every other station connected to the same hub has the ability to view the transmitted data. This possible intrusion can be mitigated by the use of switches instead of hubs in a wired network, which provide a separate collision domain for each connected host. Unlike in a wired network, in a wireless network you cannot prevent host A from looking at data sent from host B using the same access point. Wireless networks transfer data through the air, making them more vulnerable to eavesdropping techniques where an intruder listens secretly and captures private communication between two parties. The Physical and Data Link layers in a wireless network change network security that needs to be better protected to assure that any intruder listening on a private communication does not understand its context and does not capture sensitive data.

# Protecting the Data Link and Physical Layers

The Data and Physical layers of wireless networks need additional protection from unwanted intruders on its wireless LAN. Just as you would not hang your physical wire cable from your wired corporate network outside your office window for anyone to freely connect to, you need to make sure only authorized users are able to connect and use the wireless networks that beam outside of your office windows. Measures need to be taken to protect wireless Physical and Data Link layers to prevent unauthorized users outside your window to successfully connect via wireless into your corporate network.

## Tracking and Attacking Anonymity

Tracking and identifying intruders is more difficult in a WLAN environment than in a LAN environment. On a local wired network, the administrator can track an offending user by mapping the user's IP address to their network card MAC address. Knowing the MAC address of a user on a corporate LAN allows the network administrator to view the MAC address tables on a switch to find out which wired port is associated with the user's MAC. Once the port is known, the administrator can trace the physical cable connecting the port directly to the user's PC.

There are no cables in wireless networks. An offending user or intruder can use a wireless network from the company parking lot or from the building next door. This makes the user more anonymous and harder to track. To protect your wireless network from an anonymity attack, strong per-user authentication of users and integrity needs to be implemented to avoid the need for the tracking of unauthorized users on your wireless network. You should use per-user authentication on your wireless network that will aid in tracking and suspending offending users and at the same time will protect your WLAN from unwanted intruders. Per-user authentication in 802.1X and its details are discussed in the WLAN LAN Extension 802.1x/EAP section later in this chapter.

## Attacks on Wireless Networks

Attacks like DoS can theoretically be used in any wireless installation. Unlike wired LAN networks that can be isolated for corporate users only on the same LAN segment, a WLAN can be attacked by sending signals through the air and hijacking all of the available bandwidth. Doing so can block regular authorized users from using the WLAN and its resources. The unlicensed spectrum and frequency that wireless networks use to communicate make it easy for attackers to obtain cheap hardware to perform DoS-type attacks. Attacks can involve a radio device continually blasting the 2.4 or 5.2 GHz bands, taking up all of the allocated bandwidth. If FCC regulations say wireless radio equipment must operate on a lower power level to transmit data, an attacker can simply turn the power up on their signal jammer device to create greater interference on the wireless networks.

DoS attacks are not easy to overcome on wireless or wired networks. (Measures such as installing redundant wireless access points are easy to take off during possible attacks.) Further filters such as access lists on wireless radios and wired interfaces should be implemented to drop unwanted traffic.

## Authentication

Unlike in wired networks where you plug your desktop into a cable jack in your office, you are not likely to be prompted for a username and password to successfully establish connection with the network jack. Although 802.1x makes this possible on wired networks and will be the security standard for future wired networks, it is considered a must in wireless networks and a standard security measure. In order to implement a secure wireless network, users must be authenticated prior to access point association to prevent unauthorized access from distant intruders. How 802.1x makes this authentication possible is discussed later in this chapter.

## Physical Security

It is important not to ignore the physical security of your wireless equipment. Unlike wired networks, your WLAN designs may require you to position wireless equipment outside of protected data centers or outside of buildings. These devices make easy targets for gaining access to your wireless and wired networks. Cisco wireless devices include console port access that can be used to manage the device by plugging serial cable directly into the console port. Console access should always be protected. Wireless devices should be locked up and securely mounted to prevent from anyone from disconnecting the power or network cable.

You also need to protect your networks from intruders trying to connect their laptops to your access point-wired cable to gain access to your wired network. To prevent such attacks, you may need to position a firewall between the access point and the wired network to configure limited and restricted access directly from your access point's wired IP. You can also configure port security on

Cisco switches connecting the access point to the wired network. Port security can be configured on Cisco switches to restrict communication input to a port on a switch, by allowing only trusted MAC addresses on the access point. When a violation takes place and an attacker tries to use the access point's wired cable, the switch detects the unauthorized MAC device connected to the port. The switch can then restrict the new device from sending any data and send an alert to the management station of its detection, or it can shut down the switch port from any further action.

# Preventing War Driving and Unauthorized Use of Legitimate Access Points

War driving is a form of reconnaissance where an attacker gathers information and statistics on wireless networks in a local area by using a wireless device and application to detect nearby access points A beacon packet is broadcast into the air in regular intervals from the access point with information such as the SSID and other data. By listening on these beacons and other wireless data with a wireless device, access points can be mapped out.

War driving utilities such as Network Stumbler (NetStumbler) have the ability to interconnect Global Positioning System (GPS) devices into their operations to include an exact location of a tracked WLAN. It is important to understand that the location reported by the GPS is your location at the time of detection of the wireless signal, not the actual physical location of the access point. A GPS allows for the mapping and creation of detailed maps of wireless networks that an intruder can use to explore its access.

War driving can be compared to war dialing and Internet scanning. When modems were popular, scanners would write a program to scan lists of numbers or entire area codes looking for any interesting data connections that could be further explored. The same principle applies to Internet scanning, where scanners scan IP addresses for different services and possible vulnerabilities that can be used to explore a network or a system device. Wireless networks scan the air frequency for possible signs of wireless detection.

# Devices Required in War Driving

To accomplish successful war driving, the following list of equipment and application programs are needed:

- **Portable Computer** A portable computer to use while driving around detecting access points.
- **Wi-Fi Client Adapter** A supported wireless client card is required to scan and detect the air for data. NetStumbler supports a limited number of client cards. Its supported list should be reviewed prior to purchasing a card.
- **Antenna** Although a client adapter can be used alone, it is highly recommended that you have an external antenna connected to your wireless adapter for best results.
- **GPS** Optional but very useful.
- **War Driving Software** A program that you can run on your portable computer to design, analyze, and store captured data.

## Wi-Fi Client Adapter

To war drive you need a wireless card. When purchasing a wireless card for war driving, it is important to know which card is supported by the war driving software. Another important factor in a client adapter card is external antenna support. A card without an external antenna will still allow you to scan but will have much less effect than a higher quality antenna.

## Antenna

You will need an external antenna (preferably an omnidirectional vertical antenna) that can be mounted on top of your car roof while war driving. Directional antennas are used for specific point-to-point transmissions and are not widely used in war driving. The idea of war driving is to be able to detect and scan an entire area, not just in a specific direction. If you are using war driving to detect the in-house activity of unauthorized access points, your LAN adapter alone may be enough to accomplish the task. An external antenna is not required but will improve your success rate in detecting wireless networks.

To attach the external antenna to your wireless LAN card you will need a *pigtail cable*. A pigtail cable is a short coaxial cable, usually 1 to 2 feet long, with connectors on both ends that allow you to interconnect both devices.

# GPS

A GPS can be incorporated into a scanning utility such as NetStumbler to input data and found access points directly into its database. The GPS data can then be used to map out exact locations and create wireless access maps for a specific area. Figure 7.7 displays a national map of detected wireless networks. Many GPS devices connect directly into a portable laptop through a Universal Serial Bus (USB) port, making it accessible to scanning software. Note that the current version of NetStumbler does not support readings from a USB port. A port bridge is required between the USB and one of the COM ports on the laptop. A supported GPS driver is required to make this transmission bridge possible in order to send data from the USB into the COM port where NetStumbler can find it and incorporate it into its options.

**Figure 7.7** Wireless Networks Discovered with NetStumbler as of November 2003

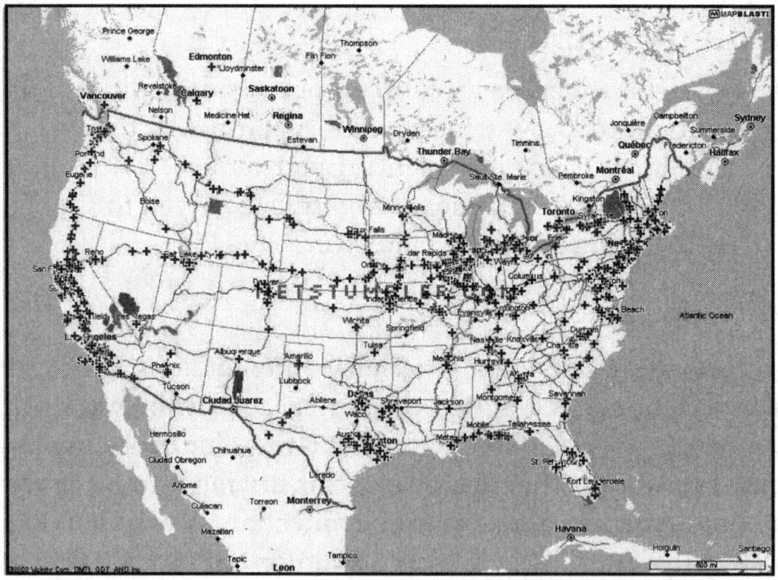

# War Driving Software

To make it all work you need a software application that will instruct and work with your client adapter, antenna, and GPS to detect wireless networks. Many different software applications are used in different operating systems. One of the most popular is NetStumbler. NetStumbler works on a Windows platform and is free for anyone to download. NetStumbler listens for wireless networks and their publicly announced beacons as well as other information sent from an access point. It then records it to a disk for later analysis with its GPS location.

Figure 7.8 shows an example of war driving in action using NetStumbler. Within 10 minutes of driving, over 60 access points were found of which 21 were unconfigured to their default settings. Less than 20 percent of found access points were using encryption for confidentiality.

When using software like NetStumbler, make sure that you do not autoconnect to access points that you are not authorized to connect to. To stop your client from trying to autoconnect and obtain IP through dynamic host control protocol (DHCP) on a wireless network while war driving, disable the TCP/IP settings for your client adapter. You do not need Layer 3 network connectivity to detect access points. As covered previously, Layers 1 and 2 are all that are required.

**Figure 7.8** NetStumbler in Action Listing Detected Default-configured Access Points

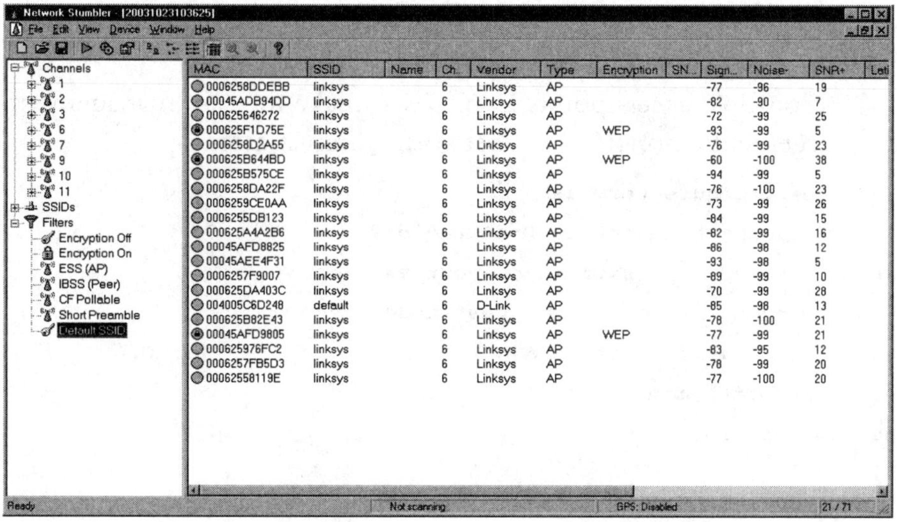

# Protecting Against War Drivers

Now that you are familiar with how war driving works, you can design security options to eliminate some of the threat of war driving. It is important to understand that eliminating the general detection of wireless networks is near impossible. As long as you are sending communication through the air, someone within reach of the signal can detect it. However, you can protect against specific war-driving utilities by knowing how they operate. In wired Internet scans, you can block ICMP on the firewalls from reaching the corporate networks, which will disable scanner-type programs that operate by sending ICMP pings to detect a device. In Cisco wireless access points, you can disable the broadcasting of beacons and their SSID information to eliminate automatic detection by some war driving programs. The only way to eliminate total detection is to turn off the access point.

## Disabling SSID Broadcasts

By default, Cisco access points send a SSID "tsunami" in its information beacon. To disable the broadcast of your SSID and to change your SSID, go to the **Security** page navigation window and select **SSID Manager**. Scroll down to where it says "Set Guest Mode SSID" and select **None**. When done press the **Apply** button to save the changes. Make sure you do this for both radios in your 1200 access points.

### *Disabling SSID in Command Line Interface (CLI)*

New IOS-compatible access points from Cisco make for easy management and familiar management among Cisco's network administrators.

```
AP1200-60a66e# configure terminal
AP1200-60a66e(config)# interface Dot11Radio 0
AP1200-60a66e(config-if)# ssid <type-your-ssid-here>
AP1200-60a66e(config-if-ssid)# no guest-mode
AP1200-60a66e(config-if-ssid)# end
AP1200-60a66e# write mem
Building configuration...
[OK]
AP1200-60a66e#
```

Make sure you apply **no guest-mode** to both radio interfaces 0 and 1 if you are running 1200 with two installed radios such as 802.11a, 802.11b, or 802.11g.

# Protecting Against Unauthorized Wireless Access

There are many different ways of protecting against unauthorized access to wireless networks. Your security policy should dictate what steps you take to implement best security measures to protect your wired and wireless networks. For the best security measures in WLAN you should implement using 802.1x protocol and its security techniques or by using IPSec VPN over wireless. Both technologies provide secure methods of authenticating on a per-user basis prior to authentication to wireless access points. IPSec and 802.1x are described in more detail later in this chapter. The following sections review some basic steps you can take to implement better authentication security to prevent unauthorized wireless access.

## Open Authentication

Open authentication allows for any device to authenticate to an access point and then communicate with it. If using an open authentication device to try to communicate and authenticate with an access point, you must use WEP and provide an identical WEP key matching the key on the access point. Devices not using WEP or unable to provide the correct WEP key will not be allowed to send data to the access point. Open authentication can be combined with additional authentication such as MAC or Extendable Authentication Protocol (EAP) (covered later in this chapter).

## Shared Authentication

A shared authentication type is provided by Cisco devices but should be avoided due to security flaws. When an end device tries to authenticate to an access point it does so by sending a request. In return the access point sends an unencrypted text (challenge) which the end device then encrypts and sends back to the access point (see Figure 7.9). An access point uses its own WEP key to decrypt the challenge and match it against the cleartext challenge previously sent. If a match is made, the access point determines that the end device has the correct WEP key and authenticates it. By sending both the unencrypted and encrypted string in the air, an intruder can capture both sequences and use them to determine the WEP key.

**Figure 7.9** Shared Key Vulnerability

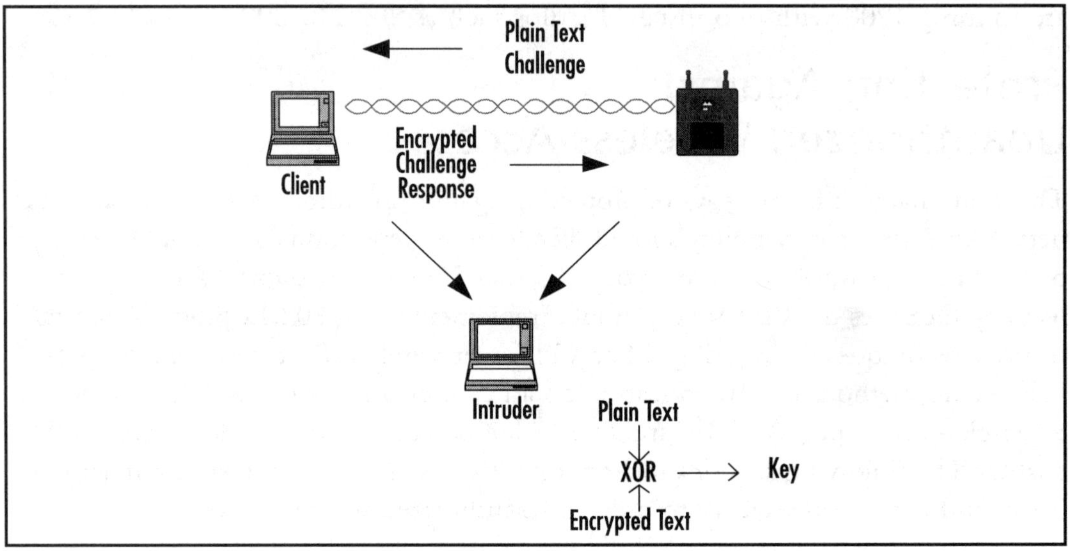


# EAP Authentication

The EAP used in 802.1x is one of the highest levels of authenticating in wireless networks. EAP interacts with the RADIUS server to authenticate the client device and in turn the client device authenticates the access point for mutual authentication. EAP provides for per-user authentication and per-user dynamic WEP key distribution. EAP authentication is discussed in more detail in the "WLAN LAN Extension 802.1x/EAP" section of this chapter.

# MAC Address Authentication

MAC is assigned a unique address that is programmed into the network adapter at the time of manufacture. It is used to communicate on Layer 2 of the OSI model. Because MAC addresses are uniquely assigned, there should never be the same MAC on two different network adapters. Every vendor card is assigned a unique MAC address.

A RADIUS server can be used to authenticate clients against a MAC address list, or a MAC address list of allowed clients can be created on the access point. Clients wishing to access and authenticate WLAN must be on the allowed list. It is important to note, however, that an intruder can spoof (counterfeit) a MAC address on a client adapter, allowing them to pick a MAC from the allowed list and bypass authentication. Some client adapters allow a configuration change of their MAC address as a supported option to users. MAC address lists can also pose manageability difficulty in large organizations. MAC authentication is less favorable than EAP, but can be used as an alternative method of authentication for clients that do not support EAP. When MAC authentication is used, it can be combined with EAP authentication and WEP authentication methods to increase its security. Figure 7.10 shows a list of supported authentication types on Cisco access points.

**Figure 7.10** Authentication Types

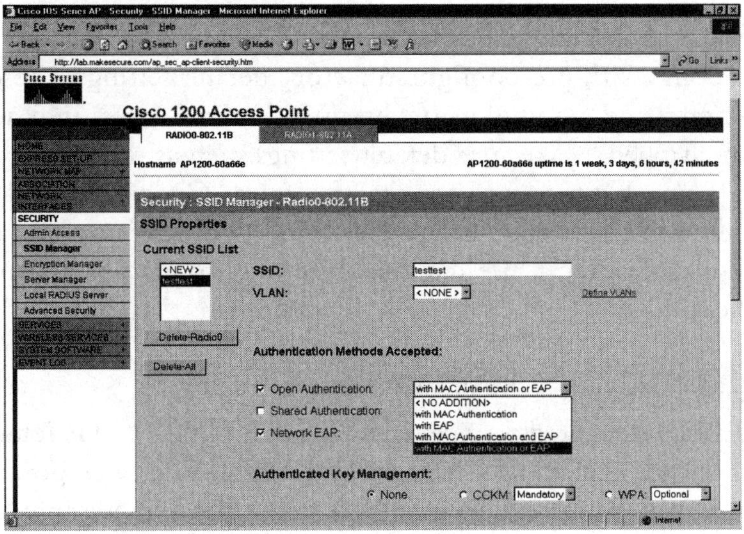


# VPN Authentication

For utmost security, you can implement IPSec VPN over WLAN. Client devices will need to successfully authenticate using their VPN client before they are allowed to pass through the access point into the wired network. Currently, Cisco access points and bridges do not support termination of the VPN tunnel. Devices such as a VPN concentrator or a Private Internet Exchange (PIX) firewall needs to be used behind the access point to authenticate and terminate a user's secure tunnel. How to implement a VPN solution over WLAN is discussed in the "WLAN LAN Extension IPSec" section of this chapter.

## Protecting Against Unauthorized Access Point Access

It is important not to forget to secure the actual access points that handle most of the security configurations. These access points are as good a target as the wired networks behind them. It is important to secure these devices to prevent intruders from trying to reconfigure their settings in order to obtain access into wired networks.

### Changing the Default Settings

Cisco devices come with pre-configured factory default settings. These settings are publicly known and are used to obtain unauthorized access into access points if they are not changed. Make sure default settings such as the default login and password and the enable password are changed from "Cisco." If you do not restrict management access and change the default settings into your access points or bridges, anyone can use the default settings to gain access and change its configuration.

### Disabling Unwanted Services

You should disable unwanted services such as NTP, CDP, HTTP, Telnet, and SNMP if you do not plan to use them. Vulnerabilities in service ports could be found and used against your access point. They can also provide information to an intruder during reconnaissance that will help them gain unauthorized access into the access point. If there is a valid need for a service, you should apply an access list to restrict its access to trusted hosts. If you are using HTTP to manage your access point through a Web browser, you should restrict its access use by using access lists. Just like FTP and Telnet, HTTP sends login information and all of its data in cleartext. Anyone listening in can use this information to gain access into an access point. It is recommended that only SSH be used, which provides for a secure and encrypted tunnel between the client and the access point to manage the access point through Command Line Interface (CLI).

# Exploring Rogue Access Points

A rogue access point is an unauthorized access point installed by an intruder or a user on a corporate wired network. When legitimate clients within the WLAN try to authenticate to this false access point, they may reveal sensitive information such as the authentication method and user credentials. In turn, rogue access points can record and later use this information to obtain access to a valid access point. Employees also may install unauthorized rogue access points for their own convenience, leaving the default settings turned on, not realizing that they have just created a backdoor into a corporate wired network that can be explored by an intruder. The wireless network administrator must recognize and mitigate the threat of rogue access points.

# Detecting and Protecting against Rogue Access Points

The following sections discuss different techniques to mitigate threats against unauthorized access points. Some are general and some use specific Cisco's solutions.

- Corporate policy and user awareness
- Mutual authentication
- Sniffers
- Cisco rogue detection by client reports
- Physical detection
- Wired detection

## Corporate Policy and User Awareness

Employees that install access points often do not understand the security risks. These rogue access points, mostly with default settings, often leave network corporations wide open to intruder attacks. To avoid and control this situation you must implement a wireless security policy that mandates that employees obey proper security measures, and coordinate with the Information Technology (IT) department to implement access points. You must audit and communicate this policy to your employees on a regular basis. A security policy only works if the employees are aware of and abide by it.

## Mutual Authentication

Mutual authentication of the user and the authenticator eliminates the potential for authenticating with a rogue access point. Users and authenticators need to authenticate against their credentials before they can associate with each other.

An access point checks the credentials of the user and the user checks the credentials of the authentication server. (This is provided in the 802.1x suite EAP in the TLS and Light Extendable Authentication Protocol (LEAP) extensions, both of which are discussed in detail later in this chapter.) Unlike in 802.11 MAC authentication where authentication is established in one way rather than being mutual, the client has no way of knowing if an access point they are authenticating with is valid.

## Sniffers

Sniffers and access point detection tools such as NetStumbler can be used to constantly sniff the air and record communications. Network administrators can then analyze the data. This requires the administrator to walk throughout the entire facility area to detect rogue access points, which can be time consuming and not always a best choice. Rogue access points may also be turned off during a sniffer scan. An employee may only turn their access point on during a conference meeting. It is a good idea to perform sniffer traces randomly during all hours of business.

When done sniffing an area for possible rogue access points, captured data such as MAC addresses, SSIDs, and authentication methods must be compared against a database of valid access points. If a new MAC address or unrecognized SSID is found, it must be red flagged, considered rogue, and further investigated.

## Cisco Rogue Detection by Client Reports

Cisco provides a feature for clients to report possible rogue access points to an access point. This feature is presented in 802.1x and its LEAP authentication protocol, in which a client running firmware 5.02.17 and later works with the access point to detect rogue access points. If a Cisco client tries to authenticate with an access point using LEAP and is unsuccessful, the client records the unsuccessful authentication information and passes it to the next access point that it is able to authenticate with. By doing so, it informs the valid access point of a possible rogue access point on the network. The access point can then store this information locally or send an SNMP trap to the management station that then sends an alert to the network administrator. The message includes the MAC address of the access point that the client was not able to authenticate with and the reason why. The message format will look like *DOT11-6-ROGUE_AP: Rogue AP [mac-address] reported. Reason: [chars].* This type of detection is a passive method of detecting rogue access points because access points do not actively scan the environment to detect unauthorized access points. Instead, they rely on users to report them.

CiscoWorks Wireless LAN Solution Engine (WLSE) is a newly developed managing solution for wireless devices. WLSE can be used to receive rogue alerts from clients and access points via SNMP. It can quickly alert the administrator and display the location of a rogue access point in its management viewer along with information such as the switch port to which the access point is connected.

### *Configuring SNMP Rogue-access Point-trap*

In the Web browser manager go to **SERVICES** and click on **SNMP**. Enable SNMP and assign a public read-only string that matches your WLSE engine's configured string. Scroll all the way down to "SNMP Trap Community" settings and fill in your WLSE IP address. Finally, place a check mark next to the **Rogue AP Trap** option that will be used to send the SNMP alert from the access point to the management server, and enable the all applied configuration by clicking on **Apply**.

**NOTE**

When configuring SNMP for management purposes, it is always a good idea to protect its access with filters, allowing only permitted management a connection. This can be accomplished with an access list.

## Physical Detection

Sometimes all it takes is a walk around the office to look for unauthorized devices and access points plugged into wired ports at a user's desk. If you find an unauthorized access point plugged into a wired network, turn it off and inform the user of the corporate security policy.

## Wired Detection

Many access points are administered by HTTP and Telnet. Since it is unlikely that an employee workstation is providing Web access through HTTP on their desktop that is connected to a LAN, a port scan of a local LAN revealing Web access or Telnet could be a sign of an unauthorized access point connected into a wired corporate LAN.

# Designing for Security

Giving the nature of wireless and its travel reach throughout the unrestricted area, you need to design secure and robust wireless networks to protect your communication and its data. First, most organizations need a security policy that meets and explains security standards, which must be audited and followed by the network architects and every employee within an organization. The following sections review what a security policy is and what should be included within it.

## Creating a Security Policy

A good security policy is a must when designing security measures for a network. A security policy will not eliminate threats in your wireless networks, but it will create a proactive approach for protecting your infrastructure with tools and procedures against security threats. It will set a guideline for current and future network implementation to ensure security.

A security policy should fit the organization and be dynamic enough to expand when new technology or vulnerabilities are discovered. It is the fundamental element of an organization's security practices, which network implementations are modeled after. You will use a security policy to design your secure networks and pick compatible hardware to fulfill all security measures outlined in your security policy.

Both networks and a security policies should be audited internally and externally to ensure they have good controls in place and are up to date with the latest threats.

### Risk Assessment

A security policy should include risk assessment. Understanding your organization's risks will aid in protecting against threats and allow for proper security measures to be implemented. You must understand what and whom you are protecting your network from and the risks and costs involved if intruders attack or a malfunction occurs. The first step of risk assessment is identifying network vulnerabilities, critical services, and data sensitivity and their impact on your organization if attacked, stolen, or disabled.

### The Big Three

Confidentiality, integrity, and availability ("the big three"), as shown in Figure 7.11, should be included in your security policy. Outlined procedures and measures of how to assure confidentiality, integrity, and availability between different devices and services need to be clear. Confidentiality should outline the type of encryption methods clients and devices must use to protect sensitive data. Integrity should specify the type of authentication methods between different user groups and devices to assure for best security. Availability must be specified for mission-critical devices that require redundant wireless devices to prevent down service.

**Figure 7.11** The Big Three

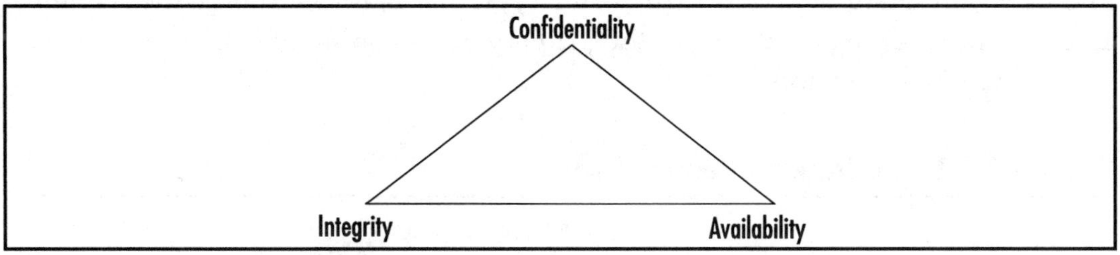


## Logging and Accounting

Logging authorized and unauthorized access into a wired network from wireless clients should be recorded. Logging services serve many beneficial purposes and therefore should be outlined in the security policy. Accounting for user activity will aid in detecting user misuse, accountability, and unauthorized access. A logging policy can include logging using access lists. Defined access lists can detect a match against unauthorized traffic on a wireless LAN and log it. RADIUS accounting services can also be used to log connecting users into a wireless network. An SNMP rogue trap can be defined as a type of logging used to detect unauthorized rogue access points. These logging options can be included in the security policy to provide standard measures.

# Hot Standby

The hot standby access point is a backup access point that assures constant availability in case of hardware failure. The hot standby access point must be configured exactly the same as the primary access point, without the interface IP.

**NOTE**

To duplicate a primary access point configuration without spending a lot of time, download a copy of your access point's configuration file via FTP, Trivial File Transfer Protocol (TFTP), or Remote Copy (RCP) and upload it to a secondary access point using the same protocols.

The standby-configured access point monitors the primary access point radio interfaces and wired interface. If the primary device fails to respond, the standby access point takes responsibility of the primary access point and takes over the wireless user's connections.

**Figure 7.12** Hot Standby Redundancy

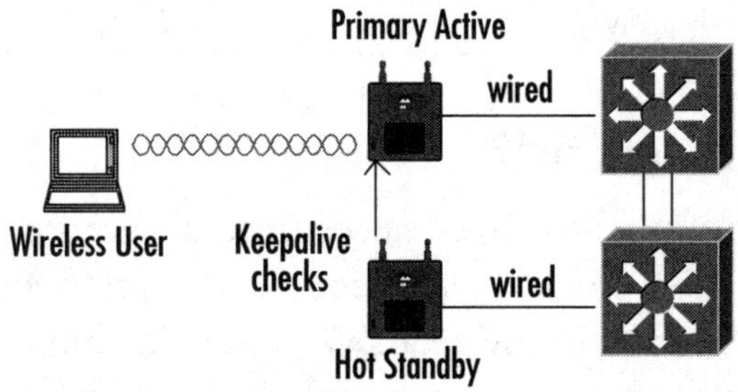


# Configuring Hot Standby

Before you configure a secondary access point as hot standby make sure it has identical configurations except for the wired interface IP address. The SSID, IP mask, default gateway, data rates, WEP settings, and authentication settings must all match.

1. On the primary access point, configure the radio interface settings to **Access Point Root (Fallback to Radio Shutdown)**. This allows the primary access point to disable radio interfaces if it loses connectivity to the wired network allowing the secondary access point to take full control.
2. Enter into configuration mode in the IOS CLI on the standby access point and configure the MAC address of the monitored unit using the **iapp standby** command. If you are configuring an access point that has

two radios, you must enter two MAC addresses. The first MAC address represents the 802.11b 2.4GHz radio and the second MAC address represents the 802.11a 5GHz radio.

**NOTE**

When using CLI in IOS instead of the Web browser to manage the access point, you can use the question mark (?) after the command to give you a list of possible options. Use the **TAB** key if you are not sure of the spelling of a command. The **TAB** key in IOS will complete the command you have started typing or will display all possible matches.

3. Configure the SSID by going into radio interface configuration mode in the IOS CLI and marking it as the infrastructure SSID. This SSID will be used by the standby access point to associate and monitor the primary access point radio interface. SSID must match on both units.
4. Configure timeouts. Use the **iapp standby pool-frequency** command in the IOS CLI followed by an integer that specifies the number of seconds frequency that the standby unit checks the primary unit interfaces. The **iapp standby timeout** command is used to specify the number of seconds a standby unit must wait for a reply from the primary access point before it determines that the monitored access point is down.
5. Save all configurations with the **copy running-config startup-config** command. All of the commands you typed in previously to enable hot standby in CLI are automatically applied but are not saved in case the access point reboots. You must always manually save all changes with the **copy** command.

After you have configured your primary and hot standby access points, make sure you test it by disabling the Ethernet cable from the primary.

# Implementing Firewalls for Additional Security

To further secure your wired network from wireless users, implement a firewall between the access point and your wired network, as shown in Figure 7.13. Instead of using filters on your access point you can offload their function and allow for better inspection, auditing, and accounting by installing a separate firewall.

**Figure 7.13** Firewall Design

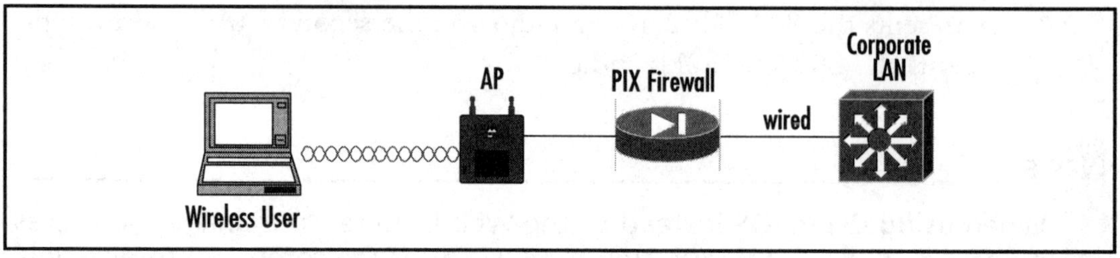


# Public Secure Packet Forwarding

Public Secure Packet Forwarding (PSPF) prevents wireless clients associated with an access point from communicating and sharing files between other clients connected to the same access point. This configuration can be useful in public access-deployed WLANs like airports and schools where you would want to restrict access between individual users. Figure 7.14 shows user1 restricted from accessing user2. Because PSPF is enabled, the access point will not allow such communication between individual clients to take place.

**Figure 7.14** User1 Blocked with a PSPF-Enabled Access Point

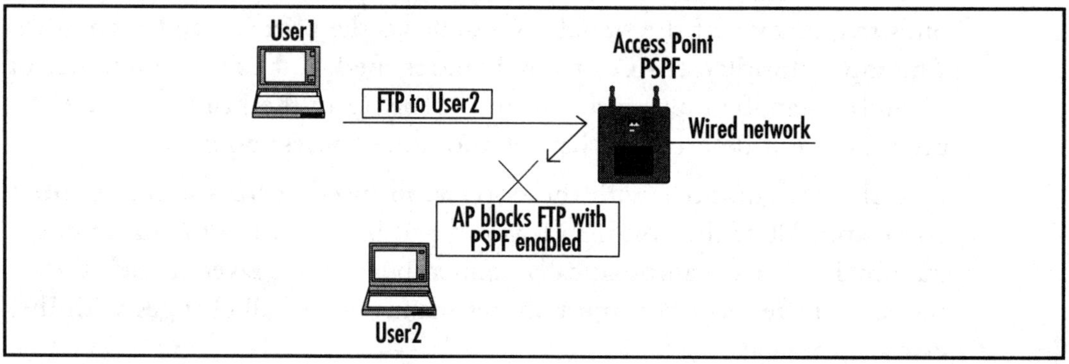


If using more than one access point on the same WLAN, in order to make sure a client from AP-A cannot communicate with a client from AP-B, you need to prevent both access points from communicating with each other across your wired network. PSPF only works on unique access points. To prevent the same LAN access points from communicating between each other on the wired network, you need to apply port access lists on compatible Cisco switches interconnecting the two access points, or use protected port settings to restrict communication between ports. PSPF can be configured under radio interface settings using Web browser management. Refer to Figure 7.15 for a PSPF configuration option.

**Figure 7.15** Configuring PSPF

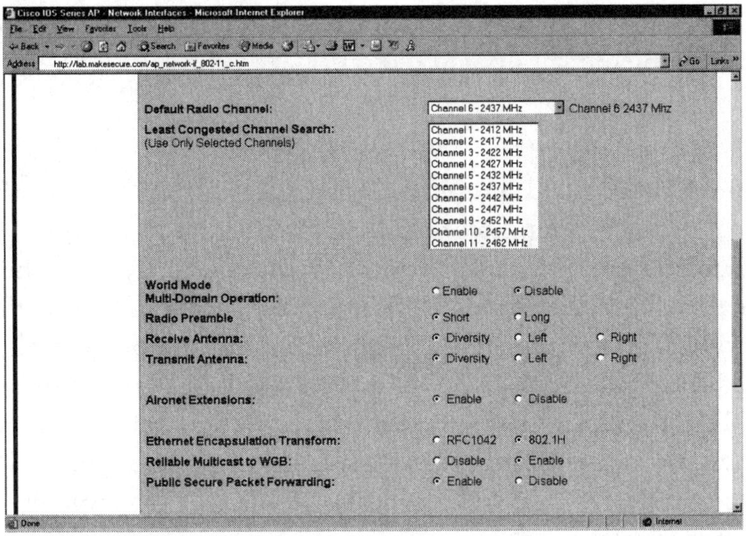


# Filters

Filters allow or restrict specific communication through an access point's Ethernet port or radio ports. Cisco WLAN devices support MAC, IP, and Ethernet-type filters. IP filters can be configured to disallow a Telnet or HTTP Web administrator access into their access point from wireless clients. IP HTTP filters can be configured to restrict administrators-only connectivity into an access point. MAC filters can be used to authenticate and restrict access to specific network client adapters. You can also use filters for services such as Quality of Service (QOS) to group different communication types into separate QOS policies.

# WLAN LAN Extension 802.1x/EAP

The original 802.11 authentication and security components proved to be a security risk in one way or another. A need for enhanced 802.11 security components was soon realized and thus the 802.1x standard was developed. 802.1x used in wireless addresses the shortcomings of the 802.11 authentication process. One of the underlying of 802.1x components is Extensible Authentication Protocol (EAP). EAP was originally created for Point-to-Point (PPP) protocol and soon found a new home in wireless networks. There are many extensions to EAP protocol. EAP was designed to provide future enhancements that leave its specific use and implementation up to the network administrator. Cisco's LEAP, EAP Transport Layer Security (EAP-TLS), and Protected EAP (PEAP) are all part of EAP and the 802.1x standards, and are used separately or together based on the required implementation and design.

# EAP

The 802.1x protocol is based on the EAP formally created and specified in RFC 2284. The EAP protocol supports multiple authentication methods. The advantage of this dynamic protocol is that it does not itself specify an authentication method, rather it allows the authentication process to request a specific authentication method before it completes. This allows administrators to pick different extensions and methods for the EAP protocol and the use of third-party devices such as RADIUS to manage such requests. Hence, the advantage of EAP is that it supports multiple authentication mechanisms. As seen in Figure 7.16, EAP can be used in many different link and physical solutions. Its dynamic use makes it popular and easy to adopt.

**Figure 7.16** EAP Dynamic Authentication Types

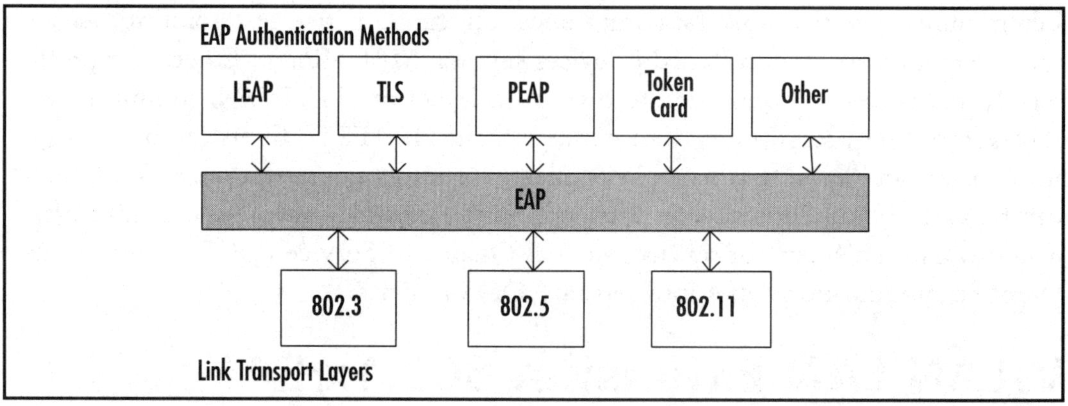


## EAP Packet Format

The fields in an EAP packet consist of code, identifier, length, and data. The **Code** field is one byte long and is the first field in packet. It consists of codes, 1=Request, 2=Response, 3=Success, 4=Failure. The **Identifier** is also one byte long and is used to match requests with responses. The **Length** field is two bytes long and represents the overall size of the EAP packet. The **Data** field is the last field that is determined by the value of **Code** identifier. Refer to Figure 7.17 for the EAP packet format.

**Figure 7.17** EAP Packet

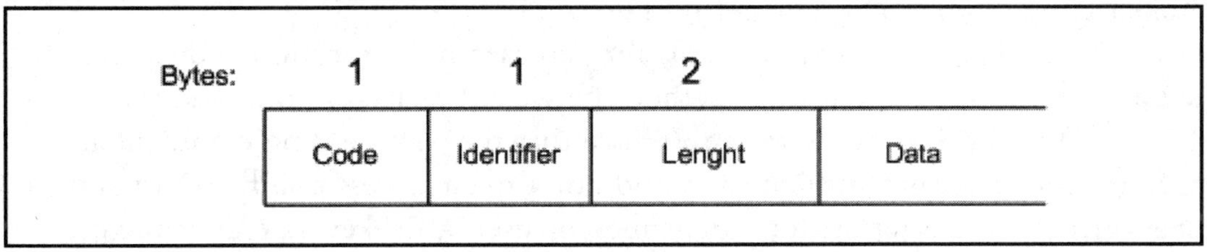


# EAP Request and Response

A request and response packet is used by the authenticator to request an authentication type based on its **Type** field. All EAP authentications must start with a **Code=1** indicating a request and finish with either **Code=3** or **Code=4** that indicates the success or failure of the supported authentication. The new **Type** field indicates the type of authentication request. This field can include a **Type=4 MD-5 challenge**, a **Type=5 OTP**, a **Type=13 TLS**, and other type codes. **Type=3 NACK** is reserved and used to indicate to the authenticator to pick a different type code. Negative acknowledgements (NACKs) are sent if a system does not support the requested authentication type. Refer to Figure 7.18 for the EAP authentication request and response packet structure.

**Figure 7.18** EAP Request and Response Packet

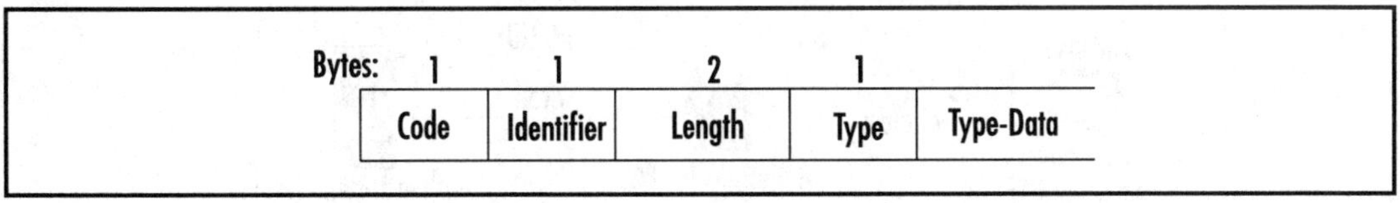


# EAP Success and Failure

If authentication is successful, the authenticator will send a success packet with code equal to **3=Success**. The access point watching the EAP transactions between a client and a third-party device such as RADIUS will open up the communication to wired networks upon identifying a successful authentication packet. If the authentication is a failure, the client failed to authenticate with the appropriate method and will be disconnected.

# 802.1x

802.1x is a port authentication control protocol used to translate messages from a variety of different authentication types into their appropriate frame formats. 802.1x can be used in other 802-based technologies such as Ethernet (802.3) or Token Ring (802.5). This chapter uses 802.11.

802.1x supports the requirements for per-user authentication and settings. It supports mutual authentication methods between the access point and the wireless user. In other words, users can validate that they are sending confidential information to a valid authenticator and not a rogue access point. 802.1x and its standards provides support for a dynamic per-user WEP key. The dynamic creation of WEP keys and per-user keys allow for better security and manageability and also allows for user accountability.

Although 802.1x does not choose what authentication and algorithm types it will use, it works with EAP to provide such information.

The three common components of 802.1x are the *supplicant* Port Access Entity (PAE), the *authenticator* PAE, and the *authentication server*. The supplicant is the client end user trying to authenticate and connect to the wireless network resources. The authenticator is normally the access point that enforces authentication before it allows access to resources. The authentication server is used to verify end-user credentials against a local or remote database. Figure 7.19 shows the relationship between all three components.

**Figure 7.19** 802.1X Components

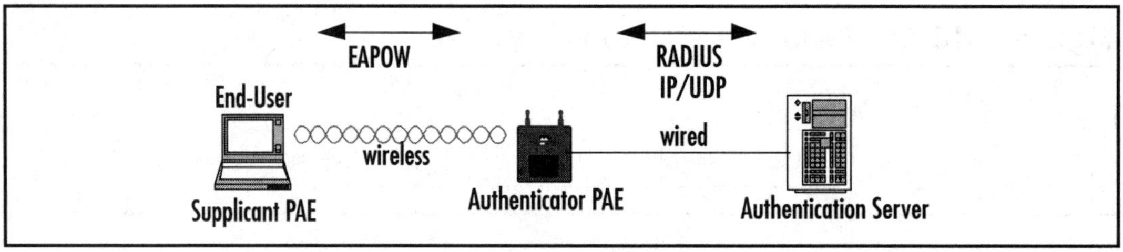


Extensible Authentication Protocol Over Wireless (EAPOW) is an EAP message that is encapsulated over wireless networks. During the authentication phase, the access point only allows EAPOW traffic through to the wired network from the end-user trying to establish connectivity. This prevents the authenticating user from sending anything but its credentials into the wired network. After successful authentication, an EAP message is sent letting the access point know that the authentication was successful. The access point then lifts the EAPOW access-only filter and opens other communication specified by the security policy for an authenticated user.

## EAP Types

There are many authentication types supported in EAP. The following sections review some of them and compare their security measures.

### EAP Message Digest 5

The Message Digest 5 (MD5) authentication type is represented by type code 4 in the EAP frame field. MD5 is a one-way hash function used in usernames and passwords to produce irreversible output. The end user uses the hash function to hash its secret and sends it to the authenticator (access point). The access point then turns to an authentication server such as RADIUS, which compares the user-produced output against its own hashed output derived from the local user database. If the hash outputs from both the user and authentication servers are identical, the end user is proven to have correct credentials and is authenticated.

### EAP Generic Token Cards

Generic Token Card authentication is represented by a type code 6 in the EAP frame field. Token cards such as RSA's SecureID are popular providers of random one-time passwords. The user is prompted to enter a token from their token card in order to authenticate. Token cards such as RSA's SecureIDs are compatible with Cisco's Access Control Server (ACS). These portable cards are time-synced with an authenticating engine that produces a unique access code every 60 seconds. The same access code displayed on the user's token card is the access code they must enter with their username and password in order to gain access. A secure token can only be used once.

### EAP TLS

Transport Layer Security (TLS) supports mutual authentication to protect from rogue access points, and uses certificates instead of usernames and passwords to authenticate wireless users. TLS takes advantage of Public Key Infrastructure (PKI). An authorized user creates a certificate and a private key that is then signed by the in-house certificate authority (CA) and returned to the user. When the user is tries to authenticate they must provide a valid signed certificate from the CA. The user authenticates the authentication server by using its own certificate to assure validity. Both the user and the authenticating server must trust the CA. Each user has a unique certificate that must be signed by the CA before they are allowed to be used as an authentication method.

EAP TLS using digital certificates is a secure way to validate users. Digitally signed certificates are almost impossible to forge. However, this design requires more time and manageability.

## Cisco EAP

Cisco wireless EAP, also known as LEAP, was developed by Cisco to further enhance EAP and its security practices. LEAP provides several enhancements and differences than those EAP authentication types discussed so far in this chapter. LEAP provides mutual authentication. However, it does not provide mutual authentication using certificates such as EAP TLS, but rather it uses usernames and passwords to mutually authenticate. This may be less secure than digital certificates but proves to be a different, faster, and less manageable option than TLS. LEAP also supports dynamic WEP implementation. WEP keys can be dynamically renewed between the client and the access point by deriving a new WEP key after the configured timeout expires. This allows for better security in case the WEP key is compromised during a client's communication period. Current static WEP keys and their vulnerabilities are covered later in this chapter.

## LEAP Authentication Process

LEAP is based on username and password authentication. The following process (shown in Figure 7.20) takes place when a wireless user authenticates to an access point.

**Figure 7.20** Leap Authentication

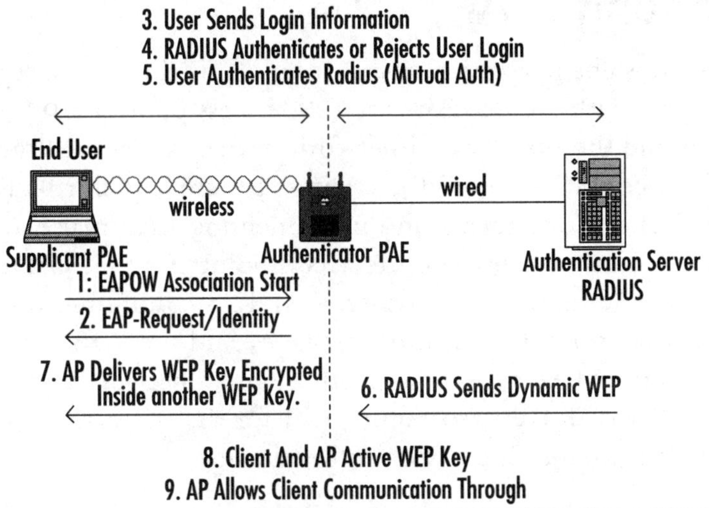


The phases for LEAP authentication include:

1. The client sends an EAPOW-Start to start the authentication process into the wireless network.
2. The access point responds with an EAP-Request/Identity to request identity from the client.
3. The client responds by sending username and password identification
4. The access point passes the client credentials to the RADIUS server where RADIUS verifies the information against a local or remote database. If authentication matches, the user has passed.
5. The user authenticates the RADIUS server to assure mutual authentication and wireless network integrity.
6. When authentication is complete, RADIUS sends a dynamic WEP key to the access point.
7. The access point sends the WEP key to the client over EAPOW.

8. Both the client and the access point install a WEP key and use it to further communicate.
9. The temporary block list used to block all communication except a client's EAP protocol, is removed to allow the authenticated client access to the wireless/wired network.

# Implementing LEAP

For purposes of this chapter, our network topology includes a compatible Cisco EAP client adapter with a Cisco Aironet 1200 access point. For routing between the access point and the other wired network devices a Layer 3 compatible switch is used. Cisco's ACS is used for storing user credentials and RADIUS authentication. Before configuring and implementing EAP, make sure you have basic network connectivity between the access point and the RADIUS server. Test connectivity by pinging one another. RADIUS communicates on IP/UDP ports 1812 and 1645 for authentication messages, and ports 1813 and 1646 for accounting messages. Make sure these ports are open between the access point and the RADIUS server. Refer to Figure 7.21 for the network topology being will be used to implement Cisco EAP authentication.

**Figure 7.21** EAP Topology

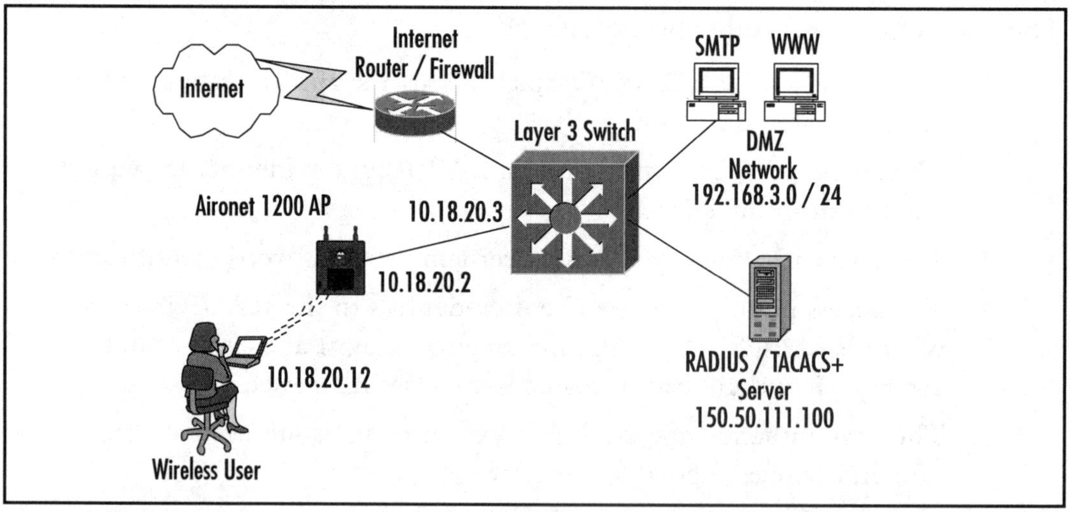

# Configuring ACS

The EAP authentication type is supported in Cisco ACS version 2.6 and higher. This section covers the configuration of how to add access points to ACS in order to communicate. For further information on how to add and define usernames into ACS, refer to www.cisco.com.

1. Add an access point to a group of allowed devices that are able to authenticate and connect to the ACS server. To accomplish this task, click on **Network Configuration** in the navigation window and then click on **Add Entry**. A window similar to Figure 7.22 will appear. In this window fill in the name of your access point, your IP address, and a shared secret that will be used to authenticate the access point to the ACS. Select **RADIUS (Cisco Aironet)** from the authentication drop-down option menu to recognize the device as Aironet.

   To configure compatible settings according to the topology in Figure 7.21, configure the IP address to be 10.18.20.2, and pick **Cisco** as the shared key. When done, click the **Submit + Restart** button.

**Figure 7.22** Adding an Access Point to ACS

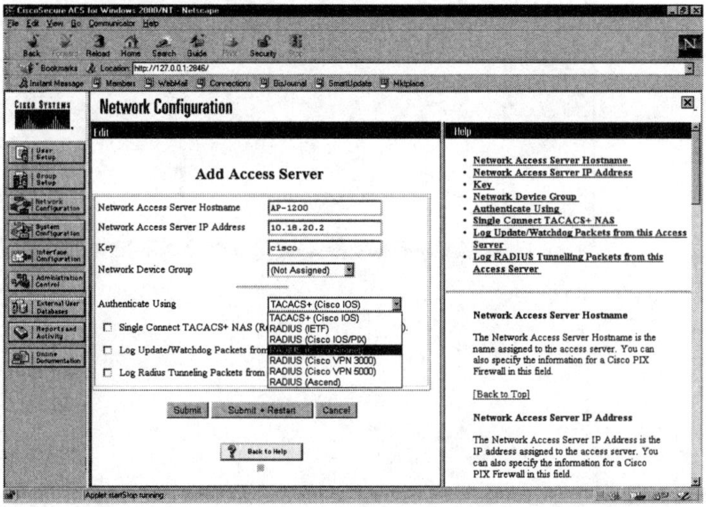


2. Enable the renewal of WEP keys. Cisco EAP utilizes this option to expire and renew a user's WEP key after a specified time has passed. This ensures the best security and protects confidentiality by always changing the encryption key that is used to encrypt data between a wireless client

and an access point. To configure a key timeout go to **Group Setup** from the navigation window (refer to Figure 7.23) where you can see group 1 is configured with one username called *makesecure*. To edit the group specifics click on **Edit Settings**. Scroll down to **IETF Attributes** and look for the value called [027] Session Timeout. Check mark the timeout value and enter a number in seconds to represent the timeout value after which the WEP key will be renewed. Figure 7.24 shows the Internet Engineering Task Force (IETF) attribute option and timeout value of 450 seconds. On a busy network, expiring and renewing WEP keys should be set to less than two hours.

**Figure 7.23** Group Setup

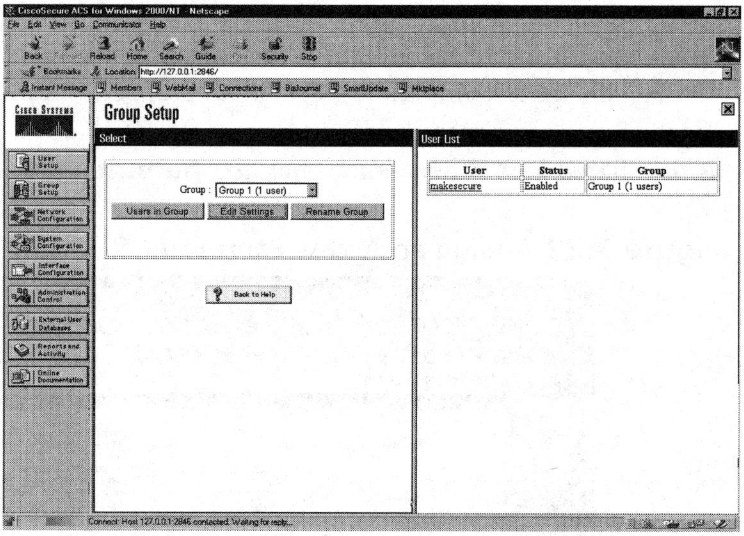

**Figure 7.24** Group Settings

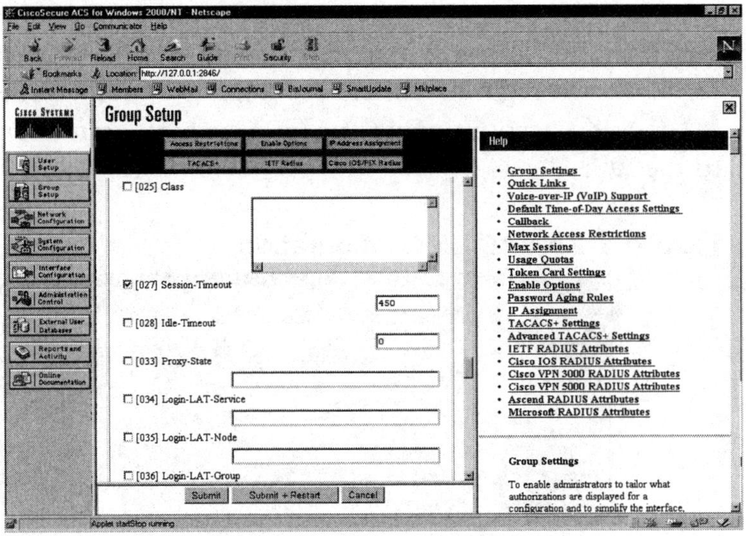

This concludes the basic configuration for allowing an access point to communicate to the ACS and pass user credentials via RADIUS for authentication. If you do not have a RADIUS server available, and have a minimum amount of users, you can configure a local user database on the access point itself. The local RADIUS feature was introduced on access points starting with IOS release 12.2(11)JA.

When configuring usernames and passwords, make sure you set your policy for hard-to-guess passwords to prevent intruders from guessing a password, and enable automatic expiration for passwords. Expiring passwords will force wireless users to change their passwords regularly.

# Configuring Access Points

Now that you have configured the ACS to allow access points to communicate with it, you must configure the access point to send authentication requests to the ACS.

1. Configure RADIUS settings in an access point. To enter the RADIUS server manager window go to **SECURITY** and click on **Server Manager** (see Figure 7.25). Fill in the IP address of the ACS server and the shared secret key that you configured previously on the ACS. Fill in the authentication port and the accounting port. Use the 1645 port for authentication and the 1646 port for accounting. If you are not running

your RADIUS server on its standard ports, this is where you need to configure them so that the access point can communicate with RADIUS. You then need to check mark **EAP Authentication** for server use. In Figure 7.25, the accounting option has been check marked to send user accounting information such as login attempts information to the ACS.

**Figure 7.25** RADIUS Configuration

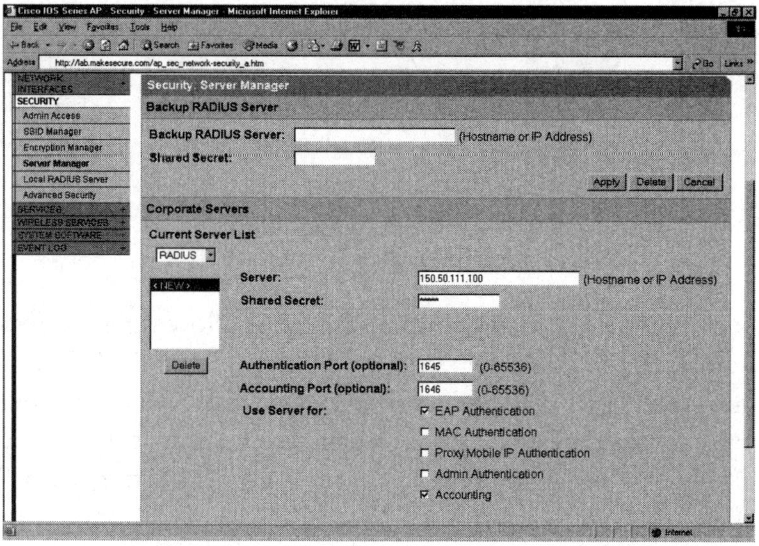


2. After configuring the access point settings to allow communication with the RADIUS server, you must configure the authentication type to be allowed for wireless clients. Go to **SECURITY** and click on **SSID Manager** (see Figure 7.26). Pick an existing SSID or create a new SSID that will be shared with the wireless users. Pick **EAP** from the list of authentication methods and apply all changes. If working with an access point that supports multiple radios, apply the changes to both radios.

**Figure 7.26** SSID and Authentication Type Configuration

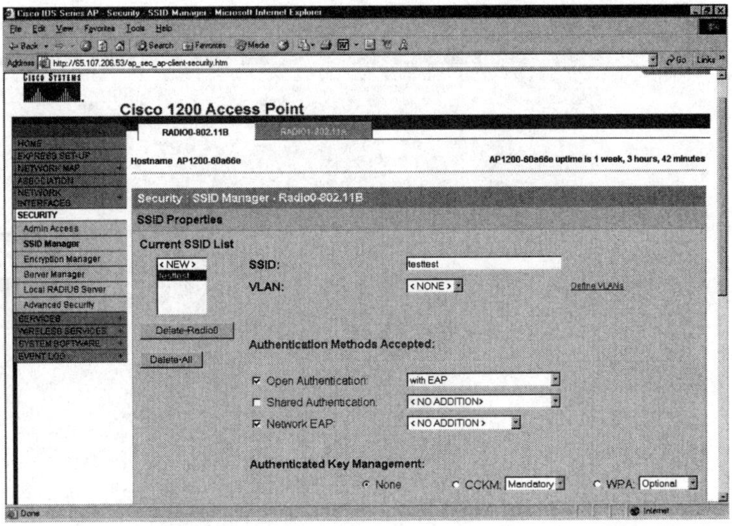


3. Configure a broadcast WEP key. Although a unique WEP key is derived dynamically from the RADIUS server to the wireless client, this unique WEP key is used only to encrypt all Unicast traffic. You still need to configure a separate WEP key for encrypting the broadcast traffic sent between the access point and the client. Refer to the "Static WEP Keys" section later in this chapter for detailed information on how to configure a WEP key. This configured WEP broadcast key will be sent to the wireless user encrypted inside the users Unicast WEP key derived from the RADIUS server. For increased security, Cisco allows you to rotate this broadcast key just as it allows you to rotate the per-user dynamic Unicast key. In EAP authentication, no WEP keys are required to be configured on the client side; the broadcast key and Unicast keys will be supplied after the authentication phase and renewed according to their timeouts. Figure 7.27 shows a configured WEP key and an enabled rotation period set to 3600 seconds.

**Figure 7.27** Encryption Manager

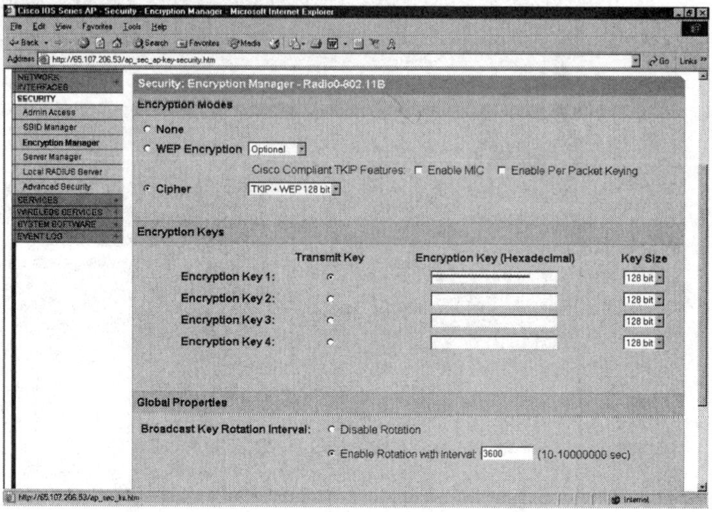


## Configuring the Client

Now that you have your access point and ACS configured, you must configure your Cisco software client to accept compatibility with Cisco EAP authentication. The first step is to match the SSID.

1. Launch the Aironet Client Utility (ACU) and select **Profile Manager** to edit an existing or create a new authentication profile. Inside the profile settings click on **System Parameters** and configure the appropriate SSID matching your configured access point (see Figure 7.28).

**Figure 7.28** LEAP Client SSID Configuration

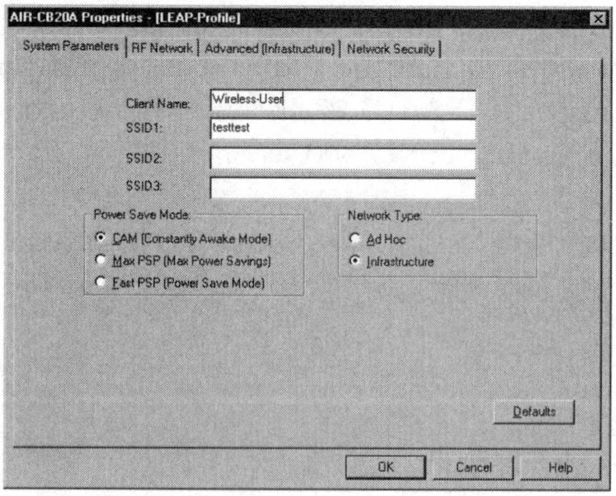

2. After configuring the SSID information, click on **Network Security** and select **LEAP** as your network security type (see Figure 7.29).

**Figure 7.29** Client LEAP Authentication Type

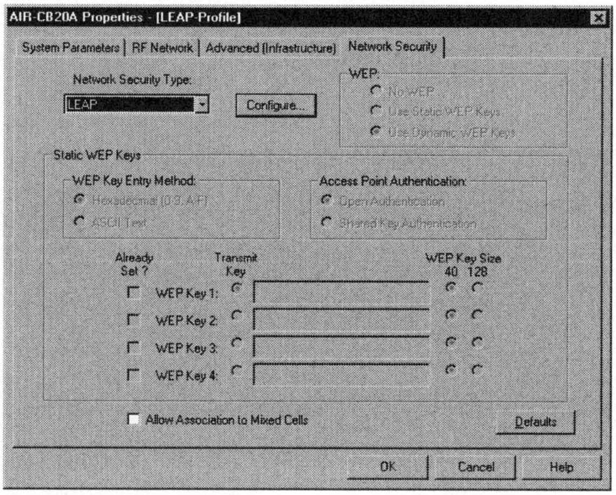


3. Click on the **Configure** icon that brings you to the detailed client LEAP configuration. In this window there are different options to choose from on how to enter your credentials for EAP authentication. (Refer to Figure 7.30 for all options.) In this scenario, we have chosen **Automatically Prompt for LEAP User Name and Password**. This option will allow for a pop-up window every time you choose to login and authenticate on a wireless network. You can also pick the option to automatically save your username and password.

**Figure 7.30** Leap Client Settings

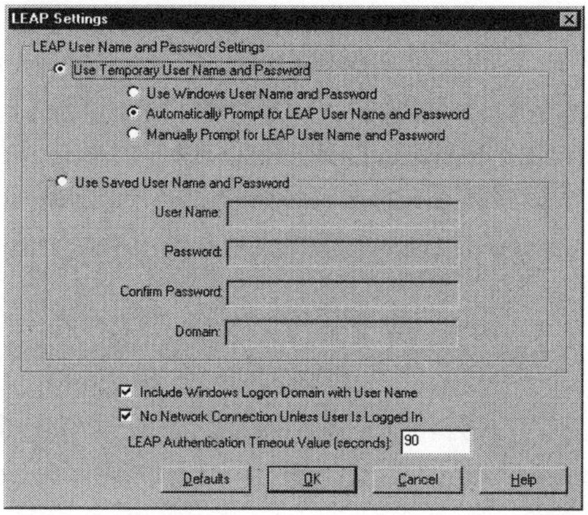

This concludes the configuration of LEAP on the client side. To start your authentication login process into your network, open your newly configured LEAP profile. If you configured to automatically prompt for your LEAP username and password, you will see a window pop up, as shown in Figure 7.31. Upon successful login and password, you will be authenticated into the wireless access point and its wired network.

**Figure 7.31** Authentication in Progress

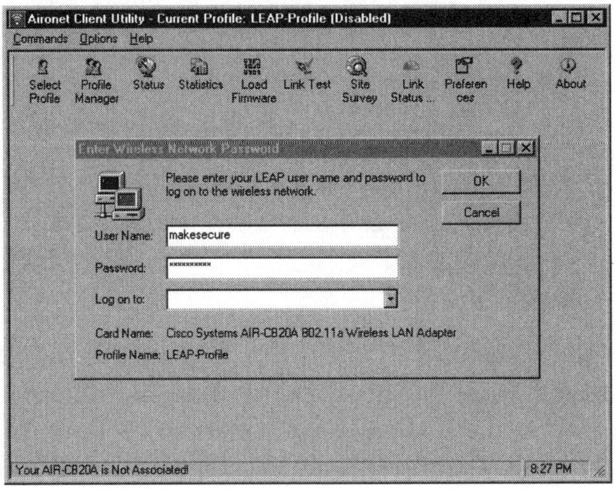


# WLAN LAN Extension IPSec

IPSec is built from a framework of open standard protocols that allows for an encrypted tunnel between two private parties or networks and allows for secure authentication of both ends of the tunnel. IPSec supports data integrity, confidentiality, and authenticity over unprotected networks such as WLAN or the Internet. IPSec was developed by the IETF to address network layer security. Overall, IPSec implementations and standards are guided by RFC 2401, "Security Architecture for the Internet Protocol."

Because IPSec is standardized, it supports vendor interoperability for easy implementation among its users. IPSec is supported in Cisco PIX Firewall, IOS, VPN Concentrator, IDS NetRanger, and many other Cisco solutions. Unlike Secure Sockets Layer (SSL) or SSH that provide security on the application layer, IPSec allows organizations to implement strong security without the need to change any of their applications. Only network layer changes such as on routers, firewalls, VPN, and in some cases, client software, are required. In most cases, the IPSec process is completely transparent from end users.

# Standards Used in IPSec

The following are the core technologies and open standards that make up IPSec:

- IKE
- Authentication Header (AH)
- Encapsulating Security Payload (ESP)

## IKE

IKE is a key management protocol used in IPSec to create a secure communication channel between two parties or networks that are securely authenticated. IKE is also used to negotiate the Security Associations (SA). SAs define how IPSec parties use their security services in the context of a particular security protocol AH or ESP to communicate securely. Both IKE and IPSec use SAs, although they are independent of one another. SAs are unidirectional and are formed per direction and protocol.

IKE requires that two parties authenticate each other prior to the exchange of key material. IKE is a hybrid protocol that uses part of the Oakley protocol and part of the protocol suite called SKEME inside the Internet SA and Key Management Protocol (ISAKMP). Oakley protocol is a key exchange that defines how to obtain keying material. Some of the basic mechanisms for Oakley include the Diffie-Hellman key exchange algorithm. Oakley is defined by RFC 2412, "The Oakley Key Determination Protocol." ISAKMP is a protocol that defines the mechanism of implementing a key exchange protocol and negotiation of SA.

IPSec can be configured without IKE, however IKE improves IPSec by providing flexibility, scalable implementations, and ease of configuration by providing automatic authentication. IKE negotiates SAs for both IKE and IPSec during two phases with different modes. These two phases are:

- **Phase 1** IKE negotiates IKE SAs
- **Phase 2** IKE negotiates IPSec SAs

In Phase 1, two peers mutually authenticate and negotiate a method of setting up a bi-directional ISAKMP SA. This SA provides a secure communication channel that will be followed and used in Phase 2 negotiations. In Phase 2, using the secure tunnel created in Phase 1, peers negotiate IPSec ESP and AH protocol compatibility. IPSec SAs are unidirectional so two different keys are used in each direction. Both of these phases use UDP protocol and port 500 for communication and negotiations.

**Configuring & Implementing…**

### Allowing IPSec through a Firewall

To allow IPSec to pass your firewall security, make sure you open these specific protocols and ports:

- **ISAKMP** UDP port 500
- **AH** Protocol 51
- **ESP** Protocol 50

If using Cisco PIX as your firewall, you can use the **sysopt connection permit-ipsec** command. This will automatically allow all IPSec-related traffic to go through without the need to configure access lists.

### *IKE Authentication*

Peers wishing to use IKE must be authenticated to each other before IKE SA can be established. IKE supports the following different authentication methods:

- **Preshared Keys** The same key is preconfigured in each IPSec device. IKE peers authenticate each other by computing and sending hashed data that includes the key. If each side can independently create the same matching hash value, the key is accepted. Since you must configure the same key on every device manually, preshared keys do not scale well in large environments.
- **RSA-encrypted Nonces** This method requires that each party generates a psuedorandom number (a nonce) and encrypt it with the other party's public key. The authentication happens when each party is able to decrypt the other party's key with its local private key.

- **Digital Signatures** In this method, each device digitally signs a set of data and sends it over to the other party. This is similar to the RSA-encrypted nonces method, except this method provides nonrepudition. The devices use CA for authentication and to derive the secret key values.

# AH

The AH is part of the IPSec security protocol that provides authentication and integrity for IP packets. AH is defined in RFC 2402 and uses protocol 51. AH does not provide data confidentiality of packets, only integrity. Its authentication is applied by a one-way hash function that is added to every packet. If any changes are made to the packet during its transmission, such as an MITM attack, the AH hash calculation will not match at the receiving end from its sender side and therefore will be rejected. Even though the AH can provide additional authentication, it has a few design issues. Refer to Figure 7.32 to note that AH encapsulates over TCP protocol.

**Figure 7.32** Using AH

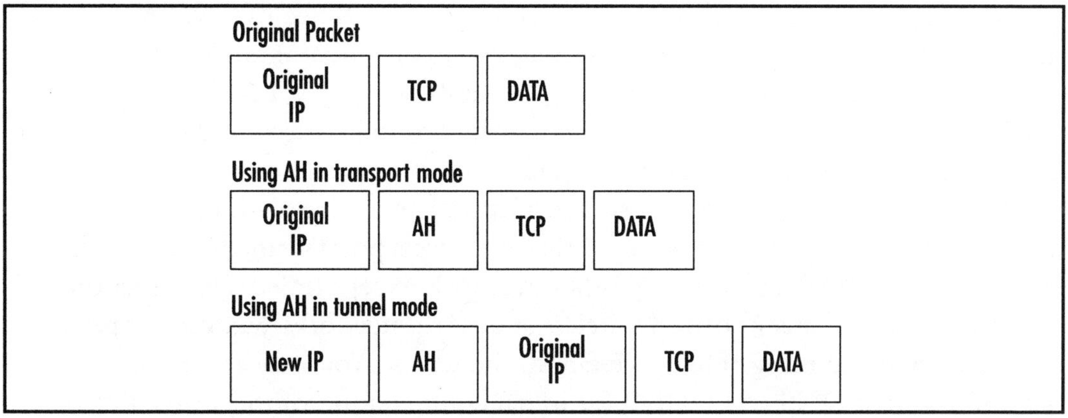


Operations such as Network Address Translation (NAT) and Port Address Translation (PAT) are required in many network environments that need to change the TCP headers of IPSec packets in order to operate. If this happens, the AH will no longer match the hash of the packet that will be performed by the receiving end from the original sender. The solution to this issue is to use ESP alone. The ESP that is used in IPSec to provide confidentiality has methods for performing additional services such as authentication, without needing to rely on the AH. In most VPN setups AH is often disabled, due to the NAT and PAT incompatibility issue. The ESP authentication service takes over for both confidentiality and integrity.

## ESP

ESP is a security protocol used in IPSec that provides confidentiality by encrypting data, authentication, and anti-reply services. ESP can be used on its own or with the AH protocol. ESP provides confidentiality by encrypting at the network layer. ESP is defined in RFC 2406 and uses protocol 50. The ESP header is inserted after the IP header and before the upper layer protocol header when used in transport mode, or it can be inserted before an encapsulated IP header if used in tunnel mode. The default algorithm for encryption algorithms provided by ESP in IPSec is 56-bit Data Encryption Standard (DES). This encryption algorithm can be increased to 3DES or AES for stronger encryption.

# Implementing IPSec over WLAN

Now that we have reviewed IPSec and its fundamental operation, lets explore the use of IPSec in wireless networks. Although Cisco's WLAN access point devices do not yet support IPSec termination, you can implement IPSec over WLAN by using a VPN device such as a PIX firewall, VPN concentrator, or IOS VPN on the back end of your access point. When deploying IPSec in a WLAN environment, an IPSec client is placed on every PC connected to the wireless network. In order for the end user to connect and route traffic into the wired network, they must establish the IPSec tunnel first by successfully authenticating with the VPN gateway. It is important to understand that the access point's function in this implementation is used to filter out any other protocols except IPSec. This function will prevent any use of unauthorized wireless and wired networks without properly authenticating and using IPSec to establish the session. You may also need to allow protocol such as DNS and DHCP for initial configuration of the wireless client. This solution will allow only authorized and authenticated users with VPN-configured access to connect to and use your wired network from wireless stations.

You should make use of IPSec when you have the utmost concern for wireless security. IPSec should not be used with 802.1x/EAP with TKIP or Cisco TKIP wireless methods. These methods should be used when you desire reasonable assurance of confidentiality and transparent user security. Refer to the previous section for more details on how to use 802.1x security features. We are using Figure 7.33 as the topology to implement IPSec over WLAN for wireless users.

**Figure 7.33** VPN Topology Implementation

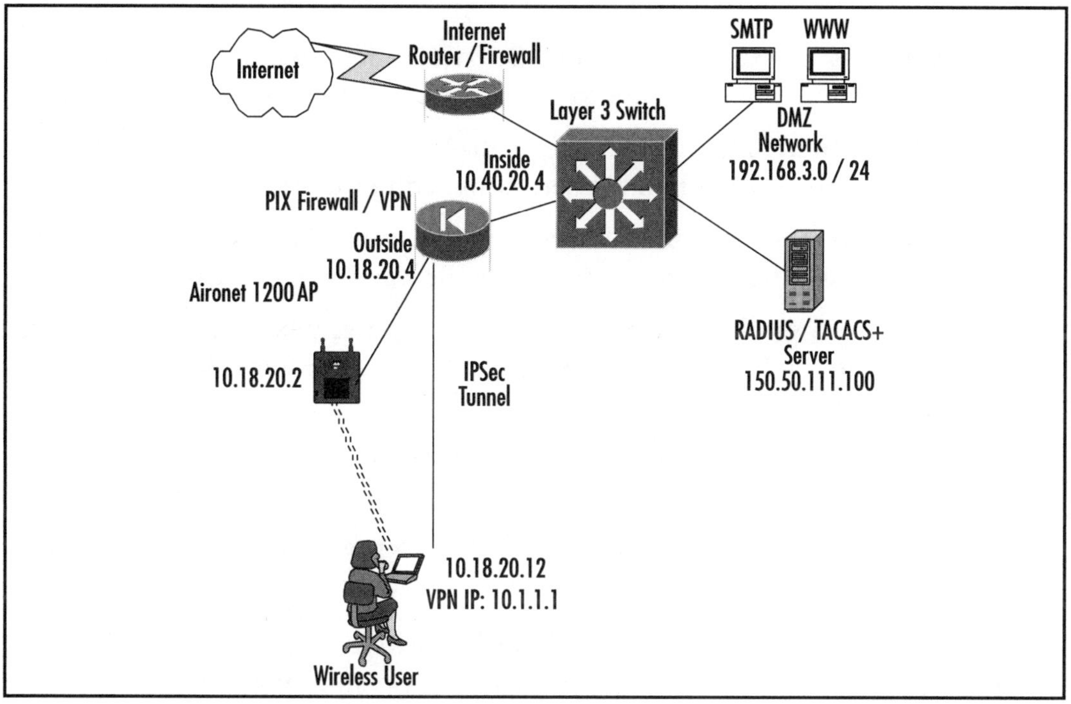


## Configuring & Implementing...

### Validate WLAN Connectivity Prior to IPSec Implementation

When configuring IPSec over WLAN, make sure your WLAN connectivity works before you add IPSec into the environment. It will ease your troubleshooting by isolating connectivity issues to IPSec implementation rather than WLAN configuration issues. Follow regular guidelines to configure wireless connectivity between your remote users and your wired network.

## VPN Device List in WLAN

- **Wireless Client Adapter and Software** The solution that provides the hardware and software to successfully connect to the access point.
- **VPN Client** The software client that provides an end-to-end VPN tunnel between itself and IPSec, terminating the VPN gateway.
- **Wireless Access Point** Wireless access points that provide the initial connection and filtering between wireless and wired networks.
- **Layer 3 Switch** Provides Ethernet connectivity between the wireless, corporate, and public networks. In addition, it can provide Layer 3 filtering for more layered security.
- **RADIUS Server** Provides for a user database and performs authentication of users connecting to the VPN.
- **VPN Gateway** Authenticates wireless users and terminates the IPSec tunnel. The gateway can also provide services such as DHCP.

You may use a modified list or additional hardware in your WLAN VPN design to best fit your network environment. For example, you can eliminate the VPN client software by providing the VPN client in a hardware device, allowing for multiple user connectivity from a branch office without any user interaction.

## Configuring VPN Gateway

Refer to Figure 7.33 as your network topology. In this scenario, you will configure secure access from wireless users into the DMZ Corporate network. You will be using a PIX firewall to terminate users' IPSec tunnels and to assume the role of the VPN Gateway. You can use Cisco's VPN concentrator, Secure IOS, or other secure devices that support termination of IPSec. To add to the IPSec authentication security, we will configure the PIX firewall to use a RADIUS/Terminal Access Controller Access Control System (TACACS) server to authenticate each user against its local user database. Using a RADIUS server as a separate user database allows for per-user authentication and accounting. If you do not have a RADIUS server in your design, and only have a small amount of wireless users, you can implement a local user database on the PIX Firewall or VPN Concentrator.

1. Configure connectivity to the radius server.

```
aaa-server myauthserver protocol radius
aaa-server myauthserver (inside) host 150.50.111.100 cisco timeout 5
```

2. Configure an IKE policy.

```
isakmp enable outside
isakmp policy 10 encryption 3des
isakmp policy 10 hash md5
isakmp policy 10 authentication pre-share
isakmp policy 10 group 2
```

3. Configure a pre-shared key as a wildcard.

```
isakmp key cisco address 0.0.0.0 netmask 0.0.0.0
```

4. Configure an access list for networks that require IPSec protection

```
access-list 120 permit ip 150.50.15.0 255.255.255.0 10.1.1.0
255.255.255.0
```

5. Configure an access list to allow and restrict VPN users access to the corporate network.

```
access-list 140 permit tcp 10.1.1.0 255.255.255.0 150.50.15.0
255.255.255.0 eq www
access-list 140 permit tcp 10.1.1.0 255.255.255.0 150.50.15.0
255.255.255.0 eq smtp
```

6. Configure NAT 0 to disable NAT operation. This step prevents our World Wide Web (WWW) and Simple Mail Transfer Protocol (SMTP) servers from being NAT'ed when returning back to the wireless user.

```
nat (inside) 0 access-list 120
```

7. Configure the transform set that defines how traffic will be protected. We are using 3DES for confidentiality and SHA for authentication in ESP.

```
crypto ipsec transform-set strong-des esp-3des esp-sha-hmac
```

8. Create a dynamic map and map it to a static map. Then apply it to a terminating IPSec interface.

```
crypto dynamic-map mydynmap 10 set transform-set strong-des
crypto map mystaticmap 20 ipsec-isakmp dynamic mydynmap
crypto map mystaticmap interface outside
```

9. Enable Xauth to support authentication of each user via a RADIUS server.

```
crypto map mystaticmap client authentication myauthserver
```

10. Configure an IP pool that will be used to assign IPs to end users.

```
ip local pool mypool 10.1.1.1-10.1.1.254
```

11. Configure a Cisco VPN client policy.

```
vpngroup mygroup address-pool mypool
vpngroup mygroup dns-server 10.0.0.15
vpngroup mygroup password cisco
```

12. Configure a PIX to permit IPSec traffic by default.

```
sysopt connection permit-ipsec
```

## Configuring an Access Point

As discussed in the previous section, in order to allow only authorized users to connect to the wired network, you need to restrict communication protocols on the access point. Restricting protocols on an access point to IPSec communication only will not allow for any other communication but IPSec and its required protocols to associate a client to an access point. If a user has not established IPSec with the proper credentials, they will not be allowed to pass traffic into the wired network.

Protocol filters used on Cisco's access point can be MAC filters, IP filters, and Ethertype filters. Filters are used to prevent or allow specific access to the access point, Ethernet wired port, or radio ports. We use an IP filter to define our filter on the access point to restrict all but IPSec. You can configure filters using a Web browser interface or by entering filters through CLI. Cisco has been converting its WLAN products into IOS-friendly devices that will make CLI management easier and well accepted among the Cisco user community. IOS is Cisco's operation software that runs on most of its core devices such as routers. IOS is supported in 1100 and 1200 access points.

## *Configuring Filters using CLI in IOS*

If you are familiar with access lists on PIX firewalls, switches, or routers, configuring an access list on an access point will not be any different. For our scenario in Figure 7.33 we will configure IP access filters to allow: ISAKMP (UDP 500), ESP (IP 50), and DHCP (UDP 67/68) protocols to pass, and deny everything else. DHCP provides initial configuration parameters such as IP and DNS information, to network users. DHCP is not part of IPSec; it allows the client to receive an IP address in order to continue with establishing communication such as IPSec. If your wireless users use static IPs configured into their wireless card, you do not need to include DHCP in your access control list (ACL).

1. After connecting into CLI using SSH or Telnet, go into configuration mode.

```
AP1200-60a66e# configure terminal
```

2. Configure Access List 102 to allow incoming traffic into the radio from wireless users.

```
AP1200-60a66e(config)# access-list 102 permit udp any host
10.18.20.4 eq isakmp
AP1200-60a66e(config)# access-list 102 permit esp any host
10.18.20.4
AP1200-60a66e(config)# access-list 102 permit udp any eq bootpc any
eq bootps
AP1200-60a66e(config)# access-list 102 deny   ip any any log
```

3. Configure Access List 103 to allow outgoing traffic from the radio into wireless users.

```
AP1200-60a66e(config)# access-list 103 permit udp host 10.18.20.4
any eq isakmp
AP1200-60a66e(config)# access-list 103 permit esp host 10.18.20.4
any
AP1200-60a66e(config)# access-list 103 permit udp any eq bootps any
eq bootpc
AP1200-60a66e(config)# access-list 103 deny   ip any any log
```

4. Apply ACLs to both radio interfaces.

```
AP1200-60a66e(config)# interface dot11Radio 0
AP1200-60a66e(config-if)# ip access-group 102 in
```

```
AP1200-60a66e(config-if)# ip access-group 103 out
AP1200-60a66e(config-if)# exit
AP1200-60a66e(config)# interface dot11Radio 1
AP1200-60a66e(config-if)# ip access-group 102 in
AP1200-60a66e(config-if)# ip access-group 103 out
AP1200-60a66e(config-if)# end
AP1200-60a66e#
```

After applying filter ACLs into the access point, make sure to test the connectivity via IPSec and confirm that ACLs are being used, by viewing its counter hits using the **show access-list** command under CLI (see Figure 7.34). For more details on using ACL in IOS and its syntax, refer to www.cisco.com.

**Figure 7.34** show access-list Command on an Access Point

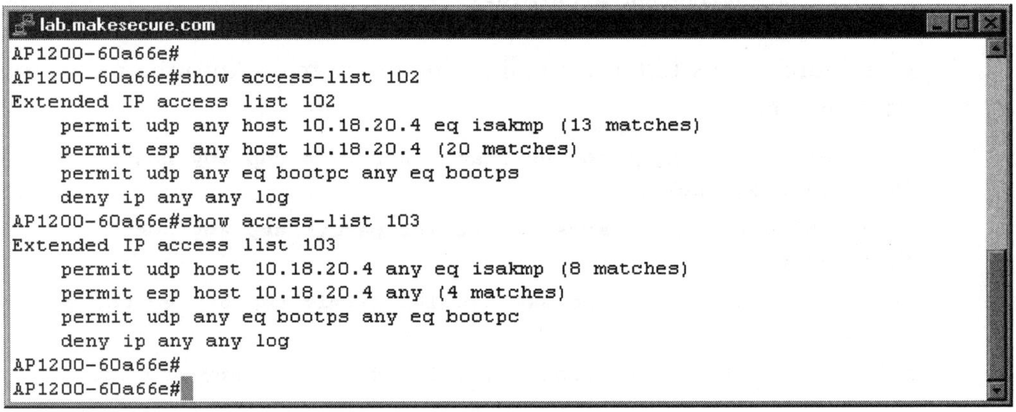


## *Configuring Filters using a Web Browser in IOS*

To configure IP filters using a Web browser, click on the **SERVICES** link in the page navigation menu. On the **Service** page click **Filters | IP Filters** to define your filters. Follow the directions on the page to create a new filter rule and all of its attributes as we have configured in the CLI using the **access-list** command. After completing the IP filter configuration, you must apply it to an interface in order to have its effect, using the **access-group** command under the CLI. After creating your filter, go back by clicking on the **Apply Filters** option (see Figure 7.35). Select your newly created IP filter to your radio interface for incoming and outgoing traffic and press the **Apply** button. If you are using multiple radios in your 1200 access point, make sure you apply your filter to both radios.

**Figure 7.35** Applying IP Filters

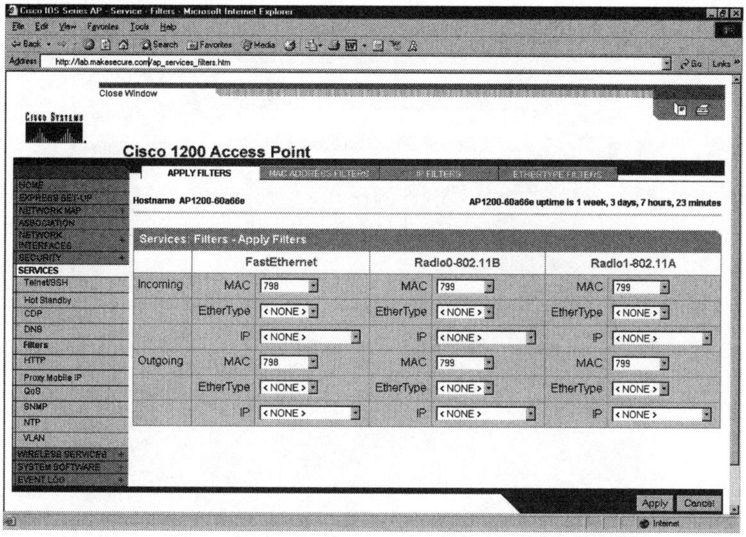


# Configuring a VPN Client

This section describes how to configure Cisco's VPN Client 4.0 to match the configuration accordingly with the VPN gateway that we configured in the previous section. Configuring Cisco's VPN client is simple and quick. In order to configure a VPN client you will need a matching group name you have configured with the **vpngroup** command and its associated password (see Figure 7.36). You will also need the IP address of the VPN gateway and the individual username and password that resides in the RADIUS server according to our scenario. For a detailed configuration of Cisco ACS software with RADIUS authentication support and user database, refer to Cisco's Web site.

For wireless users that are connecting to a wired network using IPSec VPN technology, other security technology such as WEP and other Layer 2 encryption provided by the access point itself is usually disabled. Good security is already provided by the IPSec tunnel between the wireless user and VPN gateway, assuring confidentiality and integrity for data traveling in the air.

**Figure7.36** VPN Client Configuration Property Window

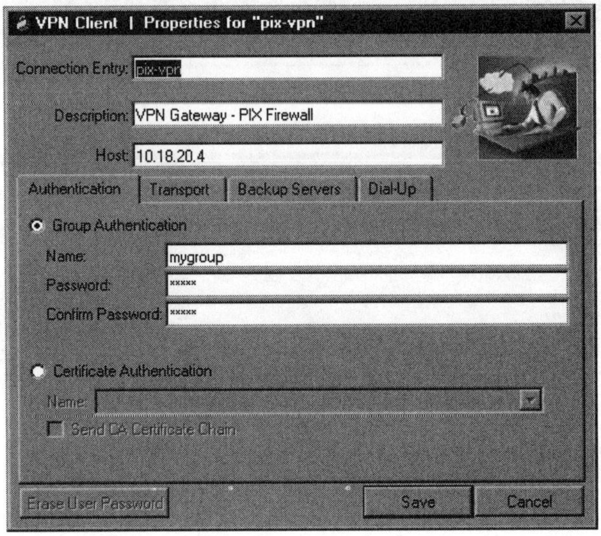


Once you configure the client and have verified that all network communication is up and working between the wireless user and the access point, you can initiate a VPN session by selecting the **Connect** icon in the VPN client software. During the connection process, the VPN client and VPN gateway will negotiate the proper IKE and IPSec SAs and form a secure tunnel. In our scenario, prior to tunnel creation, users will be prompted to enter a username and password that must match and be validated by the RADIUS server we have configured.

Once you establish an IPSec session, you can verify the connectivity by viewing the VPN statistics on your client software. Go to **Status | Statistics** to launch the **Statistics** menu. Refer to Figure 7.37 to view the established session. According to Figure 7.37, we have assumed the 10.1.1.1 IP address from the PIX-VPN gateway. If you recall, we configured an IP pool range with 10.1.1.1 through 10.1.1.254 on the PIX. You can also see that packets are being encrypted and decrypted.

**Figure 7.37** VPN Client Statistics

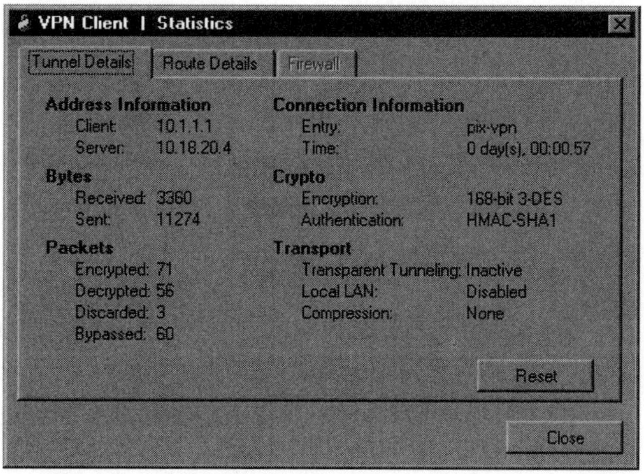


# WLAN Static WEP Keys

A static WEP key is a defined encryption key on the access point, bridge, or client adapter that is used to encrypt data. A static WEP key is composed of either 40 or 128 bits. Because it is static, it must be manually configured on the access point and every user client connecting to it. Unlike in 802.1x EAP where WEP keys are dynamically assigned on per user basis, a static key is configured once on every device. The network administrator must configure each client with the same key. If one client is compromised, every client must be reconfigured. Further, if two clients communicate on WLAN to the same access point with the same static key, they can potentially decrypt each other's communication.

## WEP

WEP is defined as the 802.11 standard for a mechanism to encrypt data moving across the air. WEP works at the Data Link layer of the OSI model. It was originally designed with a 40-bit key to avoid conflicts with U.S. control of the export of strong encryption. Many vendors now support and use 128-bit keys.

WEP static keys and their implementation have been exploited and exposed by engineers, where its vulnerability to decipher an actual key can be broken in less than 30 minutes on a busy network. WEP uses the RC4 stream cipher invented by Ron Rivest of RSA Data Security Inc., for encryption. The RC4 algorithm is a symmetric stream cipher where both parties share the same key to encrypt data. The Initialization Vector (IV) is a component used with the encryption key to create ciphered text. Added IV is used to randomize and ensure that the same plaintext data will not generate the same ciphered data. Figure 7.38 illustrates the WEP encryption process.

**Figure 7.38** WEP Encryption

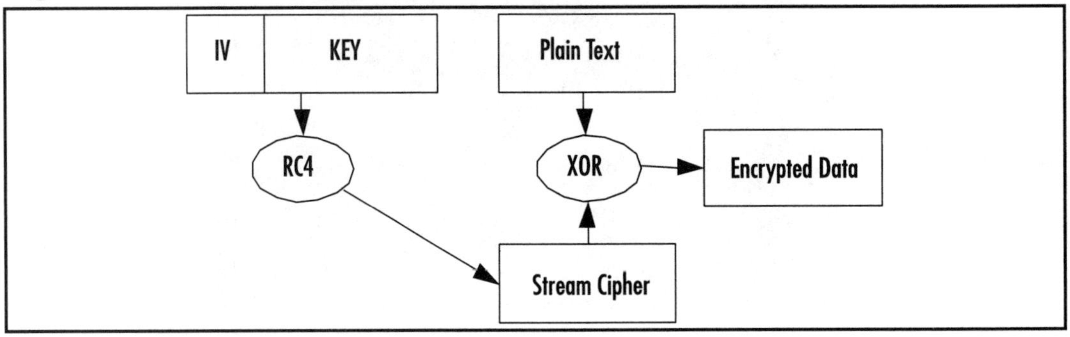


XOR is a mathematical function that combines the resulting cipher stream with plaintext to produce encrypted data. IVs that consist of a random 24-bit value must be sent along to the receiver. IV is sent in cleartext, attached to an 802.11 frame.

# IV WEP Vulnerable

The IV being sent in cleartext makes WEP vulnerable to attackers that can capture encrypted frames and derive their context. IV is 24 bits long and provides 16,777,216 possible values. A University of California, Berkeley engineer found that when the same IV is used on two packets that are captured by an intruder, this method can be used to derive the contents of the two packets. Packets using the same IV are called a collision. For further information, refer to www.isaac.cs.berkeley.edu/isaac/wep-faq.html.

# IV and RC4 Vulnerabilities

A vulnerability in the RC4 key-scheduling algorithm that can expose static WEP keys, was discovered by Fluhrer, Martin and Shamir (FMS). Due to RC4 implementation in WEP and its use of 24-bit IV, different methods can be used in this static pattern to derive the secret WEP key. This so called FMS attack uses between 100,000 to 1,000,000 encrypted packets using the same static key to derive the WEP key. For further information, refer to www.cs.umd.edu/~waa/class-pubs/rc4_ksaproc.ps.

A popular utility for automating FMS attacks is AirSnort. AirSnort operates by passively monitoring encrypted transmissions. When about 5-10 million encrypted packets have been captured, AirSnort will use the discovered vulnerability patterns to derive the secret WEP key. AirSnort is a free program that runs on Linux. For further information refer to http://airsnort.shmoo.com/.

# Mitigating WEP Vulnerability

Cisco has pioneered the development of wireless security standards that improve the implementation of WEP. WEP is still widely used and, because it is the standard, must be enhanced to protect it from known vulnerabilities.

## TKIP

As covered in the previous section, most of the attacks against WEP are due to poor implementation in IV and its use for the same key per packet. By using different keys per packet, you can eliminate some of the threat. TKIP is used to enhance WEP with per-packet-keying. As shown in Figure 7.39, the temporary key derived from a hash function is using IV and a base-configured key is being used to encrypt the packets instead of the base-configured WEP key.

By implementing TKIP, you do not eliminate the vulnerability of deriving a WEP key. Instead of the intruder deriving your base key that is used to encrypt all traffic, they only derive the temporary key that is different from every packet, thus eliminating some of the threat.

**Figure 7.39** Per-pocket-keying Feature of TKIP

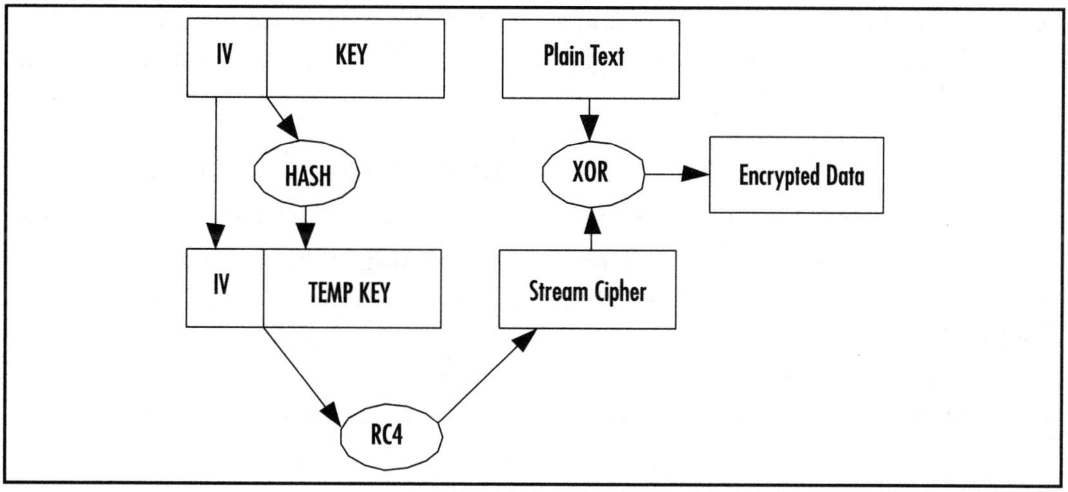

To prevent other attacks such as vector collisions, you must implement a mechanism that changes the base WEP key before the 24-bit IV value recycles and starts over. You must use a mechanism such as the EAP protocol in 802.1x, that allows not only for dynamic keying per user, but also for dynamic changing of the base key before its configured time expires. Refer back to the "EAP" section for more details.

### *Message Integrity Check*

Aside from per-packet-keying, TKIP also offers Message Integrity Check (MIC). MIC assures the integrity of data and prevents MITM attacks. MIC adds two new frames inside an encrypted frame: the sequence number and integrity check field. The sequence number is used by access points to validate packet order and to discard any out-of-order packets. The integrity check field is a hash of fields and data of the frame. Any changes made to these fields during the transmission will not match the transferred hash and the receiving device will discard the frame. It is important to know that by enabling MIC you may reduce the throughput of radio by as much as 30 percent.

# Configure Static 128-bit WEP with TKIP

For a smaller office with a lower number of users that do not require extra protection by using EAP or IPSec, a 128-bit WEP with TKIP is good security. Even if you do not transfer confidential files on your WLAN, you should still configure WEP. Put yourself into the shoes of an attacker. If an intruder wants quick anonymous access into a nearby WLAN in order to launch an attack against someone on the Internet, they would pick from the default unprotected WLAN networks rather than spend time trying to decrypt WEP keys to gain anonymous access.

## Using a Web Browser for Access Point Configuration

To configure a 128-bit WEP with TKIP, go to the **SECURITY** tab in the right navigation window and click on **Encryption Manager**. In the encryption manger window, click on **CIPHER** and use the scroll down menu to pick **TKIP+WEP 128bit** (see Figure 7.40). Go down to encryption key 1 and type in your WEP key in hexadecimal characters that will serve as your base key. Pick 128-bit from the key size scroll down menu. When finished, click on the **Apply** icon on the bottom of the page to apply all changes. If configuring a Cisco 1200 series AP with two radios, make sure you repeat the same process for the second radio and apply all changes.

**Figure 7.40** Static 128-bit WEP Key + TKIP

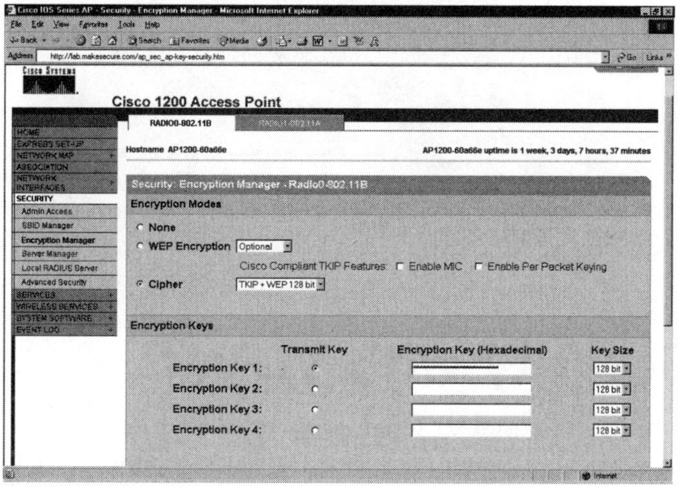


# Configuring the Client

To successfully authenticate, you must match all compatible configurations in your client adapter with those you have configured in your access point. In the Cisco ACU, go to **Commands | Profile Manager** and edit or create a new profile. To configure a WEP static key, click on the **Network Security** tab, as shown in Figure 7.41, and configure the matching WEP configuration. Make sure you chose 128-bit to match the access points configuration. When finished, click **OK**.

**Figure 7.41** Client Static WEP Configuration

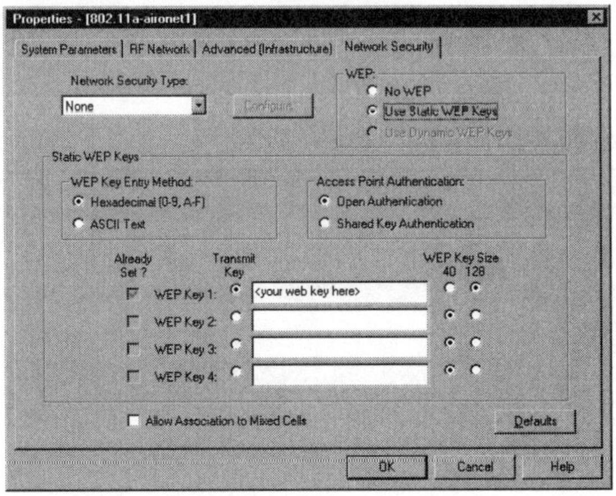


# WLAN Security with VLANs

A VLAN by definition is used to separate Layer 2 segments into multiple collision domains. Two devices connected into the same switch on a LAN can be separated in to two different collision domains by configuring their ports on the switch into different VLAN membership. By doing so, device A will not be able to send traffic to device B. A Layer 3-aware device such as a router would be required in order to route communication between VLAN A and VLAN B. VLANs have similar attributes as physical LANs, with the added capability to group stations into separate LAN segments regardless of end station location.

VLAN technology was expended from wired-aware networks into wireless-aware networks. It provides technology to separate multiple wireless users into different segments. Without VLAN technology in access points, network administrators had to install two separate wireless access points in order to separate two different groups of wireless users. For example, you may have a request from a local university to provide students and teachers with wireless access around the campus area. Students will only be allowed access to Web sites on the Internet through http, with no additional authentication or encryption. Teachers, on the other hand, will require access to the school's student databases as well as the Internet. Because confidential data and grades are stored in student databases, teachers will be required to securely authenticate and encrypt all communication they send through the air. Without the use of a VLAN, you would need to install two separate access points in every location: one access point with a student guest access policy and one access point for teachers with authentication and encryption policy rules.

VLANs in WLANs make it possible to separate different user groups on the same access point and assign different security policy configurations for each separate group. This is a big step for security as well as the manageability and acceptance of wireless LANs.

# VLAN in Aironet

Cisco supports VLANs in VxWorks Firmware Release 12.00T or later and Cisco IOS release 12.2.4-JA or later in all of its wireless devices. Cisco Aironet devices support up to 16 VLANs. Each VLAN can be configured with a separate security policy supporting different wireless groups and their restricted access into the wired network.

## VLAN by SSID

SSID is used to separate and map different users into their proper VLANs. A client configures the appropriate SSID that matches a configured VLAN SSID in an access point. The access point then knows by the SSID which VLAN security policy to associate the connecting client with. After the client is matched to the correct VLAN, the client is challenged and restricted by the matched VLAN security policy and, if successful, it will be mapped and allowed access into the appropriate mapped VLAN on the wired network.

## VLAN by RADIUS

There may be a situation where you have two separate groups using the same authentication to access the wired network. These two user groups must be configured into two separate VLANs. Because both groups use the same authentication method and only rely on the SSID for VLAN mapping, you need a way to prevent the group A client from changing its SSID to be assigned to the group B VLAN. This is where RADIUS 802.1x or MAC authentication can help. Upon successful authentication by the user with the RADIUS server, the RADIUS server sends the allowed SSID list that the user is able to connect with into the network. If a user from group A is using the SSID of group B, the user is not mapped into the VLAN and is discarded from the network.

VLANs help in separate wireless users into different security policy groups. Not only can you have different authentication and encryption measures per VLAN, but you can also apply MAC or IP filters to them. Doing so allows you to restrict specific wireless user groups from accessing restricted resources on the wired network. VLAN in WLAN operations are covered in more detail in Chapter 9.

# Summary

To fully understand and implement security measures in networks, you need to understand all of the different technologies and their functions. Security fundamentals and principles are built around confidentiality, integrity, and availability. Understanding these three fundamentals and their applications will prove to be efficient in your security designs.

Security is a never-ending cycle that needs continuous testing and constant auditing. Stay up-to-date with new releases and fixes. Use Cisco's bug notification tool that will alert you if a new bug is found in your current firmware.

It is important to recognize the threat of a wireless network as a network with no boundaries. Unlike in physically wired networks, intruders can connect to your wired corporate network through a wireless access point miles away. Data such as corporation trade secrets traveling through air waves, must be properly secured.

This chapter identified and reviewed tools an intruder may use to detect wireless networks. Knowing that there is very little to protect against wireless network detection, you must be alert and implement good security measures to protect your networks from unauthorized access after its detection. Make sure you reconfigure default settings such as passwords and authentication types.

No matter how much you protect your wireless networks from unauthorized access into your wired corporate network, one rogue access point can allow intruders in. An unauthorized access point (rogue) installed inside your corporate network by an employee or intruder must be detected and eliminated. Cisco offers solutions such as a rogue SNMP alert or mutual authentication to mitigate the threat of rogue access points.

Through careful review of a static WEP design, we identified several significant weaknesses in the algorithm and implementation used. These weaknesses can lead to unauthorized access and the viewing of confidential data sent over wireless networks. Cisco and its partners acknowledge these vulnerabilities and provide several enhancements to mitigate these threats. Enhancements such as TKIP and MIC can be used where static WEP implementations are required, such as small homes or remote office.

EAP protocol, supported in all Cisco wireless devices, helps mitigate some of the WEP vulnerabilities and allows for per-user authentication. It is now a widely accepted protocol that should be used in all serious wireless designs.

Using a VPN over a WLAN is one of the best security measures you can take to protect sensitive data. Encryption such as 3DES in IPSec can be used for stronger authentication and encryption of remote wireless users connecting into the wired network. Other encryption methods such as SSH, SSL, and PGP can be used to encrypt the application layer between the wireless user and the wired corporate device.

Wireless networks are expanding at a tremendous rate. Its popularity among home users and businesses is allowing attackers to take advantage of poor and unsecured configurations. The fact that an intruder does not need to be in your house or inside your protected data centers in order to connect into your wireless frequency wave should alert you. Knowing how the technology works and its vulnerabilities will allow you to overcome security threats and build a secure wireless network.

# Solutions Fast Track

## Network Security in General

- ☑ The big three security fundamentals are confidentiality, integrity, and availability.
- ☑ Reconnaissance includes gaining unauthorized access into a network or device. Techniques include a ping sweep, port scanning, war driving, application version code, and so on, all of which help in getting to know the network or device in order to exploit it.
- ☑ Data manipulation allows an intruder in the middle of the communication to change data without user or application consent, to benefit the intruder.
- ☑ Security is a continuous cycle that must always be improved. The cycle includes securing, monitoring, testing, and improving.

## How Wireless Technology Changes Network Security

- ☑ Physical access is not required in order to connect into a corporate network. Intruders can connect from the company parking lot into unprotected wireless access points that can lead to access into corporate network.

- ☑ Tracking and anonymity in wireless networks becomes a challenge for security administrators. Wireless connectivity and long ranges makes tracking the user difficult, unlike in LAN networks where the administrator can trace the cable.
- ☑ Due to RF reach and exposure to unwanted areas, confidentiality is required to protect sensitive data from intruder sniffing the frequency.

## Preventing War Driving and Unauthorized Use of Legitimate Access Points

- ☑ War driving is a technique that allows individuals to drive around with wireless equipment detecting transmitting access points within the area.
- ☑ The equipment required to war drive is a laptop, a wireless adapter with external antenna, and a software application such as NetStumbler that decodes and records detected wireless access points.
- ☑ Disable guest access into your access point and provide a strong authentication option such as open method with WEP and MAC or EAP authentication to prevent from unauthorized use.
- ☑ Disable unwanted services such as Telnet, SNMP, CDP, and HTTP that can be exploited to gain access into your access point, if you do not use them. Use SSH to securely connect and administer your access point.

## Exploring Rogue Access Points

- ☑ A security policy should include and prohibit employees from installing a wireless access point and connecting it to a wired network. Educating employees to this policy is a must.
- ☑ Mutual authentication allows a user to authenticate the wireless device, just as the wireless network authenticates the user. This prevents the user from connecting into a rogue access point.
- ☑ Cisco allows clients running in a LEAP environment to alert the access point of a rogue access point that it detected but could not authenticate. In turn, the access point can then send a SNMP trap to the management station alerting it to the incident.

- ☑ A port scan of a corporate-wired LAN can reveal a rogue access point by detecting port 80 for administration.

## Designing for Security

- ☑ A security policy outlines standards for new and current security implementations.
- ☑ Risk assessment identifies network vulnerabilities, critical services, and data sensitivity and their impact on an organization when an attack takes place or when a network is down.
- ☑ The PSPF option in a Cisco access point, allows you to restrict access between wireless users connected to the same access point.

## WLAN LAN Extension 802.1x/EAP 4-2

- ☑ EAP provides per-user authentication and per-user WEP keying.
- ☑ LEAP further provides for dynamic user key rotation timeout and mutual authentication between a client and a wireless network.
- ☑ EAP is a dynamic protocol that consists of multiple authentication extensions such as LEAP, TLS, Token Cards, MD5, PEAP, and others.
- ☑ The TLS EAP option makes use of PKI digital certificates to authenticate clients.

## WLAN LAN Extension IPSec

- ☑ IPSec is a framework of open standard protocols used in VPN to provide secure connectivity between peers.
- ☑ IPSec is used in wireless networks for the best security and to overcome the vulnerabilities of RC4 WEP encryption. IPSec can use 3DES for better encryption and authenticate on a per-user base.
- ☑ IKE, AH, and ESP are the three major components of IPSec.
- ☑ Client VPN software is required on wireless client's computers to connect into VPN using IPSec.

- ☑ Av access point does not support the termination of IPSec; a separate VPN device such as a Cisco PIX or Cisco VPN Concentrator is required to terminate and authenticate IPSec users.

## WLAN Static WEP Keys

- ☑ Used on its own without any enhancements, WEP is vulnerable to multiple attacks.
- ☑ TKIP allows for unique per-packet keying, where a different temporary key is derived from a base-configured WEP key on a per-packet basis and used to encrypt data.
- ☑ Keys must be manually configured and match every client and access point.

## WLAN Security with VLANs

- ☑ A VLAN allows a unique access point to define multiple user groups with different security authentication policies. Further filters can be configured to restrict or permit different access policies between different user groups.
- ☑ Cisco access points allow for configuration of up to 16 VLANs.
- ☑ Users are grouped into different VLANs by SSID. They then need to pass the required VLAN authentication policy in order to gain access.

# Frequently Asked Questions

The following Frequently Asked Questions, answered by the authors of this book, are designed to both measure your understanding of the concepts presented in this chapter and to assist you with real-life implementation of these concepts. To have your questions about this chapter answered by the author, browse to **www.syngress.com/solutions** and click on the **"Ask the Author"** form. You will also gain access to thousands of other FAQs at ITFAQnet.com.

**Q:** How do I know which WLAN security is right for me?

**A:** Every design is unique. If you are designing WLAN for a corporate network, your first step will be to consult the security policy in place. Corporate policy may require a design where per-user authentication and account must be supported, in which case you will need to use protocols such as EAP and RADIUS server.

**Q:** Access point filters or extra firewall?

**A:** Cisco access points can be configured with filters on both radio ports used as the wired interface ports. These filters can restrict or permit certain traffic. Firewalls like Cisco's PIX access point have special security features and guards as well as filters that can further safeguard a wired network. It also has greater capacity and better performance throughput than an access point. Offloading filters from the an access point can increase performance. Firewall can also be used as a VPN termination point for wireless users.

**Q:** What about IDS?

**A:** IDS, like Cisco's NetRanger, can be installed on the back end of an access point-wired network for better security awareness. It can watch all of the traffic coming from the access point and detect possible intrusions.

**Q:** I have enabled dynamic WEP with TKIP and per-user authentication using EAP. Am I now secure?

**A:** Security is a never-ending cycle. Although you are using some of the better security improvements in wireless, you still need to worry about newly discovered vulnerabilities, rogue access points, or someone simply stealing your access point. Always apply security in multiple layers. Security is a chain of layers, where you are as secure as your weakest layer.

# Frequently Asked Questions

The following Frequently Asked Questions, answered by the authors of this book, are designed to both measure your understanding of the concepts presented in this chapter and to assist you with real-life implementation of these concepts. To have your questions about this chapter answered by the author, browse to www.syngress.com/solutions and click on the "Ask the Author" form. [illegible]

**Q:** How do I [illegible] WLAN security [illegible]?

**A:** [illegible] design is [illegible]. If you are securing a WLAN [illegible] the security policy [illegible] [illegible]

**Q:** [illegible]

**A:** [illegible]

**Q:** [illegible]

**A:** [illegible] work for [illegible] access point and [illegible]

**Q:** [illegible] WEP [illegible]?

**A:** [illegible] Although [illegible] security [illegible] still need [illegible] security [illegible] access point [illegible] security [illegible] multiple layers [illegible] your weakest [illegible]

# Chapter 8

# WLAN Rogue Access Point Detection and Mitigation

**Solutions in this Chapter:**

- The Problem of Rogue Access Points
- Preventing and Detecting Rogue APs
- IEEE 802.1x Port-based Security to Prevent Rogue APs
- Using Catalyst Switch Filters to Limit MAC Addresses per Port

☑ Summary

☑ Solutions Fast Track

☑ Frequently Asked Questions

# Introduction

This chapter discusses what may be the single greatest problem of wireless local area networks (WLANs): rogue access points and unauthorized people using otherwise legitimate access points. This chapter covers wireless-aware product features that address both of these problems, as well as how to set up and use them.

This chapter also we will take a closer look and discusses how to mitigate the threat of rogue access points that pose significant security threats to businesses and their networks.

Employees install wireless devices in their offices and cubicles for their own personal use because they are convenient and inexpensive. Installing access points is as easy as plugging into an Ethernet jack. Unauthorized wireless devices can expose protected corporate networks to attackers, allowing for a security breach. In this chapter, you will learn how personal access points can introduce such threats to your networks and how you can mitigate the threat of rogue access points by using both wireless- and wired-aware devices and their techniques.

You will study traditional techniques such as manual sniffing, physical detection, and wired detection to detect rogue access points, and will also use Cisco's new centralized solutions for detecting rogue access points. In a Cisco-aware infrastructure network, all wireless devices can work hand-in-hand to detect and report unauthorized access points to the central managing station.

# The Problem with Rogue Access Points

A rogue access point is an unauthorized access point. Unauthorized access points can pose a significant threat by creating a back door into sensitive corporate networks. A back door allows access into a protected network by avoiding all front door access security measures. As discussed in previous chapters, wireless signals travel through the air and, in most cases, have no boundaries. They can travel through walls or windows, reaching long distances far outside of a corporate building parameter. Figure 8.1 shows a wireless signal from access points beaming through the air outside of a corporate building into the parking lot and nearby buildings across the street. These radio signal frequencies may represent both rogue and valid access points that carry sensitive confidential data from inside the corporation or from outside mobile workers. The difference between the radio frequencies from these two wireless access points is that the rogue unauthorized access point was installed by an employee with limited security protection, often leaving it at its default plug-and-play unsecured configuration, while the authorized access point was installed by a skilled engineer with full security support. Further, unlike authorized access points that are configured to protect radio signals confidentially with a robust authentication process, the rogue access point installed by the employee probably does not support such security options, as it does not have access to interact with third-party security servers to provide such services.

The bottom line is that rogue access points installed by employees pose a significant threat because they provide poor security measures while extending a corporate network's reachability to attackers from the outside.

**Figure 8.1** Wireless Reachability

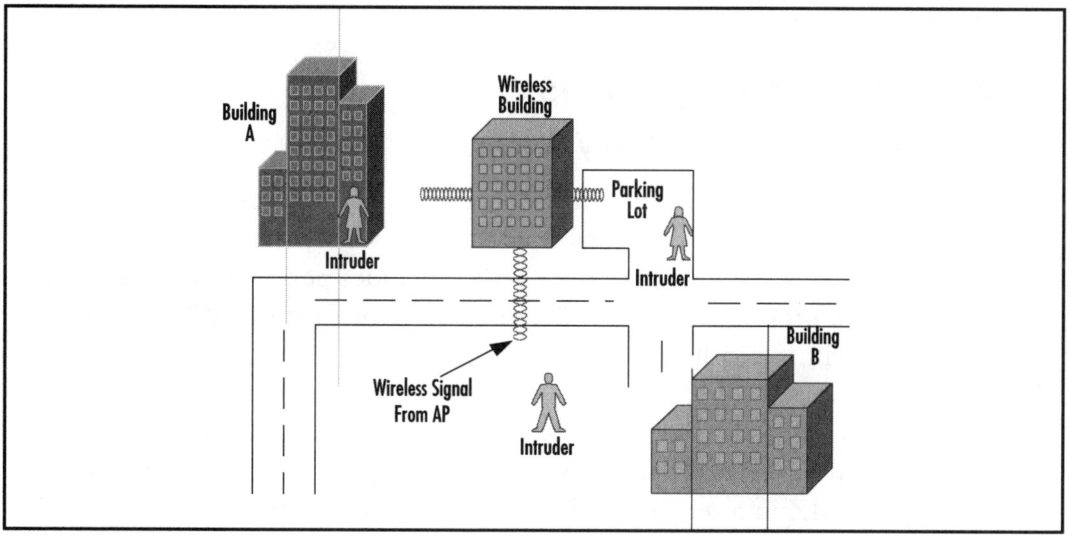


Employees usually install unauthorized access points because of poor performance of current wireless infrastructure, because they may be located in a dead spot, or simply because their company does not provide wireless access. It is important to note that a rogue access point is most likely to be installed in an organization that does not support wireless networks for its employees.

**NOTE**

Audits to detect rogue wireless access points are required in all corporate network environments, even if they do not provide wireless access.

Unauthorized installed access points are unsecured. An average employee is not an expert on wireless security and does not realize the threat they pose with their newly installed rogue access point. Most rogue access points implement a plug-and-play feature allowing for minimal configuration by the user in the order of their use. Security settings are turned off by default, and default passwords are used that need to be reconfigured to prevent from intruders.

As covered in Chapter 7, the best security is implemented using 802.1x protocol features or virtual private networks (VPNs). Both of these security solutions require a third-party device that employees would not have access to; thus, rogue access points are not secure and can be easily attacked to gain access into the connected corporate network.

# A Rogue Access Point is Your Weakest Security Link

A network is only as secure as its weakest security link. For example, consider that you have implemented a very stable and secure wireless and wired network. Your secure wireless local area network (LAN) includes per-user authentication using an 802.1x protocol, a dynamic Wired Equivalent Privacy (WEP) protocol key assignment with periodic key rotation for confidentiality, and logging for audit purposes.

Now consider that all of the time and money spent providing a secure wireless access can be diminished by a single rogue access point. Figure 8.2 represents a wireless DMZ in a secure wireless network topology. In order for valid User A to gain access onto the protected corporate network, they must go through the proper authentication process, pass the firewall and Intrusion Detection System (IDS), and use encryption. Unlike User A, User B does not need to go through any security measures in order to gain access to the corporate network. User B is simply taking advantage of a rogue access point that was most likely installed with a weak security policy and default settings.

**Figure 8.2** Bypassing Security with a Rogue Access Point

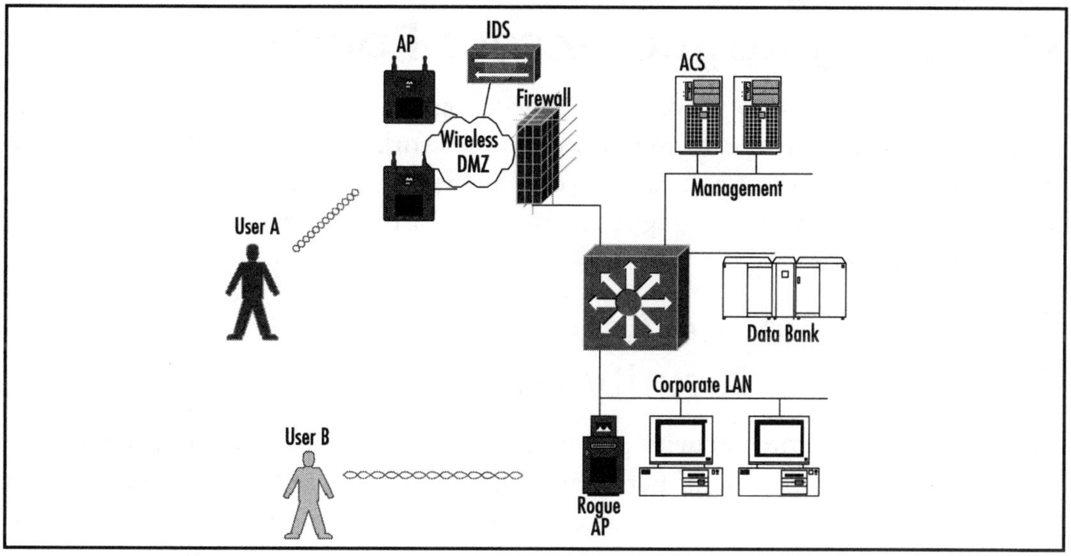


This example represents a back door into a corporation that can be used by the employee who installed the rogue access point and by an intruder that may take advantage of the poorly secured rogue access point.

# An Intruder's Rogue Access Point

An intruder can also install a rogue access point into a corporation. The difference between an intruder's access point and an employee's access point is that the intruder's is not connected to the wired network. How does this make it an unauthorized access point? It is still an unauthorized access point within the radio signal strength area that is used as the trap device to catch valid users. When a valid user tries to connect to an intruder's access point, the intruder's access point can trick the user into providing useful information such as the authentication type and credentials of the user, which can then be recorded and used later by the attacker to gain access to a valid access point.

One way to mitigate an intruder's rogue access point is to provide for dual authentication. In dual authentication, the user needs to authenticate the access point and the access point has to authenticate the user. Dual authentication is supported in the 802.1x protocol. Dual authentication allows the user to verify the validity of the access point before its use. The details of dual authentication are covered in Chapter 7 and are reviewed in the 802.1x protocol section later in this chapter.

# Preventing and Detecting Rogue Access Points

Many techniques exist to prevent and detect rogue access points. Detecting rogue access points should be performed on every network audit to avoid possible back door exposure. As mentioned earlier, your security is only as strong as your weakest link. Do not let one rogue access point dismiss your entire security-configured infrastructure.

## Preventing Rogue Access Points with a Security Policy

First and foremost, your security policy must include the use of wireless networks and prohibit the use of personal rogue access points. A security policy does not eliminate the threat of rogue access points, but it does set guidelines for current and future network installations and what steps to take if a rogue access point is detected. A security policy should mandate that all employees follow proper security measures for wireless networks and should also require written approval from the Information Technology (IT) and Security teams approving the installation of a personal access point. It is important that all employees know that freelance access points are prohibited, why they are prohibited, and what will happen if they break the rule. The risks are such that some companies will fire individuals for setting up their own access points.

For a security policy to be successful, it needs to be communicated to the users. If users are not aware of these security rules, they will not follow them. Continuous education and audits of the security policy are a must.

## Provide a Secure, Available Wireless Network

Most rogue access points are installed by non-malicious employees who simply want wireless access in their work area. One way to prevent employees from installing such rogue access points is to provide wireless access to them. Installing stable wireless access throughout meeting rooms, the cafeteria, and the outdoor campus, allows you to control its access and security implementation. Doing so does not mean you can stop auditing and searching for rogue access points within your network, but it will decrease their detection count and improve overall security.

# Sniffing Radio Frequency to Detect and Locate Rogue Access Points

Another technique for detecting rogue access points is to manually use a network sniffer to sniff the radio frequency within your organization's perimeter. A wireless sniffer allows you to capture all communication traveling through the air, which can then be used for later analysis such as Media Access Control (MAC) address comparison. Every wireless device has its own unique MAC address. If a new, unknown MAC address of an access point is detected in a wireless sniffer trace, it will be red flagged as a rogue access point and investigated further.

### Designing & Planning...

#### Finding MAC Addresses

Every manufacturer programs a unique MAC address into their network card. Every network card has its own MAC address that it uses to communicate with. A MAC address is 48 bits long. The Institute of Electrical and Electronic Engineers (IEEE) controls the first 24 bits (3 octets) of the address. These first 3 octets are called the Organizationally Unique Identifier (OUI). OUIs are given to corporations that produce network devices such as network cards. These corporations must use the unique first 3 octets assigned to them in all of their network devices. The second 24 bits of the 48-bit long MAC address are controlled by the manufacturer. If the manufacturer runs out of unique addresses for the second half of the MAC address, it requests a new 3-octet address from the OUI.

If you detect a MAC address and want to look up its manufacturer, refer to the OUI database Web site at http://standards.ieee.org/regauth/oui/index.shtml

Knowing that every network device has a unique MAC address, you can find out a lot of useful specific information about each device. In Figure 8.3, MAC address 000CCE211918 has been detected. Entering 000CCE (the first half) into the OUI online database reveals that the device detected is a Cisco device.

Tools such as NetStumbler can be used as rogue access point detection sniffers. It displays a list of detected access points within the area of signal strength that can be compared to a friendly database of access points. NetStumbler can further be used to zero in on a physical rogue access point and its location by measuring the signal strength. Figure 8.3 shows a detected access point with MAC address 000CCE211918. After checking the list of friendly access points, we have determined that this detected MAC address does not match any of the authorized access points and thus is a possible rogue access point. To locate this rogue access point, we begin searching by walking around with a laptop and the NetStumbler utility following the signal strength. Notice that the signal strength increases as we close in on the physical location of the detected access point.

**Figure 8.3** NetStumbler: Finding a Rogue Access Point with Signal Strength

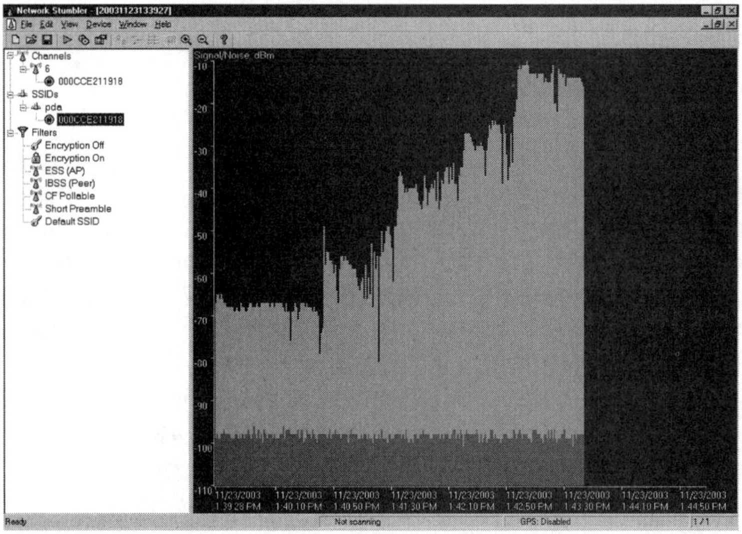


Tools such as Cisco's Aironet Client Utility (ACU) can also be used to follow the strength of a radio signal in order to find a detected rogue access point's physical location. The ACU is installed with Cisco's Aironet wireless adapter. Figure 8.4 shows the Link Status Meter tool in the ACU that displays the signal strength for MAC address 000CE211918, which was determined to be a rogue access point in the previous example. Another useful tracking tool within Cisco's ACU application is the Site Survey tool, as shown in Figure 8.5. Again, using the Site Survey tool, the closer you move to the physical location of a detected access point the higher the signal strength will be.

**Figure 8.4** ACU: Link Status Meter

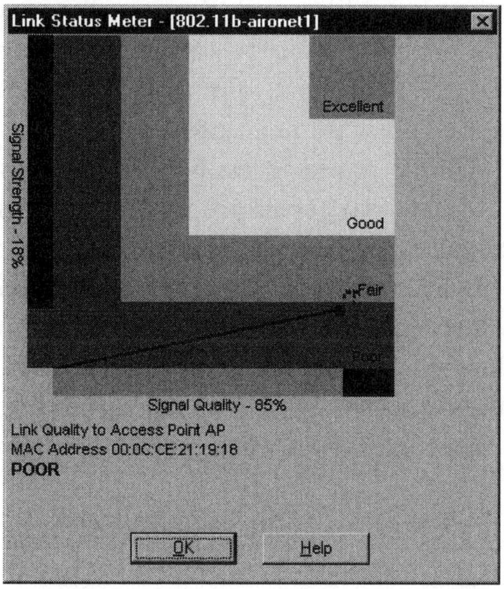


**Figure 8.5** ACU: Site Survey

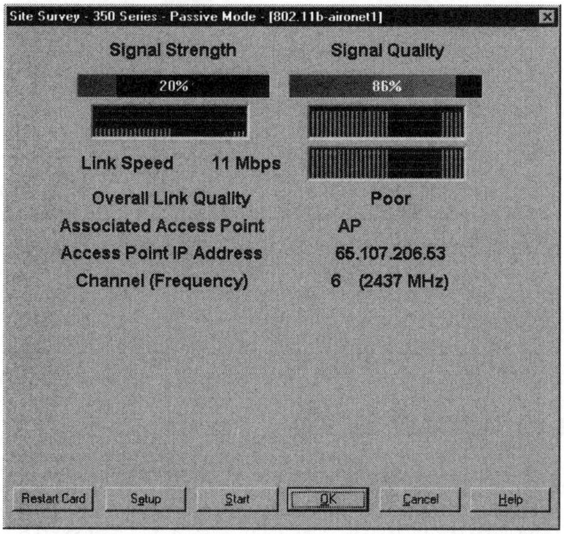


# Cisco's Rogue Access Point Detection

Detecting rogue access points with a sniffer device can be a time-consuming and almost impossible task in large-scale wireless and wired environments. The administrator must walk throughout the entire area and manually compare friendly detected access points with possible rogue access points. This task must be repeated almost daily to assure security against rogue access points.

Cisco has developed a more robust solution to overcoming the manual work effort of sniffing for rogue access points. Instead of walking around with a laptop and antenna to detect possible rogue access points, Cisco's solution allows you to turn all of the wireless clients and access points into an army of sniffers that continually analyze and monitor the radio frequencies around them (see Figure 8.6). This allows you to perform 24 hours a day/7 days per week automatic detection of rogue access points throughout all locations where authorized wireless clients and access points are located. Rogue access points detected by wireless clients and access points are then sent to the central management station where the network administrator is alerted.

**Figure 8.6** All Cisco-aware Devices Become Sniffers

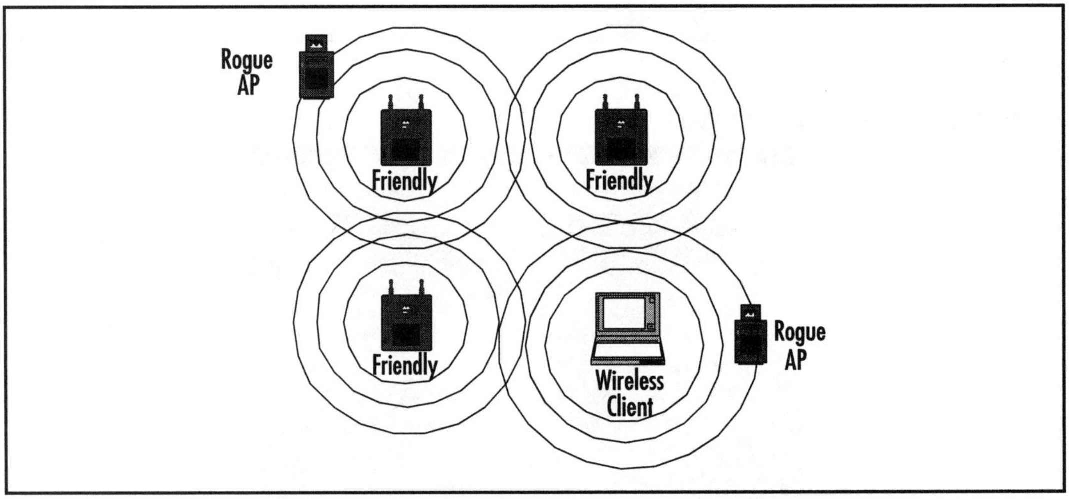


# Central Management with WLSE to Detect Rogue Access Points

The Wireless LAN Solution Engine (WLSE) is a CiscoWorks application that provides central management for all Cisco-aware wireless devices. WLSE can be used to receive rogue access point-detected information from wireless clients and access points through Simple Network Management Protocol (SNMP). When a wireless client detects a possible rogue access point, it sends the information to a friendly access point, which then sends it to WLSE engine via SNMP-trap protocol to inform the management server of its findings (see Figure 8.7). WLSE receives this information and compares it against a database of friendly access points. If the WLSE cannot find the reported access point on its friendly list of valid access points, it red flags it and alerts management that a possible rogue access point has been detected.

A WLSE centralized solution is welcomed by administrators in large- and mid-sized Cisco wireless-aware environments, as it provides scalability and central management and greatly improves the overall security against rogue access points, with its automated process.

**Figure 8.7** Rogue Access Point Detection by Client

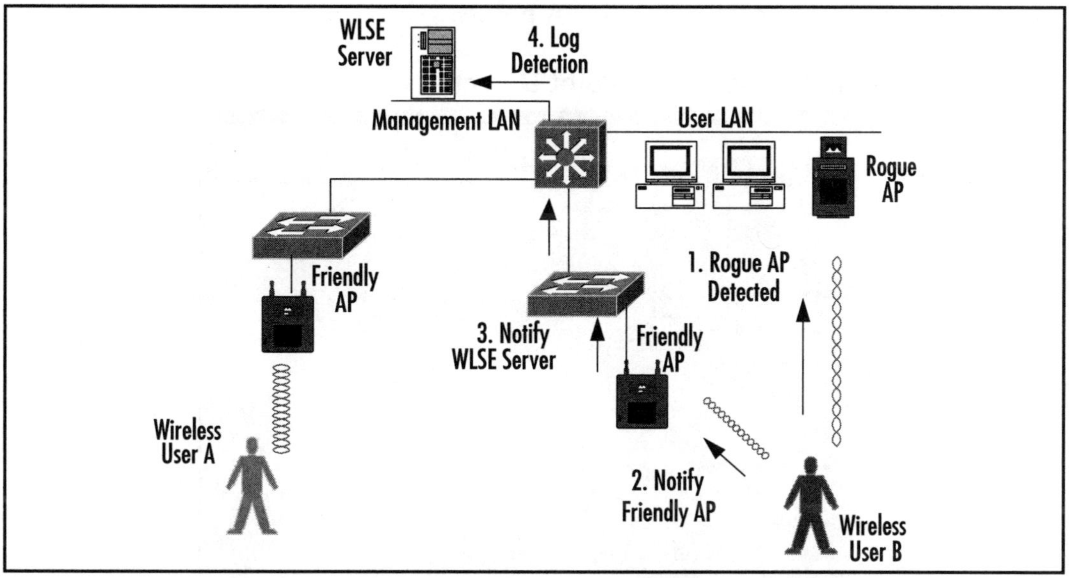


The WLSE can also use triangulation to calculate the physical location of rogue access points, by using the signal strength of multiple wireless clients and access points at the time of detection. This allows you to not only detect rogue access points, but also to know its approximate physical location. WLSE is also capable of providing the switch IP and port details into which the rogue access point is physically connected to, allowing you to quickly locate and disable the rogue access point to eliminate its security threat.

Figure 8.8 shows a rogue access point detection alert from the WLSE that reports that an unauthorized access point has been detected by four friendly access points. Further information shows that the detected rogue access point is broadcasting "ROGUE" SSID in its beacons. The Received Signal Strength Indicator (RSSI) next to each reporting access point represents the signal strength relationship between the rogue and the friendly access point, and is used to estimate the approximate physical location of the detected rogue access point.

One WLSE feature allows you to import and configure your floor blueprints, which can be used to provide a visual of the wireless clients and access points within the network wireless area. In Figure 8.9, a floor map is used along with RSSI information from friendly access points to visualize the location of a detected rogue access point. As you can see, the visual map shows four friendly access points reporting the detected rogue access points and their estimated physical location. Such automatic and detailed support from WLSE allows you to quickly find and terminate rogue access points.

**Figure 8.8** WLSE Rogue Access Point Detected

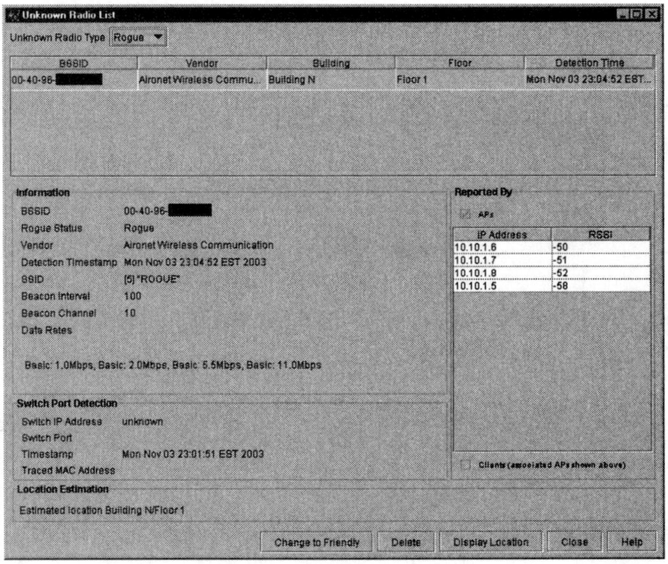

**Figure 8.9** WLSE Rogue Access Point Location Map

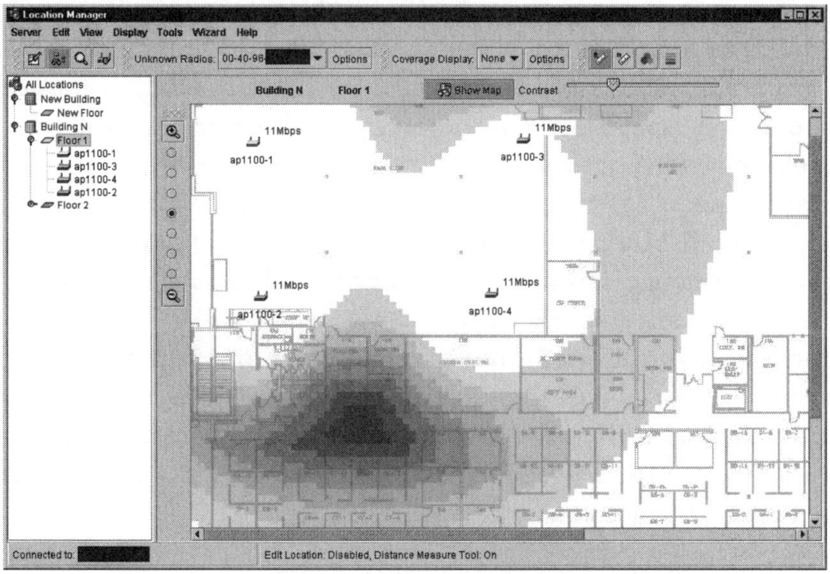


# IEEE 802.1x Port-based Security to Prevent Rogue Access Points

This section reviews IEEE 802.1x protocol, its use in wireless and wired LANs, and how it can aid in mitigating the threat of rogue access points. For further details on the 802.1x protocol and its implementation in a wireless environment, refer to Chapter 7.

As discussed earlier, there are two different types of rogue access points: one that is installed by an employee with a physical connection to the corporate LAN or one that is installed by an intruder without any physical connection to the wired LAN. An intruder's rogue access point is used to trick valid users into establishing a connection in order to obtain confidential information. A valid user needs a method of validating an access point just as the access point needs a method that validates the user, to prevent connection to a rogue access point.

## Prevent Users from Using Rogue Access Points with 802.1x

In a wireless environment, the 802.1x protocol provides mutual authentication that can be used to mitigate the threat of valid wireless users establishing a connection to rogue access points. Figure 8.10 shows a typical 802.1x Light

Extendable Authentication Protocol (LEAP) dual authentication process, where the wireless client is authenticating the RADIUS Access Control Server (ACS) server at the same time that the server authenticates the client prior to establishing a successful connection. Both challenges are derived from the user's password that only the user and a valid RADIUS ACS server have, thus providing a successful challenge response.

If the access point in Figure 8.10 were a rogue access point, it would not have access to the RADIUS ACS server because it would have failed the user's authentication challenge and in turn the user would refuse to establish connection to the access point (see Figure 8.11).

Each authorized access point must be manually configured in the RADIUS ACS server in order to access the server for authentication purposes. Therefore, unauthorized devices such as the rogue access point in Figure 8.11 would not be allowed to query or use RADIUS ACS services because it was never added to the allowed list by the administrator.

**Figure 8.10** 802.1x Mutual Authentication

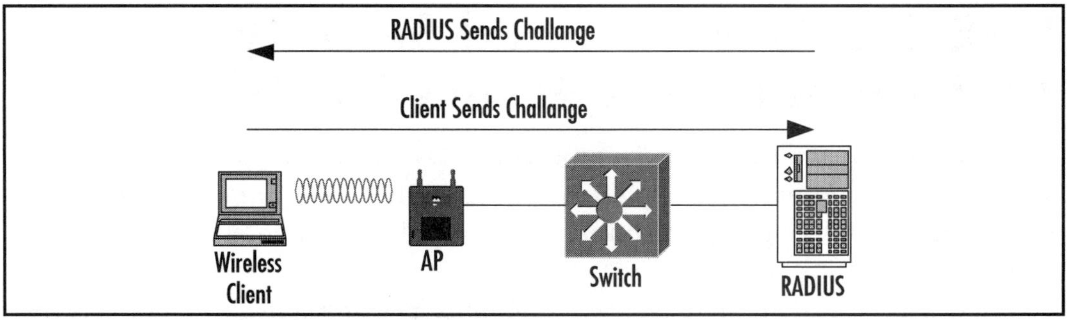


**Figure 8.11** 802.1x Failed Mutual Authentication

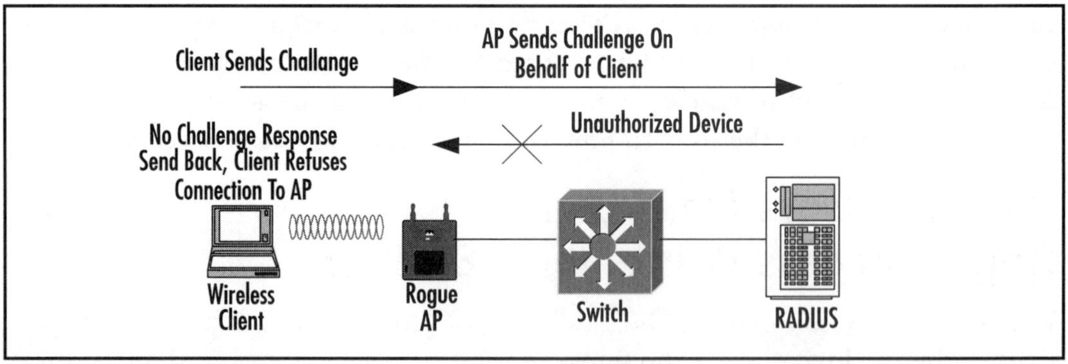

Mutual authentication is not supported in all 802.1x implementations or the Extensible Authentication Protocol (EAP). One of the supported methods of mutual authentication in EAP is Light Extensible Authentication Protocol (LEAP) and EAP-Transport Layer Security (EAP-TLS). In LEAP, authentication and challenges are derived from usernames and passwords. EAP-TLS is nearly identical to the LEAP process, but instead of using usernames and passwords it uses digital certificates. Refer back to Chapter 7 for a more in-depth review on both of these EAP types and their configurations.

# Preventing Rogue Access Point from Connecting to Wired Network with 802.1x

Now that you know how to detect and track down rogue access points and avoid using them, you must learn how to prevent them from connecting to a wired LAN in the first place. The 802.1x protocol was originally designed to control access and restrict connection to physical wired ports. This newly developed protocol allows you to authenticate a device or user prior to using a physical port on a switch. Figure 8.12 shows three workstations that are able to communicate on the wired network, and a rogue access point that is not. As soon as one of the workstations is connected to the physical port, the switch sends an authentication challenge based on a username and password from the RADIUS server that the owner of the workstation must pass in order to successfully connect to the local LAN. When a rogue access point is connected to a physical port other than a workstation, it is unable to process a challenge request from the switch and thus will not be permitted to connect to the wired LAN. This is a great step towards security that allows you to authenticate a device or users before they are allowed to connect to a physical port. This mitigates the threat of unauthorized devices and users such as rogue access points from physically connecting into the LAN and possibly creating back doors into corporate networks.

**Figure 8.12** 802.1x in Wired Network

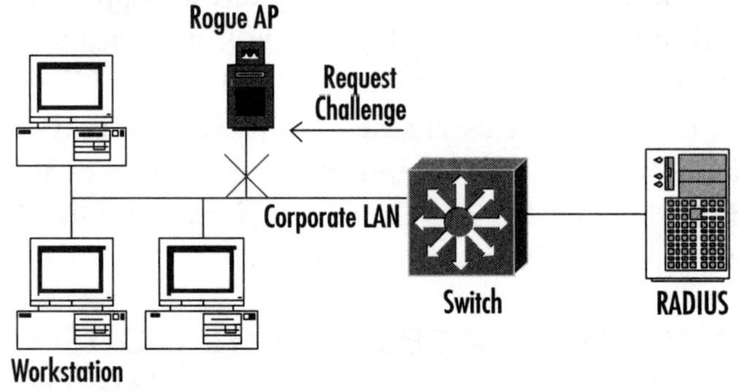


# Understanding Devices and their Roles in Wired 802.1x Implementation

Each device in 802.1x plays a specific role. Figure 8.13 includes the following three main devices:

- Client (workstation)
- Switch
- Authentication Server

The client (workstation) requests access to the LAN by sending a request to the switch. The switch can also be configured such that it automatically requests an authentication challenge from a newly connected device without waiting for a request. The client must be compatible and support the 802.1x authentication process in order to process EAP requests and its challenges.

The switch controls the physical access to the LAN based on authentication messages from the authentication server and the client. The switch acts as a proxy between the authentication server and the client. Not all Cisco switches support 802.1x authentication. The switch allows the client to only send EAP traffic in order to authenticate. After successful authentication, the switch opens its port to allow all traffic from the client to pass through.

The authentication server performs the actual authentication of users. It holds the local or external user database and its restrictions. Each authenticating user must be configured in the authentication server in order to successfully authenticate. The authentication server must support RADIUS authentication protocol and EAP extensions. Cisco ACS version 2.6 and higher supports 802.1x and RADIUS authentication.

# Configuring 802.1x Authentication on a Supported Switch

In this section you will configure 802.1x protocol on a supported Cisco Catalyst switch. Refer to Figure 8.13 for the topology. In this example, it is assumed that the client supports the 802.1x authentication process, and that the ACS – RADIUS server is configured with user database and authentication permissions.

**NOTE**

Make sure you have network connectivity between the switch and RADIUS server prior to configuring 802.1x support.

**Figure 8.13** Implementing 802.1x Topology

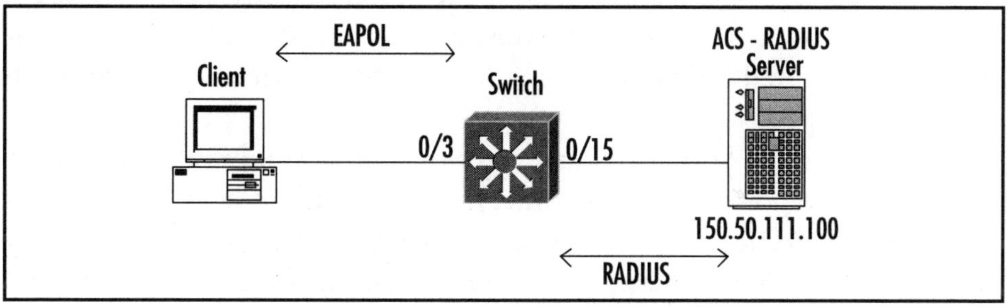


1. Configure a switch to RADIUS communication.

```
Switch3550# configure terminal
Switch3550(config)# radius-server host 150.50.111.100 key cisco
```

2. Configure 802.1x authentication.

```
Switch3550(config)# aaa new-model
Switch3550(config)# aaa authentication dot1x default group radius
local
```

3. Configure the interface to request EAP authentication when the new device connects.

```
Switch3550(config)# interface fastEthernet 0/3
Switch3550(config-if) switchport mode access
Switch3550(config-if)# dot1X port-control auto
```

4. Save all configurations.

```
Switch3550(config-if)# end
Switch3550# copy running-config startup-config
```

Now when a device connects into port 0/3 of the switch, the switch will request authentication credentials from the device. By default, all traffic but the authentication EAP protocol process will be blocked from the 0/3 port. After successful authentication the switch will allow all traffic to pass.

## *Enabling Multiple Host Authentication*

The configuration above only allows one host to connect to port 0/3 at one time. You can allow more than one device to authenticate and use the same port at one time. By default, only one host MAC address is allowed to connect to an 802.1x-configured port at one time, while other devices trying to use the same port are dropped.

Using multiple host configurations, you can have more than one host connecting to one port at the same time. In multi-host mode, it takes only one successful authentication to open up access to every other device connecting to the same port. If the multi-host port becomes unauthorized due to an EAPOL-Logoff message or when re-authentication fails, it disables access for all hosts using the same port.

Multi-host port mode may be needed when clients are not connecting directly to an 802.1x-compatible switch. Multi-mode host access can prove to be insecure as it allows for only one EAP-compatible host to successfully pass the authentication process, which could allow a rogue access point to slip by using the already authorized port with the previous user authentication.

If you need to use multi-host mode in 802.1x authentication, you should use it in combination with a port-security feature to additionally restrict and permit hosts by their MAC addresses to connect into the switch port. Using port-security features in catalyst switches is covered later in this chapter.

1. To enable multi host support:

   Switch3550(config-if)# **dot1x multiple-hosts**

2. To disable multi-host support and go back to single-host only:

   Switch3550(config-if)# **no dot1x multiple-hosts**

## *Viewing 802.1x Port Statistics*

To display the configuration and port statistics of 801.1x-configured ports, use the **show dot1x** command in main privilege EXEC mode. Figure 8.14 shows the **show dot1x interface fastEthernet 0/3** command on the Catalyst 3550 switch configured in the previous examples. The port in Figure 8.14 is currently marked as "Unauthorized," which means that all traffic is blocked except 802.1x EAP protocol. When the client is plugged in and authenticates successfully, it will change to "Authorized" mode in which the switch will allow the client to communicate freely through the port.

**Figure 8.14** show dot1x Command

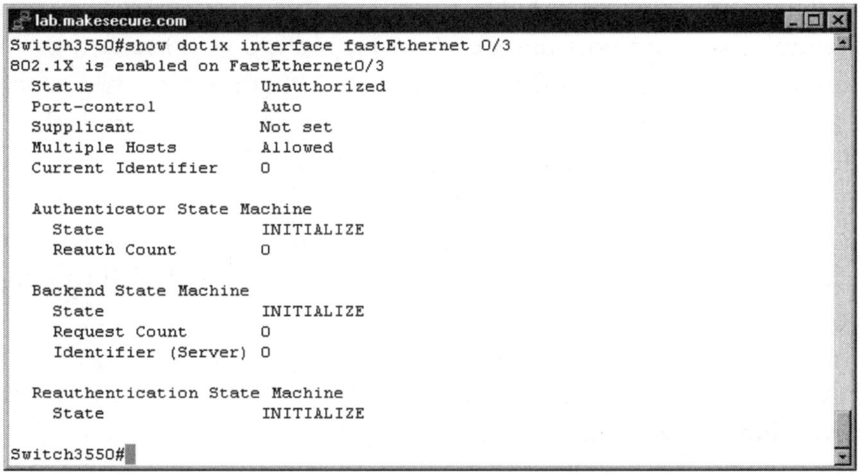


```
Switch3550#show dot1x interface fastEthernet 0/3
802.1X is enabled on FastEthernet0/3
  Status                Unauthorized
  Port-control          Auto
  Supplicant            Not set
  Multiple Hosts        Allowed
  Current Identifier    0

  Authenticator State Machine
    State               INITIALIZE
    Reauth Count        0

  Backend State Machine
    State               INITIALIZE
    Request Count       0
    Identifier (Server) 0

  Reauthentication State Machine
    State               INITIALIZE

Switch3550#
```

Fore more details on how to configure 802.1x support on Catalyst 3550 switches, refer to the documentation at www.cisco.com/univercd/cc/td/doc/product/lan/c3550/12119ea1/3550scg/sw8021x.htm.

802.1x is a dynamic protocol that can be used to accomplish mobility on wired and wireless networks. Ports can be dynamically configured and unconfigured on a per-user basis. Not only is this protocol used to restrict or permit devices based on its credentials, but it can also be used to configure per-user access lists or VLAN assignments based on individual user profiles that are stored on the authentication server.

## Detecting a Rogue Access Point from the Wired Network

Although several rogue access point detection and prevention techniques were covered in previous sections, there are still many techniques that can be used on a network to detect rogue access points. The best solution for detecting wireless rogue access points is using Cisco's centralized management solutions such as the WLSE. There may be network environments where you do not have a WLSE engine or you may have a limited number of Cisco-aware wireless devices that do not cover your entire risk area. Manual sniffing and detection can only go so far, and must be physically performed in local areas.

Detecting rogue access points from a wired network is one of the alternative techniques used to detect unauthorized access points connected into corporate networks. Detection from a wired network works by scanning the user-wired LAN and identifying rogue devices that differ from a valid user's workstation signature. This signature is based on port numbers. For example, port 80 is used on Web servers to serve Hypertext Transfer Protocol (HTTP) content to users, and is also used on most wireless access points to provide administrative access. Other ports such as Telnet (23) and SSH (22) are also opened by default on most access points for user administration. How does this help us? Normal user workstations should not have these ports open, so when performing a large port scan of your user LAN, detecting ports such as 80 or 23 may indicate that the device running these ports may be a rogue device, not a user workstation.

There are many network scanners that can be used to scan large user LANs. One of the more popular scanners is called NMAP. NMAP is a free network scanner available at www.nmap.org website.

### Detecting a Rogue Access Point with a Port Scanner

Figure 8.15 shows a typical user LAN with a large number of Windows workstations. The scanner is automatically run against these large user networks to detect any unique devices that do not match the typical workstation signature.

**Figure 8.15** Port Scanning User LAN

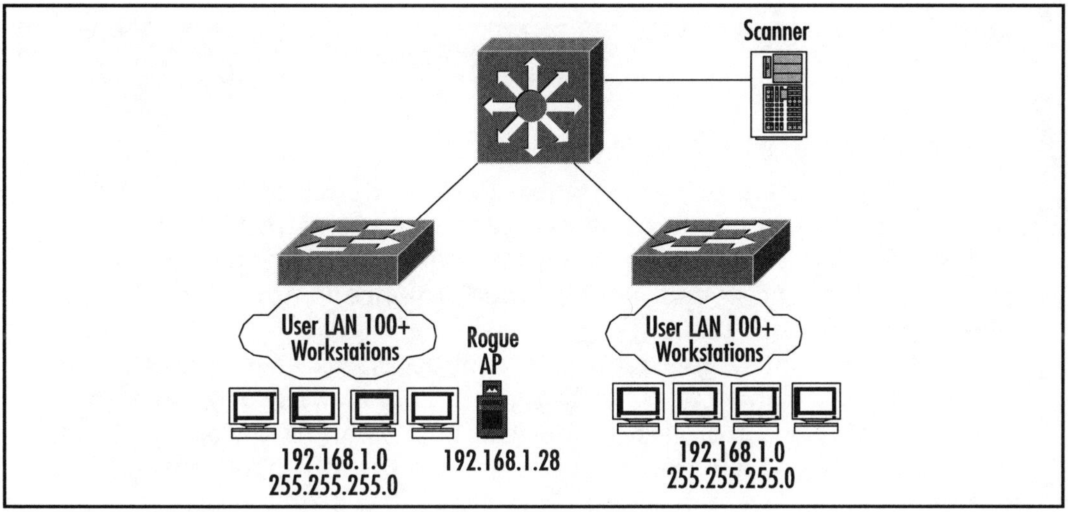


Figure 8.16 shows the actual scanner in action, scanning the 192.168.1.0 network. Notice that it found a device with IP 192.168.1.28 that has port 80, 22, and 23 open. It also detected that ports 22 and 23 are running on a Cisco device. By checking your list of Cisco network devices, you determine that 192.168.1.28 is not one of yours and thus you red flag it as a possible rogue device connected into your protected user LAN.

**Figure 8.16** NMAP Scanner in Action

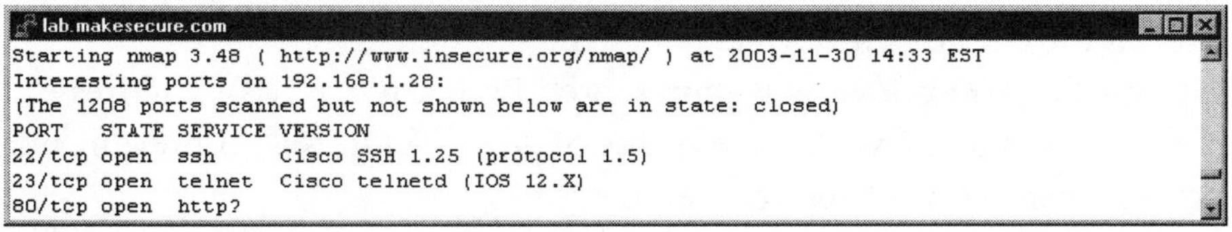

```
lab.makesecure.com
Starting nmap 3.48 ( http://www.insecure.org/nmap/ ) at 2003-11-30 14:33 EST
Interesting ports on 192.168.1.28:
(The 1208 ports scanned but not shown below are in state: closed)
PORT    STATE SERVICE VERSION
22/tcp open  ssh     Cisco SSH 1.25 (protocol 1.5)
23/tcp open  telnet  Cisco telnetd (IOS 12.X)
80/tcp open  http?
```

Once you detect a possible rogue access point on your network, you should track down its physical location by logging into the user switch and performing a reverse lookup on the detected IP to find its relative MAC address. Knowing the MAC address of the rogue device allows you to look through the MAC address table on the user switch and find out which port the device is connected to. When you know the actual port, you can trace down the physical cable to the device or disable the port.

**Designing & Planning…**

**Extra Traffic and False Alarms**

A network port scanner must connect to every device on the user LAN it is scanning, creating extra network traffic that can introduce unwanted congestion and slow performance on the overall network. You must make sure that the overall network performance is not affected when performing network scans.

Port scanners also require a connection to each device's port number. Such a connection can trigger security alarms such as personal workstation firewalls or security devices such as an IDS. Make sure network scans are coordinated between the groups that need to be aware of them order to avoid confusion and unwanted problem tickets.

# Using Catalyst Switch Filters to Limit MAC Addresses per Port

Another technique for preventing rogue access points is successfully connecting to a wired network using switch port security. Switch port security uses security features on the catalyst switch to restrict connections to a port interface based on a configured list of allowed devices.

This list of allowed devices is represented by hardware MAC addresses. Each port must be configured with its own list of MAC addresses to prevent unauthorized devices from connecting to the port.

## MAC Addresses in Port Security

There are three different types of MAC addresses that can be configured in the port security feature on a catalyst switch. These are:

- Static MAC
- Dynamic MAC
- Sticky MAC

## Static MAC

Static MAC addresses must be manually configured on each device MAC address on switch ports that are allowed to connect. Configuring a static MAC address on an IOS Catalyst switch is accomplished using the **switchport port-security mac-address <MAC>** command. By default, you are only allowed to configure one static MAC address. If you have multiple hosts using the same port, you must increase the number of allowed devices with the **switchport port-security maximum <NUM>** command. If you try to configure more than one static MAC without first increasing the number of allowed MAC addresses on a port, you will receive an error message. Static MAC addresses are saved in a configuration file so that when the switch reboots, it does not lose its MAC port security configuration.

## Dynamic MAC

Dynamic MAC addresses are learned dynamically from connected devices. If a switch port is configured to allow a maximum of three devices, it learns the first three MAC addresses dynamically and stores them in the memory table. Dynamic MAC addresses are not saved in a configuration. When the switch reboots, all dynamically learned MAC addresses are reset. Dynamic configuration is generally not used to defeat rogue access points.

## Sticky MAC

Sticky MAC addresses use a combination of static and dynamic methods to configure its list. MAC addresses are learned dynamically, but they can also be saved in a configuration file as static. This becomes useful when you have a LAN of 200 plus users. You can dynamically learn all of 200 workstation MAC addresses and then turn them into a static MAC list. Sticky port security mode is accomplished using the **switchport port-security mac-address sticky** command in the IOS catalyst switch.

# Security Violation

A port security violation occurs when an unknown device that is not on a MAC address list tries to access the switch port. Cisco Catalyst supports three different configured actions you can take when violation occurs. Each switch port can use one of the following three settings:

- Protect mode
- Restrict mode
- Shutdown mode

## Protect Mode

When a violation occurs in Protect mode, the device that is trying to gain connectivity to the port on the switch is blocked and not allowed to connect and all pockets coming from the unauthorized device are dropped. When using Protect mode, no alert message is sent out to notify the administrator of the incident.

## Restrict Mode

Restrict mode is similar to Protect mode in that all packets from the unauthorized detected device are dropped. The difference between Restrict mode and Protect mode is that Restrict mode logs the incident. It can generate an SNMP trap-to-management station alerting the administrator of a violation. It can also send a syslog message and increase the violation counter on the switch port setting.

## Shutdown Mode

In Shutdown mode, the switch port shuts down when it detects an unauthorized device trying to connect to it. The switch sends out an SNMP-trap-to-management station or a syslog message and increases the port violation counter as it would in Restrict mode. When a port is shut down it must be manually re-enabled.

# Configuring Port Security in an IOS Catalyst Switch

Figure 8.17 shows our network topology. We have user A's and B's workstations connecting to a corporate LAN. We want to make sure that only those two workstations are allowed to connect to the LAN and no other device.

**Figure 8.17** Port Security Topology

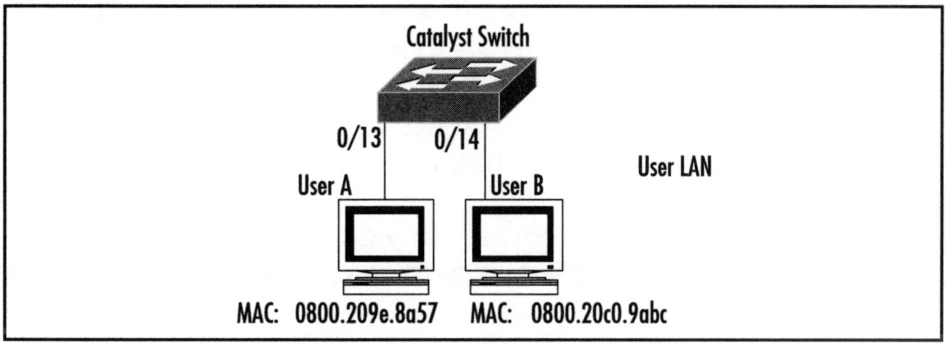


1. Configure port 0/13 on a catalyst switch with **static** MAC assignment and **restrict** violation mode:

```
Switch# configure terminal
Switch(config)# interface fastethernet 0/13
Switch(config-if)# switchport mode accesses
Switch(config-if)# switchport port-security maximum 1
Switch(config-if)# switchport port-security mac-address
0800.209e.8a57
Switch(config-if)# switchport port-security violation restrict
Switch(config-if)# switchport port-security
Switch(config-if)# end
Switch#
```

2. Configure port 0/14 on a catalyst switch with **sticky** MAC assignment and **shutdown** violation mode:

```
Switch# configure terminal
Switch(config)# interface fastethernet 0/14
Switch(config-if)# switchport mode accesses
Switch(config-if)# switchport port-security maximum 1
Switch(config-if)# switchport port-security mac-address sticky
Switch(config-if)# switchport port-security violation shutdown
Switch(config-if)# switchport port-security
Switch(config-if)# end
Switch# copy running-config startup-config
```

3. After configuring port security, verify your settings with show commands such as **show run**, **show port-security**, **show port-security address**, and **show port security interface**. Figure 8.18 shows switch **show port-security** and **show port-security address** commands. Notice that the port 0/14 MAC address was learned dynamically by using a sticky MAC list. Both ports show no violations thus far.

**Figure 8.18** show port-security Commands

```
lab.makesecure.com
Switch#show port-security
Secure Port  MaxSecureAddr  CurrentAddr  SecurityViolation  Security Action
                (Count)       (Count)        (Count)
---------------------------------------------------------------------------
     Fa0/13          1             1              0                Restrict
     Fa0/14          1             1              0                Shutdown
---------------------------------------------------------------------------
Total Addresses in System : 2
Max Addresses limit in System : 128

Switch#
Switch#show port-security address
          Secure Mac Address Table
-------------------------------------------------------------------
Vlan    Mac Address       Type                Ports   Remaining Age
                                                          (mins)
----    -----------       ----                -----   -------------
   1    0800.209e.8a57    SecureConfigured    Fa0/13        -
   1    0800.20c0.9abc    SecureSticky        Fa0/14        -
-------------------------------------------------------------------
Total Addresses in System : 2
Max Addresses limit in System : 128

Switch#
```

Now look at what will happen if user A using switch port 0/13 tries to unplug their workstation and use the cable to connect to a rogue access point or a personal hub instead (see Figure 8.19).

**Figure 8.19** Connecting Rogue Access Points

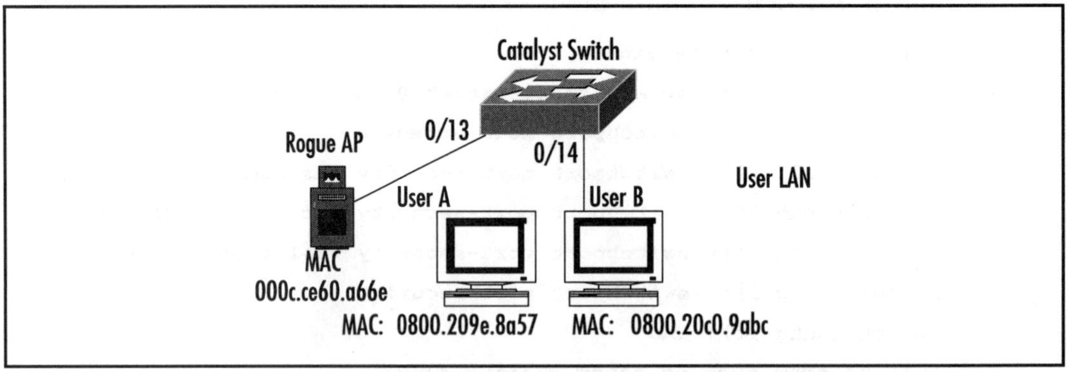

Port 0/13 goes down and up as the user unplugs and plugs the cable from their workstation to the rogue access point. As soon as the switch detects a new device that is not allowed on the configured MAC address list , it will act based on its port's security configuration. In the port 0/13 configuration the switch restricts and drops all packets coming from the unauthorized rogue access point, and sounds an alarm by sending an SNMP alert to the management station. See Figure 8.20 for the actual violation error messages taken from the switch when connecting a rogue access point to port 0/13.

**Figure 8.20** Violation Error Messages

```
lab.makesecure.com
Switch#
Switch#
Switch#
00:17:53: %LINEPROTO-5-UPDOWN: Line protocol on Interface FastEthernet0/13, changed state to do
wn
00:18:55: %LINK-3-UPDOWN: Interface FastEthernet0/13, changed state to up
00:18:56: %LINEPROTO-5-UPDOWN: Line protocol on Interface FastEthernet0/13, changed state to up
00:19:16: %PORT_SECURITY-2-PSECURE_VIOLATION: Security violation occurred, caused by MAC addres
s 000c.ce60.a66e on port Fa0/13.
00:19:26: %PORT_SECURITY-2-PSECURE_VIOLATION: Security violation occurred, caused by MAC addres
s 000c.ce60.a66e on port Fa0/13.
00:19:36: %PORT_SECURITY-2-PSECURE_VIOLATION: Security violation occurred, caused by MAC addres
s 000c.ce60.a66e on port Fa0/13.
00:19:46: %PORT_SECURITY-2-PSECURE_VIOLATION: Security violation occurred, caused by MAC addres
s 000c.ce60.a66e on port Fa0/13.
Switch#
```

Figure 8.21 shows the **show port-security interface fastEthernet 0/13** and **show port security** commands. Notice that the "Security Violation" number count from the previous example has increased from 0 to 60. This means that the rogue access point is trying to gain access but the switch keeps denying it.

**Figure 8.21** Security Violation Counter Increase

```
lab.makesecure.com
Switch#
Switch#show port-security interface fastEthernet 0/13
Port Security : Enabled
Port status : SecureUp
Violation mode : Restrict
Maximum MAC Addresses : 1
Total MAC Addresses : 1
Configured MAC Addresses : 1
Sticky MAC Addresses : 0
Aging time : 0 mins
Aging type : Absolute
SecureStatic address aging : Disabled
Security Violation count : 60

Switch#show port-security
Secure Port      MaxSecureAddr  CurrentAddr  SecurityViolation  Security Action
                    (Count)        (Count)        (Count)
-------------------------------------------------------------------------------
     Fa0/13           1              1              60             Restrict
     Fa0/14           1              1              0              Shutdown
-------------------------------------------------------------------------------
Total Addresses in System : 2
Max Addresses limit in System : 128

Switch#
```

# Summary

Throughout this chapter you learned different techniques for detecting and preventing rogue access points. Rogue access points are unauthorized access points that are installed by employees without approval from the IT and security departments, or by an intruder trying to trick valid users in order to gain sensitive information.

It is important to mitigate the threat of rogue access points as they have the ability to demolish an entire security architecture. A single rogue access point installed by an employee can create a back door into a corporate network, ignoring and bypassing all border security architecture such as firewalls and IDS engines.

Several manual techniques exist for detecting rogue access points, such as wireless sniffers and wired network scanners. A wireless sniffer has the ability to detect wireless radio signals and access points within a reachable signal area that can then be compared against a list of known authorized access points. If the new MAC address of an access point that does not match any of authorized access points is detected within the radio signal area, it must be tracked down as a possible rogue access point. Several wireless application tools exist that can measure the signal strength of an access point, which can then be used to locate an access point's physical location.

Wired network scanners can be used to scan user LANs to detect possible rogue devices. A rogue access point-scanned signature differs from a user's workstation. Signatures such as a workstation's open port numbers differ from access point port numbers. Ports such as 80 (HTTP) on a detected device or 23 (Telnet) can reveal that a device is something other than a user workstation.

Among these manual rogue access point detection techniques, Cisco offers a robust centralized detection solution. A management station WLSE can be used to control all Cisco-aware devices such as access points and wireless clients to perform automatic and periodic scans of radio signals. They then report any findings back to the central management engine that are then matched against a database of authorized access points.

Preventing rogue access points from connecting into a protected wired LAN is as important as its detection. Throughout this chapter you learned several techniques that can be used to prevent an unauthorized access point from connecting to a wired network. A good prevention technique should eliminate detection. Techniques such as 802.1x port-based authentication and MAC port security features can be used on Cisco switches to control which device can and cannot establish physical connection to the network.

It is important to note that rogue access points are most likely found in organizations where no wireless network is supported. Every network and security administrator needs to worry about rogue devices that can dismiss the entire implemented security measures.

# Solutions Fast Track

## The Problem with Rogue Access Points

☑ A rogue access point is an unauthorized access point installed by an employee without permission from the IT or Security departments.

☑ One rogue access point can dismiss an entire security architecture.

☑ Employees install rogue access points for their own benefit without realizing that they have created a back door to the corporate LAN.

## Preventing and Detecting Rogue Access Points

☑ The first step in protecting against rogue access points is having a security policy. A security policy should outline the rules against unauthorized wireless devices and employees must be educated about the policy.

☑ A wireless sniffer can aid in the detection of wireless access points throughout an area that can then be compared against a list of authorized access points.

☑ Cisco offers a centralized solution with a WLSE engine where all Cisco-aware wireless devices work together to detect possible rogue access points and report them to the central management station.

☑ Rogue access points can be detected from the wired network by using a network port scanner. Unlike a user's workstation, rogue access points usually have port 80 (HTTP) and 23 (Telnet) open for administration purposes.

- ☑ A port scanner can trigger false alarms and extra traffic on already congested traffic by scanning every device. Coordinated scanning should be performed to avoid confusion.

## IEEE 802.1x Port-based Security to Prevent Rogue Access Points

- ☑ The 802.1x protocol allows mutual authentication where the access point authenticates the user and the user authenticates the access point, to ensure that the user is connecting to a valid, not a rogue, access point.
- ☑ In 802.1x protocol, users are prompted for authentication credentials as soon as they plug their workstation into the switch port. Devices such as rogue access points that do not support such authentication will not be allowed to connect to the wired port.
- ☑ A third-party authentication server that supports RADIUS protocol is required to store all user credentials and perform the actual authentication. The access point or the catalyst switch is used as a proxy server between the authenticating client and the RADIUS server.

## Using Catalyst Switch Filters to Limit MAC Addresses per Port

- ☑ Port security in catalyst switches allows you to restrict devices that can physically connect to the port by their MAC addresses.
- ☑ The three types of MAC addresses in port security feature are static, dynamic, and sticky.
- ☑ When an unauthorized device connects to a secured port, a violation occurs. The three configurable reactions to a violation are protect, restrict, and shutdown modes.
- ☑ In shutdown violation mode the port is shut down and requires the administrator to manually bring it back up.

# Frequently Asked Questions

The following Frequently Asked Questions, answered by the authors of this book, are designed to both measure your understanding of the concepts presented in this chapter and to assist you with real-life implementation of these concepts. To have your questions about this chapter answered by the author, browse to **www.syngress.com/solutions** and click on the **"Ask the Author"** form. You will also gain access to thousands of other FAQs at ITFAQnet.com.

**Q:** Can 802.1x protocol be implemented in wired network devices just as it can be in a wireless network?

**A:** Yes. 802.1x was originally designed for a wired network. Cisco supports 802.1x protocol in many of their hardware devices and is expected to expand its collection in the future. There are, however, a few differences in how 802.1x is implemented that you should review prior to its implementation.

**Q:** I have over 1,000 users that I need to move over to a Cisco ACS RADIUS server. What is the best way to do this?

**A:** Cisco ACS supports multiple external databases. If your user database is one of them, you can link it up to provide user authentication. Refer to Cisco's ACS product details for a list of supported external databases.

**Q:** Is a once-a-week detection scan sufficient to mitigate the threat of rogue access points?

**A:** Rogue access point detection and awareness should be performed constantly to protect your networks from intruders.

**Q:** What is the best way to protect against rogue access points?

**A:** The best protection should be a combination of multiple techniques. One of the best techniques is to use Cisco's centralized WLSE engine solution along with its wireless devices to perform continues rogue access point detection scans.

# Frequently Asked Questions

The following Frequently Asked Questions, answered by the authors of this book, are designed to both measure your understanding of the concepts presented in this chapter and to assist you with real-life implementation of these concepts. To have your questions about this chapter answered by the author, browse to www.syngress.com/solutions and click on the "Ask the Author" form. You will also gain access to thousands of other FAQs at ITFAQnet.com.

**Q:** Can 802.1x protocol be implemented in wired network devices that [illegible] wireless networks?

**A:** 802.1x was originally designed for a wired network. [illegible]

**Q:** [illegible] that I need to move over to a Cisco ACS RADIUS server. What is the best way to do this?

**A:** Cisco ACS supports [illegible] database. [illegible] ACS [illegible] external databases.

**Q:** Is it often a good [illegible] rogue access points?

**A:** Rogue [illegible]

**Q:** What is the best way to prevent rogue access points?

**A:** The best protection should be a combination of [illegible] techniques. One of the [illegible] centralized WLAN rogue [illegible] wireless devices [illegible] scans.

# Chapter 9

# Wireless LAN VLANs

## Solutions in this Chapter:

- Understanding VLANs
- VLANs in a Wireless Environment
- Wireless VLAN Deployment
- Configuring Wireless VLANs in IOS
- Broadcast Domain Segmentation
- Primary (Guest) and Secondary SSIDs
- Using RADIUS for VLAN Access Control

☑ Summary

☑ Solutions Fast Track

☑ Frequently Asked Questions

# Introduction

Virtual local area networks (VLANs) represent the logical separation of physical LANs. A VLAN allows you to split up your physical network devices such as Cisco's switches and access points into different virtual local area networks (LANs) in which each VLAN takes on its own unique characteristics. Up until now only one group policy has been supported by Cisco's wireless access. Due to this one access point/one policy limitation, any wireless client group not compatible with the main policy settings have to use their own separate compatible access points. With the introduction of VLANs into a wireless network, you can define multiple compatible group policies such as voice and data groups that allow you to use one access point for all of your unique wireless client groups. VLANs can also be characterized and used to represent a group of devices on different physical LAN segments that can then communicate with each other as if they are on the same physical LAN.

WLANs have been widely used in the wired LAN industry since the 1990s. The VLANs proven support of scalability and cost savings are required by network administrators and architects in LAN network deployments.

Cisco has taken VLAN technology from wired LANs and its standards and incorporated it into its wireless devices to offer some of the advantages of a WLAN such as scalability, security, and per-VLAN policy, thus making wireless networks more scalable and appealing to corporations and more cost effective.

This chapter reviews the basic workings of the VLANs used on wired networks. It covers specific protocols and functions that make up a VLAN and its technology. You will learn how VLANs help overall network design scalability, security, availability, and cost savings.

Wireless VLAN deployment and configuration differs a bit from a wired LAN. This chapter takes a closer look at these differences and similarities. You will learn how to deploy and configure VLANs in wireless networks using Command Line Interface (CLI) in IOS and using a Web browser.

This chapter covers broadcast domain segmentation and its advantages for overall performance on the LAN and WLAN networks. A broadcast domain benefits from performance and also from certain security aspects.

You will review techniques on how users are assigned into allowed VLANs by using Service Set Identifier (SSID) and Remote Authentication Dial-In User Service (RADIUS). You will learn the differences between the two and how to configure them.

VLAN support is found mostly in multi-group support designs. It allows for differentiation in policy such as security or Quality of Service (QOS) among multiple devices such as wireless users using laptops, IP phones, or personal digital assistants (PDAs). The introduction of VLANs in wireless technology makes wireless networks more intercompatible with wired networks and more appealing to corporations.

# Understanding VLANs

This section reviews VLANs and some of their standard protocols such as VLAN Trunking Protocol (VTP) and trunk ports.

VLANs are incredibly flexible due to their logical rather than physical implementation. Logical implementation can be used to split one physical switch device into multiple Layer 2 segments representing different domain groups (see Figure 9.1). Logical separation is done by configuring VLANs. These different Layer 2 segments may then further span across multiple switches allowing one or more segments to coexist in different geographical areas (see Figure 9.2).

**NOTE**

**Layer 2 segmentation is synonymous with multiple VLANs. By creating VLANs on a switch, you logically separate them into multiple Layer 2 domains.**

A VLAN is a logical separation of a LAN. One separated segment is restricted to interact between the other segments unless a Layer 3-aware device such as a router is used to route and restrict traffic between them. In Figure 9.1, all three computers are connected to the same switch: two are configured in VLAN A and one is configured in VLAN B. Communication between VLANs A and B is restricted, even though they are connected to the same switch.

**Figure 9.1** VLAN Logical Segmentation

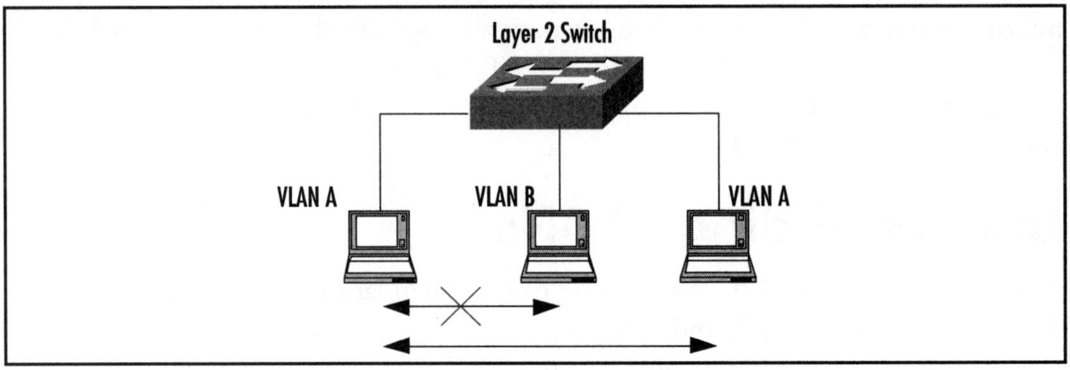


**Figure 9.2** VLANs across Multiple Switches

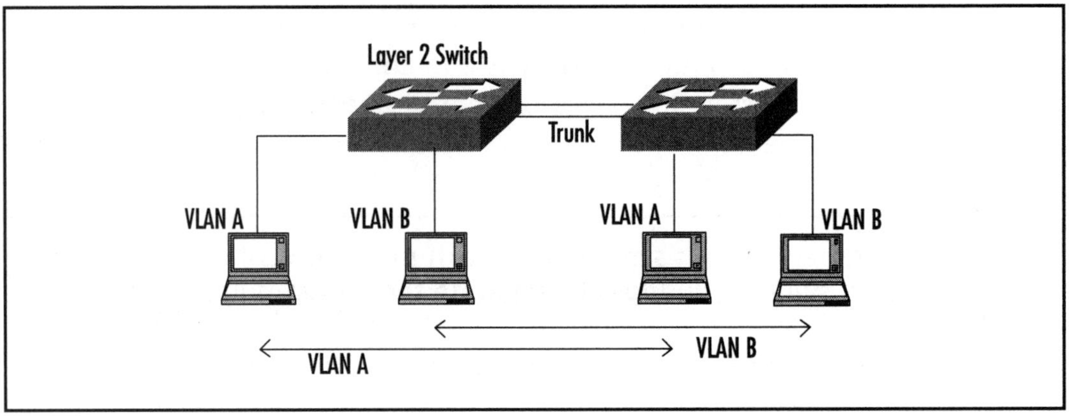


Before VLANs were introduced, two separate switches were required in order to create two separate LAN segments. If you needed to separate segments into more than two segments, it came with a heavy price tag. By creating VLAN segments, you also separate each logical LAN segment into its own broadcast domain. (Broadcast domains and their benefits are covered later in this chapter in the "Broadcast Domain Segmentation" section.) Other benefits of logical separation are per-VLAN compatibility requirements such as a security policy. For example, you may need to separate employees into multiple groups throughout multiple LAN locations and restrict specific access between them. Different VLANs can have different QOS policies configured. A voice-critical application

VLAN may have a higher priority QOS set over a user VLAN group. A slight delay in sensitive Voice over IP (VoIP) traffic may be unacceptable and should be prioritized over user traffic such as Internet browsing, which can bare some delay during congestion.

VLANs are usually associated with different IP subnetworks. For example, all devices in the same VLAN usually belong to the same subnet, which differs from other VLAN subnets. You need a router or Layer 3-aware device to route between these VLAN subnetworks. In Figure 9.3, the PC configured in VLAN A needs to pass through the router (Layer 3) in order to reach VLAN B. New devices used today to route between VLANs include Layer 3-aware switches. Layer 3-aware switches have the capability to route and thus eliminate the need for an external router. In Layer 3-aware switches, the router is built inside the switch itself.

**NOTE**

The International Standard Organization (ISO) has created a layered model called the Open System Interconnect (OSI) model. The purpose of the OSI model is to describe and define each layer in the network system. The seven OSI layers are: (7) Application, (6) Presentation, (5) Session, (4) Transport, (3) Network, (2) Data Link, and (1) Physical.

A network switch works on Layer 2 (Data Link) of the OSI model. It uses frame data such as MAC addresses to direct traffic. A network router works on Layer 3 (Network) of the OSI model. It uses packet data such as IP addresses to direct traffic.

Cisco has combined both switch (Layer 2) and router (Layer 3) technology so that one device can both switch and route. These are called Layer 3-aware switches. Layer 3-aware switches include routing capabilities and can be used to route between VLANs.

**Figure 9.3** Using a Router To to Route Between between VLANs

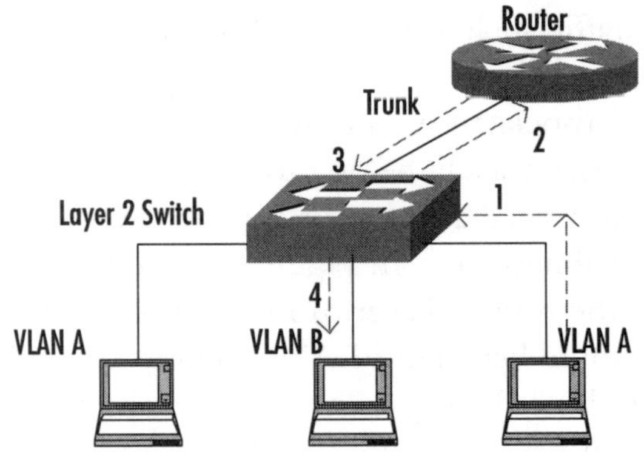


By default, every port on a Cisco catalyst switch is assigned to VLAN 1. VLAN 1 is called the native VLAN. In most devices VLAN 1 poses configuration limitations such as that it cannot be deleted. For this reason, when configuring VLANs try to avoid adding devices to VLAN 1, as it can cause potential network issues and security leaks. VLANs may be numbered from 1 through 4096 on a Cisco catalyst switch.

## Configuring & Implementing...

### VLAN Numbers

Although VLANs 1 through 4096 are supported in Cisco devices, they are further grouped and restricted.

- **VLAN 1** The default VLAN for every port on a switch.
- **VLANs 2 through 1001** These VLANs are used for Ethernet and can be deleted or created at the will of the network administrator.
- **VLANs 1002 through 1005** These VLAN IDs are used for Fiber Distributed Data Interface (FDDI) and Token Ring in Cisco devices. They cannot be deleted and are restricted.
- **VLANs 1006 through 4096** The VLANs in this range are called the extended VLANs. Additional requirements must be enabled in order to use extended VLANs.

## VTP in a Wired Network

VTP is a Layer 2 management protocol that allows the administrator to create, delete, or modify VLANs on a server switch, that are then populated throughout the client network of switches in the same VTP domain. VTP allows for better central management compatibility and avoids problems such as configuring inconsistencies of duplicate VLAN IDs, names, and types. In Figure 9.4, the administrator modifies VLAN A or B on the master switch after which new configurations will be propagated to its clients. Both the master and clients need to share identical VTP settings such as the domain name, version, and authenticating password to operate correctly.

**Figure 9.4** VTP in Virtual LAN

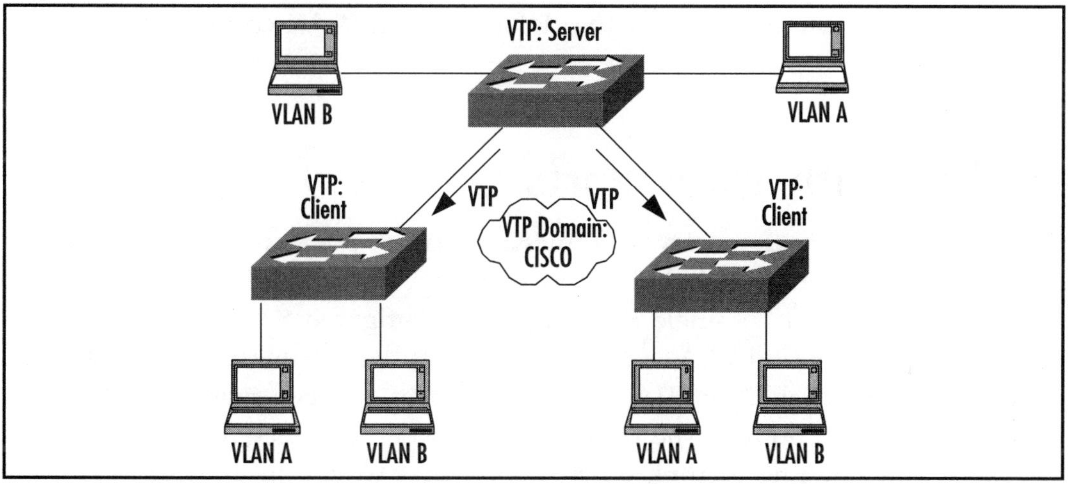


## VTP Modes

After configuring a VTP domain name, you must choose from the following three different VTP modes:

- **Server Mode** Server mode is the default configuration mode on a switch. When in server mode, the administrator has the ability to create, delete, or modify VLANs. Server mode devices advertise their VLAN configurations to other connected network devices.
- **Client Mode** When in Client mode, the administrator cannot create, delete, or modify VLAN settings.

- **Transparent Mode** A device in Transparent mode can create, delete, and modify VLAN settings. It does not, however, participate in a VTP domain nor does it advertise or sync up VLAN configuration from server-configured devices. A transparent configured device will, however, accept VTP advertisement from servers and pass it along.

**WARNING**

Use caution when adding new switches to your environment. Always check the VLAN revision number of the switch. Devices in the same VTP domain and configuration always use configurations from the highest revision number server switch. If you add a new default server configured switch with a high revision number, you risk having your other switches automatically copy the default switch VLAN configuration.

# Dealing with Trunk Ports

Trunk ports are ports that carry VTP messages. Trunk ports are also used to send multiple VLANs over a single link across two devices. In Figure 9.4, all switches are interconnected with a single Ethernet cable configured as a trunk port to carry both VLAN A and VLAN B information. Without trunk ports, you would need one Ethernet cable per VLAN to interconnect between switches. A trunk port is considered a point-to-point type link between switches, routers, and wireless access points. Without trunk ports, subnetworks would not be able to be partitioned across multiple switches or other devices. Trunks extend LAN domain where users connected to VLAN A can communicate to users on different switches that are also connected to VLAN A.

Two following Cisco-supported trunking encapsulation protocols are available for configuration:

- Inter-Switch Link (ISL)
- IEEE 802.1q

ISL is a Cisco proprietary trunking encapsulation protocol that supports spanning tree on a per-VLAN basis (PVST). PVST can block or forward based on the specific VLAN. It treats each VLAN separately. By treating each VLAN on a per-VLAN basis, the administrator can load-balance traffic by forwarding different VLANs over different trunks. ISL supports Ethernet, FDDI, and Token Ring frames with up to 1000 VLAN configurations. One implication of ISL is that it includes 30 bytes of encapsulated header on top of the network packet and also includes an additional Frame Check Sequence (FCS) that is added at the end of the ISL-encapsulated frame. When an ISL header is added to an already large Ethernet packet (1518 bytes), the packet will go over its Ethernet standard allowed limit size and will ask to be fragmented into smaller packets, thus possibly lowering the overall performance and requiring fragmentation support. The FCS is used to ensure that the frame has not been damaged during transmission and that it is added at the end of every frame. With the additional FCS around the ISL encapsulation, the switch must check the frame twice, which in a switch environment should not have any noticeable impact, but may be more difficult on routers if used for trunking.

The IEEE 802.1q is the industry standard trunk encapsulating protocol. If you are interconnecting vendor devices other than Cisco, you must use 802.1q to have the compatible standard. IEEE's 802.1q is also capable of supporting PVST in the same Cisco environment as ISL. However, unlike ISL, it only supports Ethernet frames. The header overhead in 802.1q is much smaller than in the ISL protocol, as it only includes 4 extra bytes. Further, the 802.1q does not include the extra FCS at the end of the frame. 802.1q simply recalculates the existing FCS of the frame after it adds its 4 extra bytes.

# VLANs in a Wireless Environment

VLANs in a wireless environment have the same purpose as in a wired network. They are used to separate devices into groups of specific services. Whether trying to accomplish better security, performance, or scalability, a VLAN can help.

In a wired network you statically configure VLAN settings on the switch port that the user is connected to. In a wireless network there are no cables or ports that users must connect to. Therefore, you need a mechanism to separate and identify wireless users or devices belonging in different VLANs. This fundamental of identifying to which VLAN user a device needs to be mapped to in a wireless environment is accomplished using SSIDs. A different and more secure way of identifying a user or device VLAN assignment on a wireless network is with the use of a RADIUS server (covered later in this chapter in "Using RADIUS for VLAN Access Control" section).

SSIDs are used to map to unique VLAN IDs to help the access point recognize and connect users to its proper VLAN assignment. SSIDs are not used for security purposes, rather their purpose is to separate users into groups so that access points can recognize and match an individual device into the properly configured VLAN. After separation is recognized and the user is mapped to the proper VLAN, the device must pass the VLAN-configured security policy (that may mandate for EAP, WEP, or MAC authentication) before the user is allowed to fully use its mapped VLAN on the wired side.

Cisco access points have the ability to support up to 16 different SSIDs; therefore, you can configure up to 16 different VLANs. Each VLAN can have its own policy, allowing the network administrator to use one access point to configure up to 16 different groups of devices, each with its own unique compatible service policy. In the past, without VLAN support you would need a new separate access point for each different group with unique policy settings.

## Per-VLAN Settings

As mentioned previously, an access point has the ability to support up to 16 VLANs that can be configured with different policies and restrictions. These individual policies may include:

- **Authentication Type** Open, Shared, and EAP.
- **MAC Authentication** In, Open, Shared, or EAP.
- **Maximum Associations** The maximum number of clients allowed to be connected at one time.
- **Encryption Key** Every VLAN must have a unique WEP key that is used for broadcast and multicast traffic. Broadcast domain segmentation is discussed later in this chapter.
- **Enhanced Message Integrity Check (MIC)** Used for WEP packet verification.
- **Temporary Key Integrity Protocol (TKIP)** A WEP per-packet keying mechanism.
- **Filters** Allows different filters per VLAN.
- **QOS** Allows for a different Class of Service (COS) priority per VLAN.

By allowing for such dynamic and specific settings on a per-VLAN basis, you can support multiple unique groups of users and devices on one access point, as shown in Figure 9.5. As shown in the drawing, a wireless PDA, a wireless IP phone, and different groups of users can all use the same access point but have unique configuration policies and restrictions to fit their compatible requirements. Each wireless client device must have a proper SSID configured in order for the access point to recognize its VLAN assignment. If you have two VLANs configured with identical security authentication policies but a unique restrictive access policy exists for each VLAN, you must ensure that the client device from one VLAN does not change its assigned SSID in order to be mapped by the access point into the other VLAN. To mitigate the threat of unauthorized VLAN hopping between wireless client devices, use a RADIUS server to monitor and assign a VLAN SSID other than relying on the clients to have the proper configured SSIDs. RADIUS servers and their use in VLAN assignments are covered later in this chapter in "Using RADIUS for VLAN Access Control" section.

**Figure 9.5** Multiple Devices Connected to One Access Point

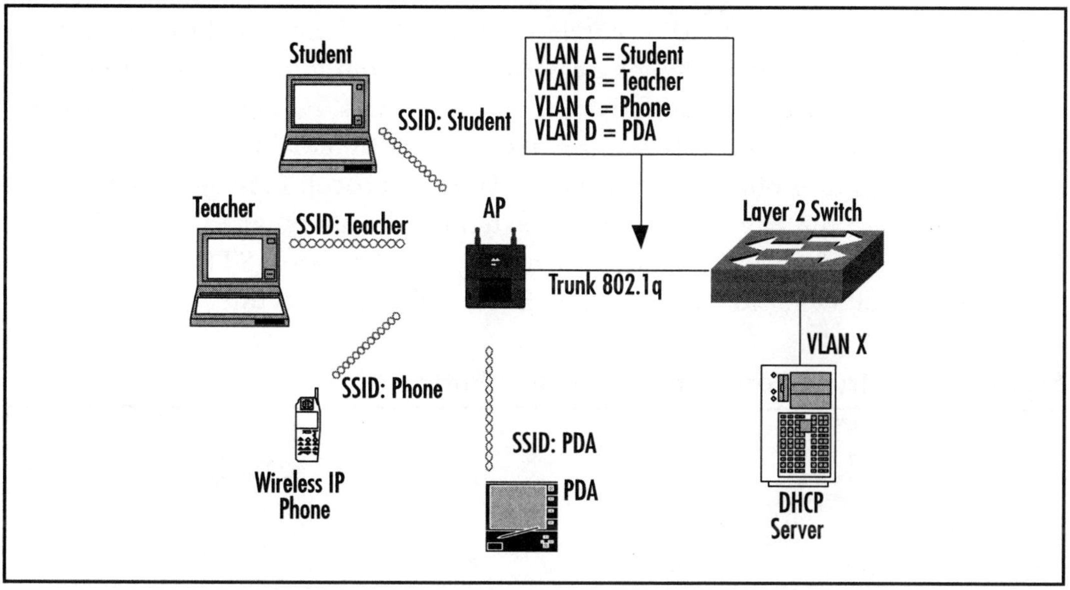

## VTP in a Wireless Network

The dynamic VTP protocol used to manage dynamic VLAN assignment settings is not supported in wireless access point devices, but is supported in wired-switch network devices. Wireless devices work as stub devices and require manual static VLAN configuration from the wired and wireless network. Currently, the server/client VTP relationship is not supported in wireless devices.

## Trunk Ports

Trunk ports are used to carry multiple VLANs over a single communication line. This allows the access point to support and map multiple devices from a wireless environment into a wired LAN using just one interface. In Figure 9.5, the trunk is configured between the access point and the switch using 802.1q encapsulation. It allows for multiple wireless VLAN devices to map into wired VLAN domains.

### Trunk Ports between Bridges

Trunk ports can be configured in wireless bridges to extend wired or wireless LANs across two or more different areas. A trunk is configured on both the radio and Ethernet interfaces for transferring multiple VLANs. Figure 9.6 shows a trunk implementation between two wireless bridges, supporting multiple VLAN communications over 802.1q trunk encapsulation protocol. This allows wireless networks to have the same advantages to extend and scale as wired networks. Additionally, wireless bridges support Spanning Tree Protocol (STP) as regular wired bridges, allowing for scalable loop-free wireless architecture.

**Figure 9.6** A Trunk Port across a Wireless Bridge

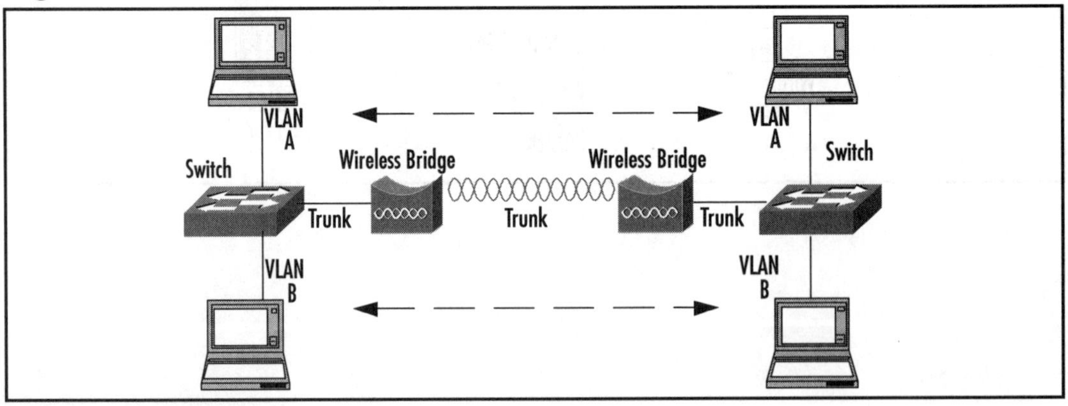

Keep in mind that, just as in a wired network, if VLAN A wants to talk to VLAN B on its local or extended LAN, you will need a Layer 3-aware device such as a router to route between VLANs.

# Wireless VLAN Deployment

Wireless VLANs are supported in Cisco 1200,1100,350,340 access points and Cisco 350,1400 wireless bridges. The supporting versions required are VxWorks Firmware release 12.00T or later and Cisco IOS release 12.2.4JA or later.

## Native VLAN

Native VLAN is the default configured VLAN. An access point must match the configuration of the native VLAN of the opposite connected device over the trunk port. If you have multiple access points or bridges in the same wireless LAN Extended Sub System (ESS) that need to communicate with each other, they must match their native VLAN configuration. All wired Cisco switches and bridges use VLAN 1 as their default-configured VLAN, therefore you will need to configure the same value on your access points to allow for interaction compatibility with connecting switches.

Any administration traffic such as Telnet, SSH, or RADIUS, directed from the access point or to the access point IP address, will be tagged using the native VLAN. An IP filter list is recommended to restrict and allow only authorized network administrators.

## Routing between VLANs

Figure 9.1 shows that two VLANs cannot talk to one another unless they use a Layer 3-aware device such as a router. Can you see a possible problem in Figure 9.5? Wireless users that are required to use services such as dynamic IP assignment will not be able to reach their dynamic host control protocol (DHCP) server located on VLAN X because they are configured in different VLANs. To overcome this problem, you must use a Layer 3-aware switch or external router that allows for inter-VLAN routing.

If you are using services such as RADIUS in your design to allow for EAP authentication, the access point is the one initiating authentication connection to the RADIUS server on behalf of the users. With that in mind, you must configure your RADIUS server to be in your native VLAN, or allow native VLAN-tagged traffic from the access point to reach RADIUS's VLAN using routing.

## Per-VLAN Filters

As discussed earlier, one of the many benefits of creating multiple VLANs for different classes of users is that you have the ability to apply different security filters. Filters restrict or allow communication to enter from a wireless LAN into a wired network.

In Chapter 7, you learned how to apply an access list filter into Ethernet and radio interfaces and what its benefits are. Each VLAN works as if it has its own interface. The main interface is split into several subinterfaces. Each subinterface can have specific settings configured to it.

**Figure 9.7** Per-VLAN Filters

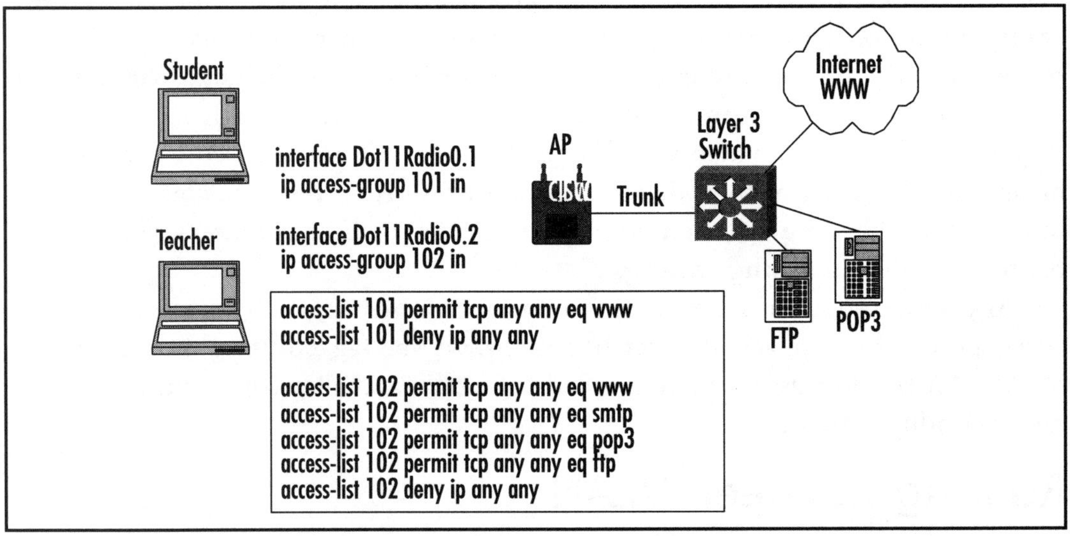


As shown in Figure 9.7, radio interface "0" has been split into "0.1" and "0.2" sub-interfaces in which unique access groups 101 and 102 have been applied. The dot "." in the interface represents a sub-interface. Sub-interfaces are used to accomplish multiple VLAN configurations with unique policies such as filters. According to the drawing, the Student group is bound to the interface with access list 101, which is only permitting HTTP access to be sent to the wired network from the Student wireless VLAN. The Teacher group with filter list 102 is allowed to access the World Wide Web (WWW), mail, and the File Transfer Protocol (FTP) on the wired network.

## Per-VLAN QOS

QOS policies can be applied on a per-VLAN basis. For example, you may want to give a higher priority to the wireless IP phone's traffic VLAN than to the student VLAN. VoIP may not work properly during congestion, therefore it is important to prioritize it. Or you may want to prioritize teachers' communication over students or guests when an access point becomes congested. You can specify different QOS policies on a per-VLAN basis where different groups are mapped. Detailed information on QOS and its implementation is found in Chapter 8.

## Per-VLAN Authentication and Encryption

Each VLAN can have its own authentication and encryption policy. You can support a guest network for your students without an authentication or WEP encryption policy, while at the same time use Cisco EAP authentication with WEP+TKIP policy for teachers. Also, your PDA devices may not support the same authentication policy as the teachers, and will require a compatible policy of its own. Just like filters and QOS, these settings are configured on per sub-interface VLAN basis.

If you need to support two different groups that share identical authentication types but require different restrictions on the wired network, you need a way to prevent the wireless user from simply changing its SSID in order to be mapped into the restricted VLAN after passing authentication. How to mitigate such a threat is discussed later in this chapter.

# Configuring Wireless VLANs Using the IOS: A Case Study

A local university has asked you to implement wireless technology for its faculty, students, and maintenance workers. After conducting a site survey and developing security policy requirements for the university, you have came up with a solution. Since students, faculty, and maintenance workers require different security policies and restrictions, your design will include three different VLANs in every access point. Refer to Figure 9.8 for part of the network topology map used in this scenario.

**Figure 9.8** School Topology

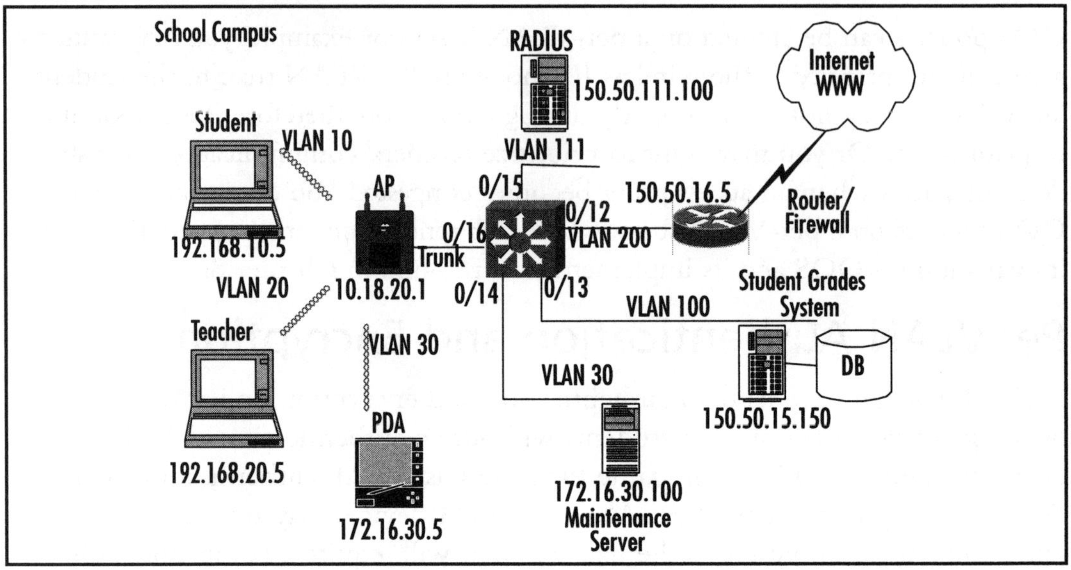


Faculty and students require strict per-user authentication in order to map into their specified VLANs. The faculty needs to access the Internet to surf the Web and access the student grades system to update records. Students will only be allowed to surf the Web. The maintenance workers will take advantage of the new wireless design to allow communication and report back to the maintenance server using wireless PDA devices. Refer to Table 9.1 for a listing of the requirements.

**Table 9.1** Table of Requirements

| | Teacher | Student | Maintenance |
|---|---|---|---|
| SSID | Teacher | Student | PDA |
| VLAN ID | 10 | 20 | 30 |
| Authentication | LEAP | LEAP | MAC/WEP |
| Encryption | Dynamic 128-bit WEP | Dynamic128-bit WEP | Static 40-bit WEP |
| Filter List | Yes #101 | Yes #102 | Yes #103 |

The following steps are required to configure the access point to support the network topology from Figure 9.8.

1. Configure SSIDs for all three groups and their authentication types. The first two authentication types for VLANs 10 and 20 are configured using the EAP method. VLAN 30 is authenticated using an open static WEP and MAC address list. (Refer to Chapter 7 for details on authentication types.)

```
AP# configure terminal
AP(config)# interface DotRadio 0
AP(config-if)# ssid teacher
AP(config-if-ssid)# vlan 10
AP(config-if-ssid)# authentication open eap eap_methods
AP(config-if-ssid)# authentication network-eap eap_methods
AP(config-if-ssid)# exit

AP(config-if) ssid student
AP(config-if-ssid)# vlan 20
AP(config-if-ssid)# authentication open eap eap_methods
AP(config-if-ssid)# authentication network-eap eap_methods
AP(config-if-ssid)# exit

AP(config-if) ssid pda
AP(config-if-ssid)# vlan 30
AP(config-if-ssid)# authentication open mac-address 798
```

2. Configure the native VLAN interface. You can configure the native VLAN only on the Ethernet interface to avoid administration access directly to the access point's IP address from wireless clients. We configure native VLAN on both the radio and Ethernet interfaces. The VLAN number is followed by the key word **native**.

```
AP(config)# interface DotRadio0.1
AP(config-if)# encapsulation dot1Q 1 native
AP(config-if)# bridge-group 1
AP(config-if)# exit
AP(config)# interface FastEthernet0.1
AP(config-if)# encapsulation dot1Q 1 native
AP(config-if)# bridge-group 1
```

3. Configure VLANs 10, 20, and 30 by creating sub-interfaces and enabling encapsulation on radio and Ethernet interfaces.

```
AP(config)# interface DotRadio0.10
AP(config-if)# encapsulation dot1Q 10
AP(config-if)# bridge-group 10
AP(config-if)# exit
AP(config)# interface FastEthernet0.10
AP(config-if)# encapsulation dot1Q 10
AP(config-if)# bridge-group 10

AP(config)# interface DotRadio0.20
AP(config-if)# encapsulation dot1Q 20
AP(config-if)# bridge-group 20
AP(config-if)# exit
AP(config)# interface FastEthernet0.20
AP(config-if)# encapsulation dot1Q 20
AP(config-if)# bridge-group 20

AP(config)# interface DotRadio0.30
AP(config-if)# encapsulation dot1Q 30
AP(config-if)# bridge-group 30
AP(config-if)# exit
AP(config)# interface FastEthernet0.30
AP(config-if)# encapsulation dot1Q 30
AP(config-if)# bridge-group 30
```

4. Configure WEP keys. Two 128-bit WEP keys will be used for VLANs 10 and 20. These two keys will be used for broadcast and multicast traffic only, as unicast WEP keys are dynamically derived on a per-user basis in the 802.1x EAP authentication process. There will be one static 40-bit WEP key to support the maintenance worker's wireless PDA compatibility. This key will be used for unicast encryption between PDAs and access points. For security purposes, the broadcast key is rotated in VLANs 10 and 20 using the **broadcast-key** command. (Refer to Chapter 7 for details on 802.1x and the rotation of broadcast keys.) Broadcast key rotation is currently only supported in LEAP authentication.

```
AP(config)# interface DotRadio 0

AP(config-if)# encryption vlan 10 key 1 size 128bit <key-here> transmit-key
AP(config-if)# encryption vlan 10 mode ciphers wep128
AP(config-if)# broadcast-key vlan 10 change <# of seconds>

AP(config-if)# encryption vlan 20 key 1 size 128bit <key-here> transmit-key
AP(config-if)# encryption vlan 20 mode ciphers wep128
AP(config-if)# broadcast-key vlan 10 change <# of seconds>

AP(config-if)# encryption vlan 30 key 1 size 40bit  <key-here>  transmit-key
AP(config-if)# encryption vlan 30 mode ciphers wep40
```

5. Configure filter lists to restrict the types of communication accepted from wireless groups into the wired network. Part of the campus requirement is to restrict student access to surf the Internet only and prevent them from accessing the student grades database. A unique filter list can be applied on each VLAN radio sub-interface. Filter lists and its configuration have been covered. (Refer to Chapter 7 for how to configure and apply filter lists to restrict or permit traffic.)
6. Apply identical configurations to the secondary radio interface. If you are using access points such as the 1200 series that support up to two installed radios such as 802.11b, 802.11g, or 802.11a, you must repeat all of the configurations for interface "DotRadio 1" as you configured for interface "DotRadio 0." This includes SSIDs and the creation of sub-interfaces, WEP keys, and IP filters.

**NOTE**

In a Web-based access point administrator graphical user interface (GUI) you can use the "Apply-all" button in the interface configuration menu to apply your settings to both of the installed radios at once. The 1200 series access point supports up to two installed radios including 802.11a, 802.11b, and 802.11g. Each radio can have unique or identical settings.

There is one big security concern and risk in the current school campus design called *VLAN hopping*. To mitigate VLAN hopping you must use a RADIUS server to authenticate VLANs. This concept is covered later in this chapter and must be considered in the design to prevent students from accessing their confidential records.

In Figure 9.8, a Catalyst 3550 Layer 3-aware switch with IP routing was enabled. Part of the switch configuration is displayed below for reference purposes. Notice that the trunk port configured under the FastEthernet 0/16 interface only allows VLANs required on the wireless side. Also, access filters can be configured that can be applied on the switch VLAN interfaces to restrict traffic communication between VLANs.

As shown in Figure 9.7, topology map Interface 0/12 is configured to be part of VLAN 200.

```
interface FastEthernet0/12
description Port to Internet Router
switchport access vlan 200
switchport mode access
no ip address
```

Interface 0/13 is part of VLAN 100 and is used as a student records server.

```
interface FastEthernet0/13
description Student Records Server
switchport access vlan 100
switchport mode access
no ip address

interface FastEthernet0/14
description Maintenance Server
switchport access vlan 30
switchport mode access
no ip address

interface FastEthernet0/15
description Radius Server
switchport access vlan 111
switchport mode access
no ip address
```

Interface 0/16 is used to establish a trunk port to carry multiple VLANs between the access point and the switch connection. The trunk is encapsulated with 802.1Q protocol to support access point compatibility. Further, VLANs that are allowed to pass the trunk with the **allowed vlan** command have been restricted. This will ensure that only required VLANs from the switch are allowed to cross to the wireless side.

```
interface FastEthernet0/16
description Trunk Port to AP
switchport trunk encapsulation dot1q
switchport trunk allowed vlan 1,10,20,30
switchport mode trunk
no ip address
```

Logical VLAN interfaces are assigned with IP addresses that are used for Layer 3 routing between the different VLANs. They are also used as default gateways for devices on each VLAN.

```
interface Vlan1
ip address 10.18.20.3 255.255.255.0
interface Vlan10
ip address 192.168.10.1 255.255.255.0
interface Vlan20
ip address 192.168.20.1 255.255.255.0
interface Vlan30
ip address 172.16.30.1 255.255.255.0
interface Vlan100
ip address 150.50.15.1 255.255.255.0
interface Vlan111
ip address 150.50.111.11 255.255.255.0
interface Vlan200
ip address 150.50.16.1 255.255.255.0
```

The default gateway is configured with the **ip route 0.0.0.0 0.0.0.0** command to match and route all traffic not directed to any specific VLAN on the switch, such as Internet browsing towards the Internet router.

```
ip classless
ip route 0.0.0.0 0.0.0.0 150.50.16.5
```

# Broadcast Domain Segmentation

Broadcast domain segmentation prevents broadcast and multicast traffic from one group from entering other segmented groups. One of the advantages of separating LANs with VLANs includes the creation of separate broadcast domains. A broadcast domain assures performance and scalability and prevents users from different logical domains from exchanging broadcast or multicast traffic.

## Traffic Types

There are many different traffic types. To understand broadcast domain segmentation and its benefits, a review of the three fundamental traffic types—unicast, broadcast and multicast— is required.

### Unicast

Unicast traffic is when traffic is directly directed to one individual. An example of this one-to-one relationship can be found at www.cisco.com. Only the client and the Web site are involved in receiving and sending traffic.

### Broadcast

In a broadcast network, the client sends only one packet that is directed to everyone. This is a one-to-all relationship. As shown in Figure 9.9, one server sends a broadcast message and everyone on the LAN receives it. A broadcast can be stopped by logically separating the LAN with VLANs, or by a Layer 3 device. Every client receiving broadcast messages must process them, thus lowering the overall performance of a LAN.

Broadcast frames contain the broadcast MAC address (ff:ff:ff:ff:ff:ff). When the switch sees this address it forwards it out of every LAN port. Servers make use of broadcast traffic to announce information services they provide. The broadcast domain is the group of logical network devices where broadcast messages are flooded.

**Figure 9.9** Broadcast Traffic

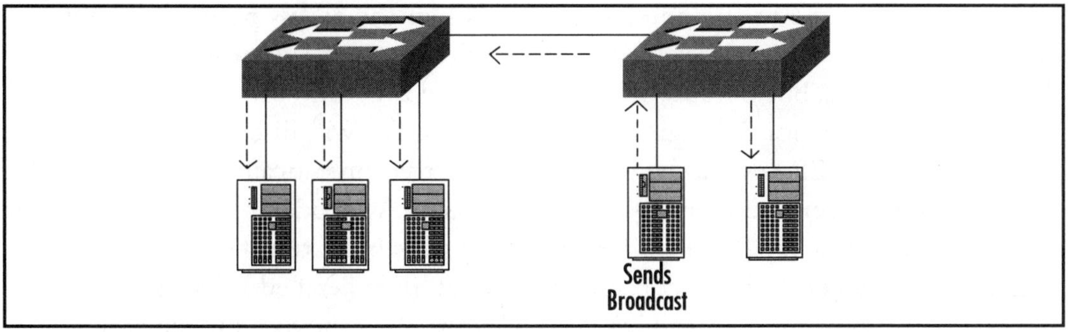


# Multicast

Multicast traffic is similar to broadcast traffic. Its intentional relationship is one-to-many. Unlike broadcast traffic, multicast traffic is sent to a set of users in a group. It is still forwarded like broadcast traffic, however, unlike in a broadcast environment where each device must process the broadcast, multicast devices that are not listening in to the specific multicast group being advertised will disregard the multicast traffic. How can multicast benefit your network? Unlike in unicast traffic where the server is required to send a copy of the same packet to every server it needs to communicate with, in multicast it only needs to send one multicast packet that will reach all of the users listening in on a specific multicast group.

# Broadcast Domain in Wireless

Now that you understand the different types of traffic and benefits of broadcast domain segmentation in wired networks, we will take a closer look at broadcast segmentation in wireless networks. In a wired network, VLANs are used to separate broadcast domains.

As discussed earlier, every packet traveling through the air can be seen by its neighbors as long as they are within signal reach. Thus, for this reason, every wireless client regardless of VLAN assignment will receive broadcast and multicast traffic. This is the difference between a wired and wireless network and their treatment of broadcasts in VLANs. You cannot prevent broadcast messages from reaching other VLAN segments on the wireless side because no physical separation (such as an Ethernet cable) exists.

Not being able to prevent broadcast messages from reaching multiple wireless users from different VLANs requires a workaround solution. Cisco wireless access point devices allow you to configure a different WEP key for the broadcast traffic for each unique VLAN. This WEP key differs from the unicast traffic key and is communicated to the wireless clients. When the access point sends out a broadcast message on its wireless side, other wireless users will still receive those broadcast messages, but because they do not share the same broadcast WEP VLAN key, devices not belonging to the same VLAN will discard them.

A broadcast WEP key can be dynamically derived or statically configured and is synced up between the users and the access point. A broadcast key shares some of the same ability as a WEP unicast key, including the ability to rotate when used with LEAP protocol within a configured timeout. (For detailed LEAP standards and their configuration refer to Chapter 7.) Figure 9.10 shows a broadcast sent from the access point to the teachers VLAN. Anyone not on this broadcast VLAN will still receive the packet but will discard the broadcast traffic because they do not share a common broadcast WEP key. If this was a wired network, the students would never receive the broadcast from the teacher, as it is in different VLAN.

**Figure 9.10** Wireless Broadcast

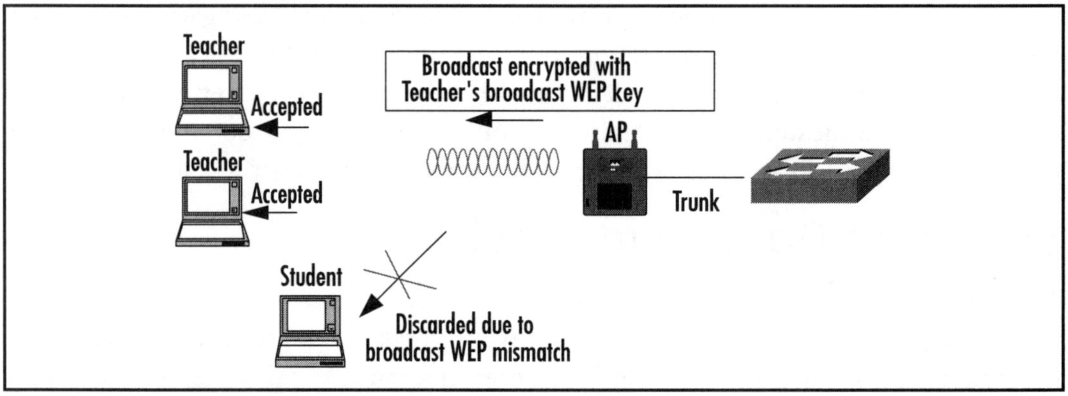


# Primary (Guest) and Secondary SSIDs

The SSID is a unique case-sensitive 32-alphanumeric character used in VLAN mappings. Up to 16 SSIDs can be configured. Hence, the limit of 16 VLANs is due to the limit of the SSID, as each VLAN must contain a unique SSID.

Each SSID can be configured with different policy characteristics. All SSIDs are active, allowing clients to use and pick from all 16 SSIDs at once. Some of the characteristics that can be configured based on a unique SSID include the authentication type, VLAN, guest mode, and RADIUS accounting among others. SSIDs are not used for any type of security purpose. SSIDs travel in cleartext through radio frequency (RF), which anyone can capture. Its use is purely to separate and recognize multiple group policy requests.

## Guest SSID

Guest SSID allows wireless users without any configured SSID to associate with the access point. Guest SSID is also used to broadcast unsolicited beacons from the access point to advertise its presence to the wireless community. The default configured SSID is **tsunami** on Cisco wireless devices and is enabled as a guest SSID. Broadcasting beacons should be disabled if you do not plan to use the access point for guest network access.

Only the primary SSID in multiple VLAN configurations can be included in broadcast beacons. Clients will still be allowed to request all different SSIDs from the access point, and the access point will respond with the proper SSID. However, in environments such as guest access networks where clients do not know the SSID, only one SSID can be used as the primary that is advertised in broadcast beacons. Figure 9.11 shows how to enable SSID as guest mode in a Web administration interface.

**Figure 9.11** Enabling Guest Mode SSID

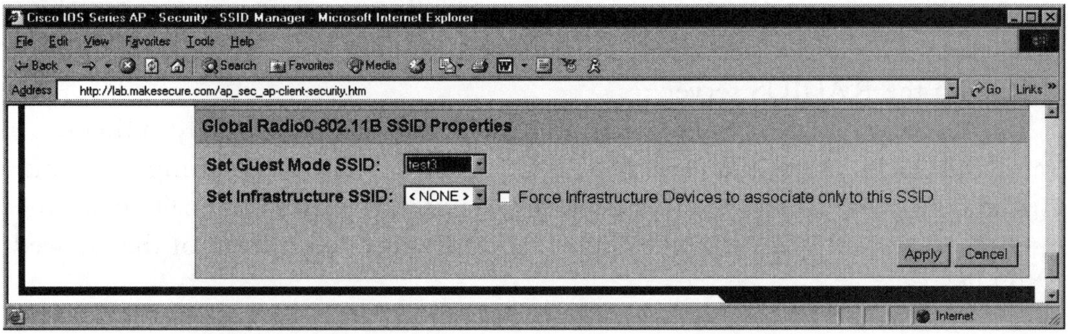


# Using RADIUS for VLAN Access Control

A RADIUS server can be used to control VLAN and SSID assignments. In previous examples, all SSIDs were configured on the access point. These SSIDs are used to map wireless devices into certain policy groups, whether it for security or QOS requirements.

Refer back to Figure 9.8 for the school campus implementation. Students and teachers share an identical authentication type. Both of these groups will require to authentication using LEAP protocol in order to be mapped to the proper VLAN base on the SSID. Further, each VLAN in this scenario has a unique access filter that allows teachers greater access on the wired network.

What will happen if a student decides to configure his adapter with the teacher's SSID? It will still be mapped to the VLAN with the LEAP authentication policy, which the student passes, after which the student will be mapped into the teacher's VLAN using the teacher's SSID. This is called VLAN hopping. VLAN hopping happens when an identical authentication type is used in multiple VLAN groups, where two or more groups can pass the identical authentication process.

To prevent VLAN hopping, a third-party service such as a RADIUS server is required to perform SSID or VLAN check assignments based on a user's record. It can be accomplished in two methods:

- RADIUS-based SSID
- RADIUS-based VLAN

In a RADIUS SSID-based verification, after a user successfully authenticates, the RADIUS sends a list of SSIDs that the user is allowed to use. If the SSID that user is using matches the list, the user is mapped into its proper VLAN. If it does not match, the user is not mapped into the VLAN and is disconnected. In Figure 9.12, student John Doe tries to access the network with teacher SSID. Student John Doe is rejected because it does not match the allowed SSID list profile on the RADIUS server.

In RADIUS VLAN-based verification, after the user successfully authenticates, RADIUS assigns the user to a VLAN based on its profile settings. For this method, no SSID is required to be sent by the user. RADIUS statically maps the user to its allowed VLAN. VLAN information is sent back instead of the allowed SSID list.

**Figure 9.12** Radius VLAN Control

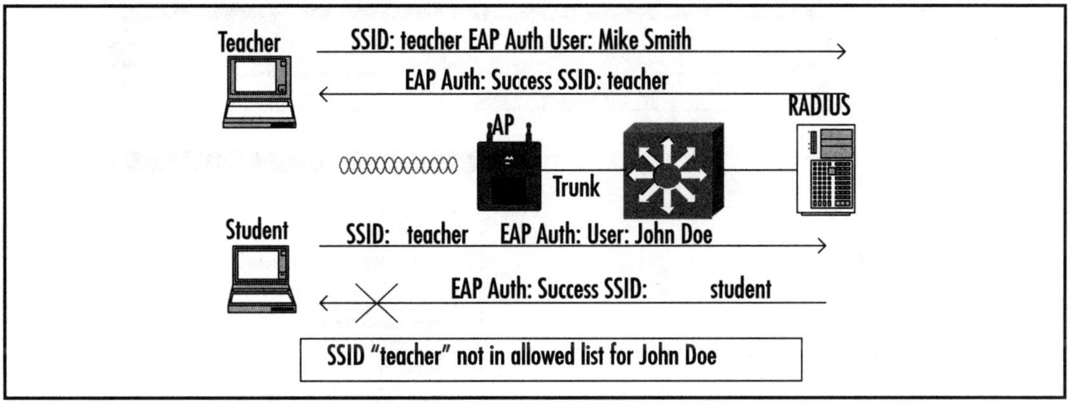


RADIUS verification can only be used when using protocols such as EAP for authentication. You need a per-user authentication method where VLAN restrictions can be verified. If you rely on static WEP key authentication only between multiple VLAN settings, each device or user can hop VLAN by changing the clients SSID.

# Configuring RADIUS Control

The RADIUS user attributes used for VLAN-based assignments are:

1. IETF 64: set this to "VLAN"
2. IETF 65: set this to "802" as the tunnel mode type
3. IETF 81: set this to the VLAN ID number you want the user to assume

For a RADIUS SSID control list configure the Cisco's 009/001 **cisco-av-pair**. This Vendor Specific Attribute (VSA) allows you to enter a list of SSIDs that the user is allowed to use in order to authenticate.

To enable and configure a list of allowed SSIDs in a Cisco ACS RADIUS server, go into User Settings and scroll down to "Cisco IOS/PIX RADIUS Attributes." Figure 9.13 shows the enabled attribute with the **ssid=student** value. This will prevent this particular student account from choosing any other SSIDs other than **student** and thus mitigate the VLAN hopping threat. You can add multiple allowed SSIDs per user.

**Figure 9.13** Configuring an SSID List in ACS

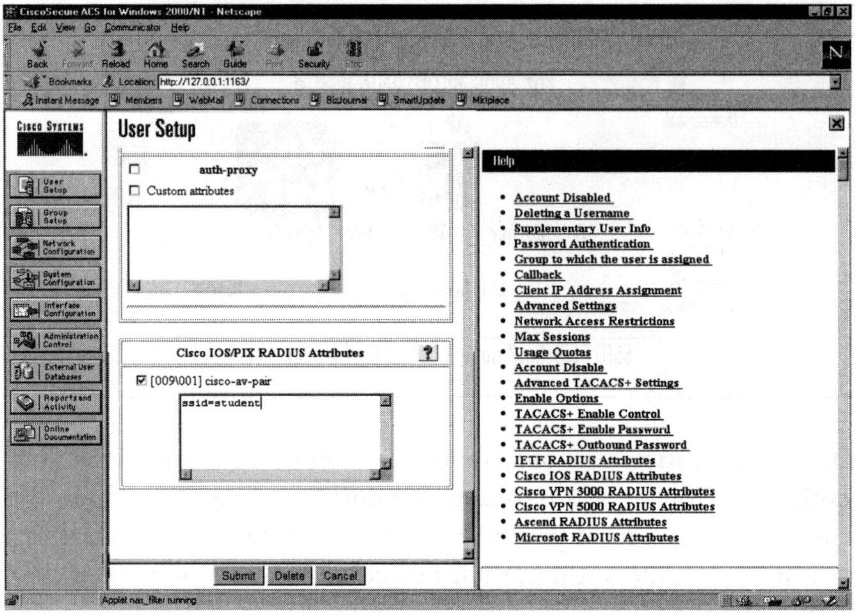

# Summary

Wireless VLANs and its technology bring wireless technology closer to acceptance with wired networks. Its integration ability with wired networks allows for scalable wireless solutions. This chapter covered the basic fundamentals of wired and wireless VLANs.

The creation of a VLAN allows you to logically separate network devices into multiple domains. These domains are unique because they work independently from other VLANs, which allow you to configure each of them with a unique characteristics policy. Some of the characteristics you can configure for per-VLAN in wireless network are an authentication method, security filters, and an encryption method. You can configure up to 16 different VLANs with unique characteristics. Each VLAN is represented by a unique SSID. In the past, without VLAN technology, there was only support for one static policy. This prohibited different devices or groups of users not compatible with the static policy from connecting. Administrators needed to purchase extra equipment if they wanted to support multiple groups with different policies.

Access points or bridges with multiple configured VLANs require to be connection to a trunk port to the wired side. A trunk port is an interface port configured to transfer more than one VLAN. Since there are multiple VLAN mappings from the wireless users, the access point or bridge needs a way to communicate with the wired network on all of the VLANs. A trunk port uses the 802.1Q encapsulation standard to communicate VLAN information between access points and switches. The access point must also include a native VLAN. The native VLAN tag is used for all traffic coming directly from the access point or to the access point IP address such as SSH, Telnet, or RADIUS administration.

When designing VLANs it is important to remember that you need a Layer 3-aware device such as a router to route between VLANs. For example, you may have a DHCP server that all wireless users need to connect to on the wired network regardless of the VLAN settings.

Each VLAN has its own broadcast domain. A broadcast sent from VLAN A cannot reach users on VLAN B on a wired network. Although this concept is applied to wired networks, it works differently in wireless communication. You cannot prevent a broadcast sent out through the air from reaching a group of users configured on a different VLAN. In wireless networks, you need to configure a unique WEP key for each VLAN to protect your broadcast and multicast traffic. When a broadcast is send out, it is encrypted with the VLAN broadcast

WEP key, so that only users belonging to that broadcast domain will recognize its content.

A RADIUS server is used to support and assign users to the proper VLAN. It is required when using an identical authentication policy in more than one VLAN. A RADIUS server prevents users from changing their SSID and hopping to an unauthorized VLAN. RADIUS works only when per-user authentication is used, such as in EAP. It verifies the user's SSID credentials that are used to map VLAN.

# Solutions Fast Track

## Understanding VLANs

☑ A VLAN is used to define the logical separation of a LAN network into multiple broadcast domains.

☑ Two configured VLANs cannot interact with each other unless they are routed with a Layer 3-aware device such as router.

☑ A trunk port is a configured interface port that allows for multiple VLAN communications. A trunk port is used between the access point and the switch to transfer multiple VLANs using the 802.1q encapsulation standard.

## VLANs in a Wireless Environment

☑ SSID is used to bind a wireless user to the proper VLAN.

☑ Each VLAN can have unique characteristics such as the authentication method, IP filters, and the encryption method. This allows one access point or bridge to support multiple groups of users and devices.

☑ A native VLAN is used to tag traffic originating and directed to the IP address of the access point or bridge, such as SSH and HTTP administration.

## Wireless VLAN Deployment

☑ Currently you can configure up to 16 VLANs. You can only configure up to 16 SSIDs on Cisco's wireless devices.

- ☑ VLANs are supported in VxWorks 12.00T release and IOS 12.2.4-JA release and later.
- ☑ Av 802.1q trunk port must be configured between two bridges supporting multiple VLAN communications.

## Configuring Wireless VLANs in IOS

- ☑ Multiple SSID configurations using the **ssid** command are configured under interface configuration mode.
- ☑ Radio and Ethernet interfaces are split into logical sub-interfaces to represent each VLAN configuration.
- ☑ You should always copy the running configuration and startup configuration to save your configuration in case the device reboots.

## Broadcast Domain Segmentation

- ☑ A broadcast domain segmentation prevents broadcast-directed traffic from one VLAN reaching other VLANs that are considered to be in a separate broadcast domain.
- ☑ Unlike in wired broadcast segmentation, in 802.11 all broadcasts are seen and processed by every wireless user, even if they are in a different VLAN.
- ☑ To overcome the differences between 802.11 and a wired network, a broadcast WEP key configuration is required per VLAN. This still does not prevent broadcasts from reaching every wireless user, but it allows only specific VLAN users who know the broadcast key to read its content.

## Primary (Guest) and Secondary SSIDs

- ☑ A guest mode SSID allows users without any SSID to associate to the access point.
- ☑ The access point sends out a guest SSID in its broadcast beacon to announce its presence.
- ☑ Only the primary (Guest) SSID can be used in beacons.

## Using RADIUS for VLAN Access Control

- ☑ RADIUS can be used to verify user VLAN mapping and prevent VLAN hopping using unauthorized SSIDs.
- ☑ RADIUS can either send a list of SSIDs to the user that they are allowed to use, or statically assign a user to a specific VLAN without the need for an SSID.
- ☑ You can only use RADIUS in a per-user authentication environment such as EAP.

# Frequently Asked Questions

The following Frequently Asked Questions, answered by the authors of this book, are designed to both measure your understanding of the concepts presented in this chapter and to assist you with real-life implementation of these concepts. To have your questions about this chapter answered by the author, browse to **www.syngress.com/solutions** and click on the **"Ask the Author"** form. You will also gain access to thousands of other FAQs at ITFAQnet.com.

**Q:** Why is there a limit on the number of VLANs in wireless networks?

**A:** Because each VLAN must be represented by a unique SSID and Cisco's wireless devices only support 16 SSIDs.

**Q:** Why use VLANs if I only have one group of users that share identical policies?

**A:** VLANs are an optional configuration, and even though you may not require one now, it allows for a future growing scalable environment without the extra expense.

**Q:** How can I block traffic between wireless users in the same VLAN connecting to the same access point?

**A:** You can configure Public Secure Packet Forwarding (PSPF) on a per-VLAN basis. PSPF prevents wireless clients in the same VLAN from communicating with each other through the access point.

**Q:** In multiple VLAN EAP authentication, do I need to make sure that all wireless VLANs can reach the RADIUS server through a Layer 3-aware device?

**A:** No. The RADIUS authentication that you provide for authentication is between you and the access point. The access point then initiates the RADIUS request to the RADIUS server on behalf of the client, using its native VLAN tag over the trunk port. The only requirement is that your native VLAN can reach RADIUS server.

# Chapter 10

# WLAN Quality of Service (QoS)

## Solutions in this Chapter:

- The Requirement for Service Quality
- QoS in General
- How Wireless Changes QoS
- WLAN QoS Design Guideline
- Configuring for Wireless QoS in IOS
- Using Existing Network QoS Configuration

☑ Summary

☑ Solutions Fast Track

☑ Frequently Asked Questions

# Introduction

As more Wireless Local Area Network (WLAN) deployments emerge in residential broadband and commercial networks, it is becoming more common for WLAN to support bundled Internet Protocol (IP) services. If a WLAN is going to implement telephony and other sensitive services, it must implement a Quality of Service (QoS) scheme. This chapter discusses setting up an Internet Operating System (IOS) to meet QoS requirements in a WLAN environment.

Throughout this chapter, MetroWiFi is used as a multi-service WLAN deployment example. By partnering with other Internet Service Providers (ISP), MetroWiFi uses WLAN technology to provide broadband network services to both residential and enterprise customers. Figure 10.1 depicts MetroWiFi's overall network architecture.

**Figure 10.1** MetroWiFi WLAN Network Architecture

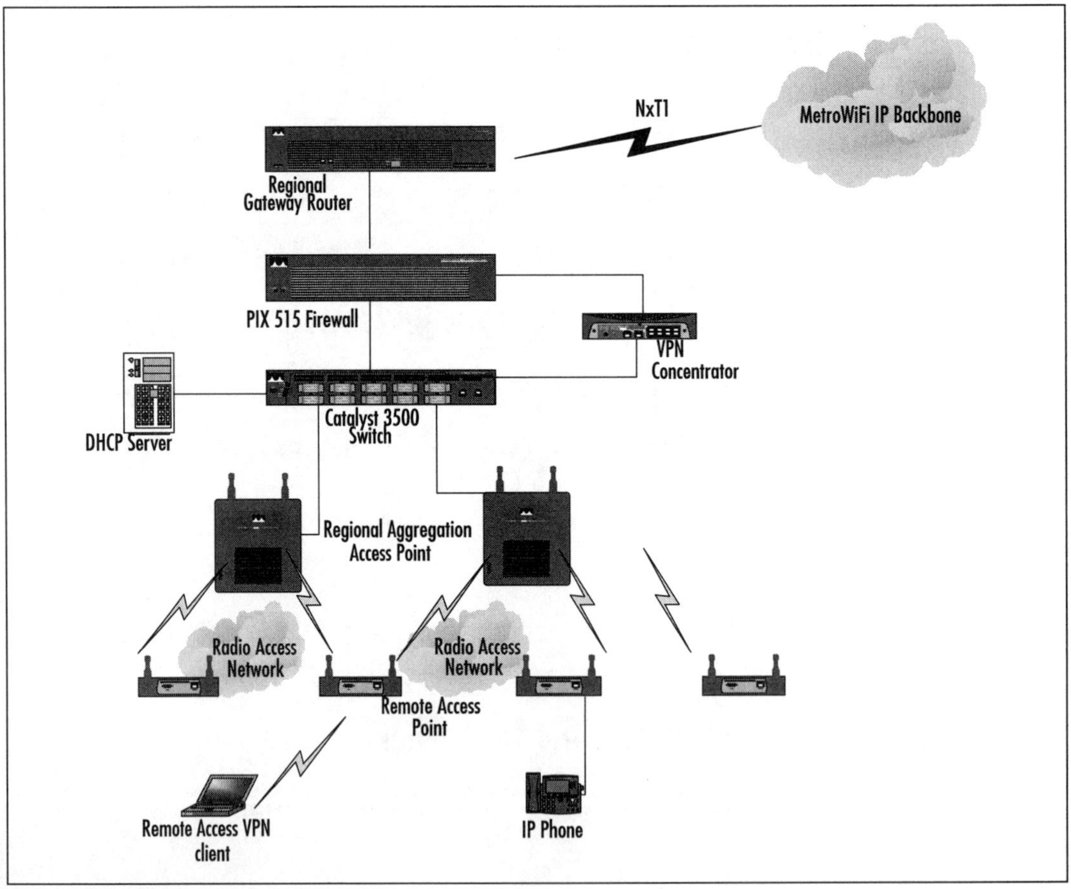

MetroWiFi provides four types of services:

- Premium Voice over IP (VoIPoWLAN)
- Premium II multimedia applications including Video (VoWLAN)
- Premium I mission critical applications such as Medical applications (MoWLAN)
- Standard best effort

The following sections use a MetroWiFi WLAN network as an example when discussing wireless QoS issues and illustrating Cisco IOS QoS configurations for WLAN.

# The Requirement for Service Quality

Studies of real-time applications and Transmission Control Protocol (TCP)/User Datagram Protocol (UDP) protocol behaviors have identified a set of parameters that are critical for providing a satisfying multi-service-over-IP network. The list includes available bandwidth, latency, jitter, packet loss, and network availability.

## Bandwidth

Bandwidth is measured by the ratio of the total number of packets transmitted vs. the time it takes to transmit them. Real time multimedia applications typically produce relatively constant bit streams and require a fixed amount of bandwidth end-to-end. On the other hand, most data applications tend to exhibit bursty traffic patterns and typically require an average bandwidth and a burst window.

## Latency

Latency (or delay) is the amount of time a packet takes to reach a destination endpoint after being transmitted from a source endpoint. One-way end-to-end delay is formulated using the following equation:

$$D_{total} = T_f + \Sigma(T_i + P_i + Q_{max})$$

- $T_f$ The packet formation time such as video or audio codec processing time.
- $T_i$, $P_i$ The transmission time and the propagation time.
- $Q_{max}$ The maximum queuing delay.

Among these four latency components, only is variable depending upon network load conditions; the rest are fixed values once a path is determined. Real-time applications normally have very low latency requirements. For example, the International Telecommunication Union (ITU) G.114 specifies that a one-way maximum delay should be less than 150 milliseconds (ms) for a toll-quality call. Voice over Internet Protocol (VoIP) requires that a round-trip delay be in the range of 0 to 300 ms.

## Jitter

A jitter is the variation of packet arrival times due to end-to-end delay variations when packets traverse from a source endpoint to a destination endpoint. A jitter is a random function of the network path that packets traverse, the traffic load, and the characteristics of scheduling algorithms on each hop along that path. A wide distribution of jitter can have a noticeable effect on time-sensitive applications such as VoIP.

## Packet Loss

Packet loss is the measure of packets that failed to be delivered to a destination endpoint after being transmitted from a source endpoint. Packet loss is normally expressed as the percentage of lost packets to the total packets transmitted.

Most real-time applications use UDP as the transport protocol, which does not tolerate packet loss as well as TCP. For instance, the VoIP application generally requires less than 1 percent packet loss to avoid quality degradation.

## Network Availability

Network availability is the ability to reroute the packet-forwarding path in the event of link or node failures. It is typically measured by either total network downtime for a specified period, or network recovery speed upon a network failure. The low latency and jitter requirement for real-time applications dictates high network availability. For VoIP service, the network recovery speed is expected to be in the range of 50 to 100 ms.

In summary, different applications require different value ranges for these parameters. This commonly takes the form of Service Level Agreement (SLA) specifications. For example, in the MetroWiFi example, VoIPoWLAN is characterized by small packets that are delay, jitter, and packet-loss sensitive. Premium II VoWLAN is delay sensitive. Premium MoWLAN is packet-loss sensitive and traffic can be bursty.

# QoS

QoS refers to the ability of a network to guarantee to meet application quality requirements for various services while transporting application traffic end-to-end. Achieving QoS is a fundamental requirement for any network that supports multiple services that have distinct quality requirements and traffic characteristics. For example, QoS helps manage time-sensitive voice and video application traffic to ensure that it receives higher priority and less delay and packet loss than best-effort data traffic.

**NOTE**

**Sometimes Class of Service (CoS) and QoS are used interchangeably. While for the most part they are the same, CoS emphasizes providing differentiated treatment to various services based on a set of pre-defined service classes. 802.1p Ethernet frame prioritization and Asynchronous Traffic Mode (ATM) traffic management functions are good examples of CoS.**

In practice, QoS commonly refers to a set of enforcement techniques that are used to cope with congestion and meet the service quality requirements of various applications in multi-service IP networks. In an IP-centric networking paradigm, transient congestion is inherent due to the packet multiplexing nature of IP networking and the packet buffer memory design commonly found in IP networking devices. Poorly designed networks with mismatched link speeds or equipment with sub-par performance can also cause persistent congestion. QoS provides enhanced and predictable network performance by managing and minimizing network congestion. For example, a stub network connected to a backbone by a link of fixed capacity is a common building block of many IP networks such as an enterprise branch office or a WLAN site network. If one user on the stub network wants to use a VoIP phone to place a call while all of the other users are busy downloading large File Transfer Protocol (FTP) files or playing online games, a QoS mechanism is needed to ensure that the VoIP user's packets get priority over the rest of the less time-sensitive packets on that link.

For wired IP networking, two main QoS schemes—Integrated Service (IntServ) and Differentiated Service (DiffServ)—have been proposed to try to provide guaranteed or differentiated treatments for IP applications. The IntServ model is based on the Resource Reservation Protocol (RSVP) and aims to provide end-to-end QoS guarantees by pre-establishing a suitable path for every flow in the network. The connection-oriented, per-flow nature of the IntServ model requires every network element in the path to maintain a reservation state of all of the flows. For a large IP network with thousands of hosts, the number of flow states can far exceed what today's network equipment can handle. Because of this inherent scaling limitation, QoS deployment based on the IntServ model in large-scale IP networks is not practicable.

In recent years, the DiffServ scheme has become widely accepted and supported by both wired and wireless service providers. It is also becoming the central element in the Cisco IOS QoS feature set. Figure 10.2 shows an end-to-end QoS architecture defined by DiffServ model (RFC2745).

**Figure 10.2** DiffServ QoS Architecture and Functional Components

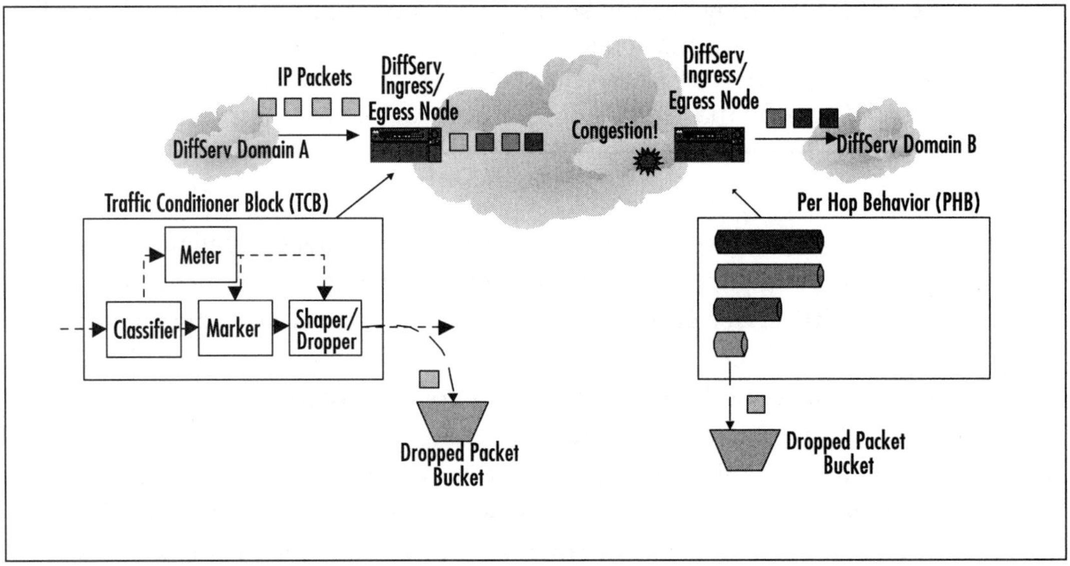


As shown in Figure 10.2, there are two key components in DiffServ architecture: Traffic Conditioner Block (TCB) and Per-Hop Behavior (PHB). TCB provides QoS functions such as traffic classification and rate limiting based on either IP type of service (TOS) bits or Differentiated Service Code Point (DSCP) bytes. In the Cisco IOS QoS toolkit, Committed Rate Limiting (CAR) is a good example of TCB. PHB represents a combination of queuing algorithms, queue management, and congestion control techniques. Cisco's Priority Queuing (PQ), Class Based Weighted Fair Queuing with Low Latency Queue (CBWFQ-LLQ), and Modified Deficit Round Robin (MDRR) are examples of PHB queuing techniques while Weighted Red Early Drop (WRED) is commonly used for PHB congestion control.

Unlike the IntServ scheme, the DiffServ approach is a CoS technology that relies on consistent traffic classification and queuing provisioning in each hop along an end-to-end traffic path to provide granular QoS support to different traffic classes. Each traffic class comprises all of the individual flows that require similar QoS treatment. By aggregating flows into a limited number of traffic classes, DiffServ obviates the scalability constraint that IntServ is facing with its per-flow based connection-oriented signaling approach.

In DiffServ QoS implementations, TCB is typically an integral part of the edge QoS strategy while PHB is essential for core QoS. A common QoS design based on a DiffServ scheme consists of a set of TCB and PHB instances mapped to a set of SLAs for supported traffic classes. For each traffic class, the corresponding TCB and PHB must be consistent across the entire network to ensure a desired end-to-end result. In other words, in an QoS-enabled network, all of the switches and routers must be configured to classify and recognize the various types of important traffic, and must treat those packets with sufficient priority.

# How Wireless Changes QoS

As shown in Figure 10.1, a WLAN network adds a new media edge network—a wireless network. The VoIPoWLAN, VoWLAN, and MoWLAN services inevitably drive new edge QoS requirements for this new sub-network. As a result, multi-service WLAN deployments need a QoS extension to the wireless portion of the WLAN network and must have seamless integration of QoS provisioning in both the wireless and wired networks. Figure 10.3 illustrates an integrated QoS architecture for WLAN.

**Figure 10.3** Integrated QoS Architecture for WLAN

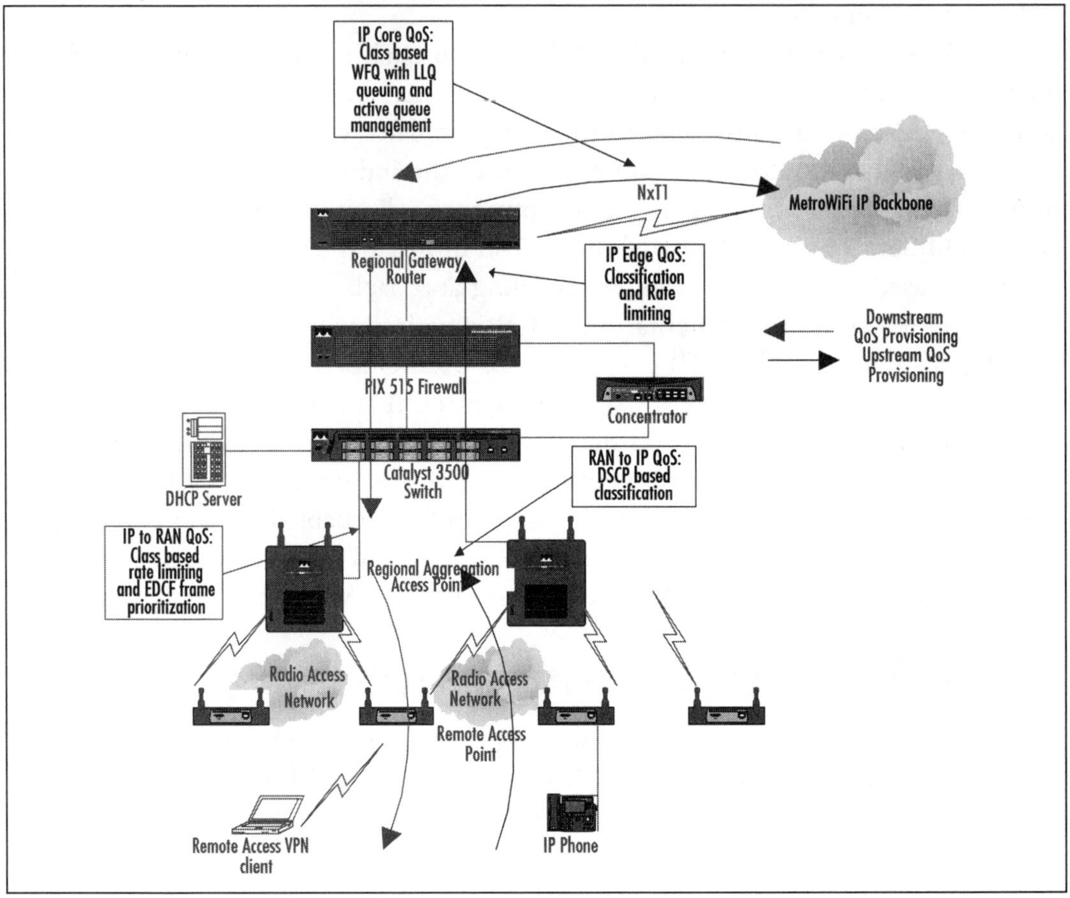


# Extending QoS Support to WLAN Wireless Network

As IP packets travel from the WLAN client through the wireless network to the IP backbone, the IP datagrams are first encapsulated into radio frames then decapsulated back to IP datagrams at the access point. IEEE 802.11 specifies the Distributed Coordination Function (DCF) as the Layer 2 media access control (MAC) mechanism for WLAN wireless networks. The DCF is composed of two main components:

- Interframe space (IFS)
- Random backoff (contention window)

IFS allows 802.11 to control which traffic gets first access to a radio channel once the carrier declares that channel to be free. High priority 802.11 management and control frames use Short IFS (SIFS) for the fastest delivery. All other data frames use Distributed IFS (DIFS) to gain radio access for transmission.

DCF uses a random backoff algorithm to manage collisions over a radio channel. The value of the random backoff timer is controlled by a contention window (CW), which is defined as ($CW_{min}$, $CW_{max}$). Initially, the backoff timer is a random number between 0 and $CW_{min}$. It decrements for every 20 μs during which the radio channel remains free. A data frame can be sent only when the available radio channel remains free after the backoff timer reaches zero. If, however, the data frame does not get sent before the initial random backoff timer expires, the WLAN client or access point will increment the retry counter and restart the process with a new random backoff window, doubled in size. This doubling in size will continue until the final window size equals $CW_{max}$. The retries continue until the maximum retries or time-to-live (TTL) have been reached.

DCF mainly defines MAC for WLAN wireless networks. Other than 802.11 management and control frames, DCF does not provide traffic prioritization directly for all other data frames transmitted over a WLAN wireless network. To fully support QoS at Layer 2, IEEE 802.11e added Enhanced Distributed Coordination Function (EDCF) as an enhancement to DCF. EDCF allows for the adjustment of the variable $CW_{min}$ and $CW_{max}$ random backoff values based upon traffic classification. Table 10.1 shows the different settings for the $CW_{min}$ and $CW_{max}$ of three common traffic class based on 802.11e.

**Table 10.1** Average Default $CW_{min}$ and $CW_{max}$ Values of Different Traffic Categories

| Category | $Cw_{min}$ | $Cw_{max}$ | Average Minimum | Average Maximum |
|---|---|---|---|---|
| Interactive Voice | 3 | 31 | 1.5 | 15.5 |
| Interactive Video | 15 | 63 | 7.5 | 31.5 |
| Best Effort | 31 | 255 | 15.5 | 127.5 |

With EDCF, all traffic waits for the same DIFS, but the $CW_{min}$ value used to generate the random backoff number is directly associated with the traffic classification. High priority traffic for real-time applications uses a small $CW_{min}$ so that it will be served faster than low priority best-effort traffic where a larger value of $CW_{min}$ results in a larger random backoff timer. As shown in Table 10.1, an interactive voice frame would have an average random backoff time of 30 μs, whereas the average random backoff time for the best-effort frame would be 300 μs. If interactive voice and best effort WLAN clients tried to transmit at the same time, the interactive voice frame would be transmitted first with a small delay.

The EDCF prioritization mechanism discussed so far is applicable to traffic leaving an access point to a WLAN client, or *radio downstream*. For *radio upstream* transmission, in current EDCF implementations, all WLAN clients are treated equally unless a client implements a proprietary mechanism of obtaining the channel more quickly than the others. That said, however, 802.11e indirectly provides radio upstream QoS support via a QoS advertisement mechanism by WLAN infrastructure devices such as access points. The 802.11e draft specifies a QoS Basis Service Set (QBSS) information element that access points can advertise to the WLAN clients indicating its key QoS parameters such as channel utilization and frame loss rate. This allows WLAN clients to select the best access point and radio channel to meet their QoS requirements. Cisco Aironet Software Release 12.00T for VxWorks access points and bridges and Cisco IOS Software release 12.2(4)JA for Aironet 1100 Series Access Points support both QBSS information elements and a proprietary mechanism based on Symbol Technologies, Inc. extensions to advertise QoS parameters. The latter can only be used for Symbol NetVision WLAN VoIP handsets.

# Integrating QoS in Wireless and Wired Networks

In the previous section you learned that EDCF allows frames in different traffic categories to receive different forwarding priority over WLANs. As shown in Figure 10.3, when an IP packet enters and leaves an aggregation access point, its QoS setting must be consistent. 802.11e also provides support for 802.1p and DSCP in access points, which greatly enhances the ability to leverage the existing wired QoS scheme and allows a seamless integration of QoS configurations over WLAN wireless and wired networks. In particular, 802.1p support in 802.11e enables transparent Layer 2 bridging between access points and aggregation Ethernet switches on a per-QoS service class basis.

Within aggregation access points, various classification schemes are available to mark IP packets or frames under a particular service category before the access point forwards them over a radio interface to a downstream WLAN client or over an IP interface to the upstream LAN switch and gateway router. This handoff between the access point and the connecting LAN switch serves as the hook for the integration of the WLAN wireless network QoS and the wired IP network QoS. Figure 10.4 illustrates the available classification schemes and the order they are applied in Aironet 1200 series AP:

**Figure 10.4** QoS Classification Precedence on an Aironet 1200 Access Point

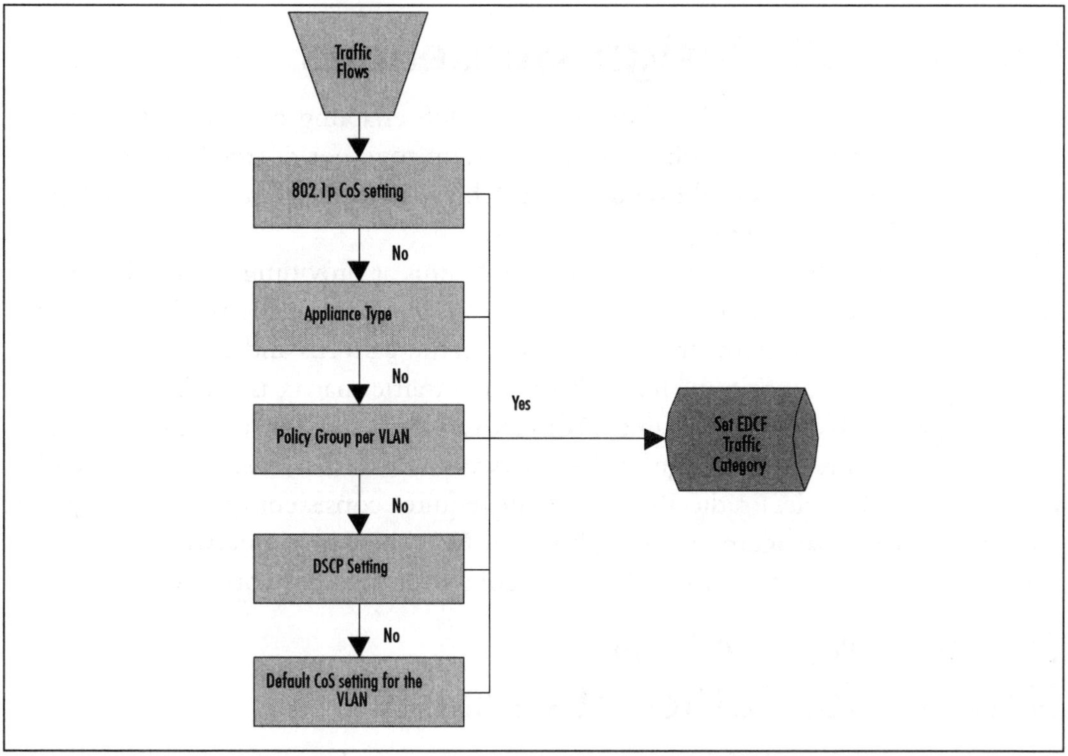


As shown in Figure 10.4, these classification rules are evaluated sequentially. When a match is made, the evaluation process stops and the corresponding QoS settings are applied. In particular:

- If a frame arrives at the access point with a CoS setting via IEEE 802.1p, select and use it.
- If a WLAN client has identifies itself as a particular CoS via per-appliance QoS mechanism such as Symbol VoIP device, select and use it.
- If a policy group is defined on a VLAN or interface, the CoS defined by that policy group is selected and applied to the matched traffic flow.

- If DSCP is marked within the IP packet, it is converted to the appropriate CoS as defined by the DSCP-to-CoS mapping table on the access point.
- If none of the previous mechanisms are viable, the default CoS setting for the VLAN is selected and used for all traffic.

# WLAN QoS Design Guidelines

Previous sections covered SLA requirements, QoS enabling components, and common QoS deployment schemes. To put them together correctly, however, a sound design principle should be followed. This is particularly important when designing a WLAN network.

You have also learned that, in essence, QoS aims at providing predictable network performance during congestion conditions. The predictable performance of a multi-service network stems from predictable traffic patterns and predictable treatments. Appropriate capacity planning based on a traffic matrix is the key to ensuring predictable traffic. Recently, WLAN service providers started deploying rate-limiting features at the edge of their networks to further ensure that network traffic stays predictable. Predictable treatment requires consistent queuing discipline and active queue management throughout the WLAN wireless network and the aggregation local area network (LAN) and core wide area network (WAN).

## Dimensioning WLAN Network for Sufficient Capacity

The basic principle of WLAN network planning is to determine the theoretical maximum number of video, voice, or data clients on a per-access point and per-Layer 2 subnet (i.e., per VLAN) basis, and to apply the appropriate traffic engineering model for over-subscription (such as Erlang tables for voice) to derive a practical deployment size limit for wireless application clients based on each individual application traffic type and usage pattern.

Take VoIP as an example to illustrate how to apply this principle. Different VoIP codec produces different VoIP packet size. For uncompressed G.711 with a 20 ms sampling rate, the actual VoIP packet size is calculated as follows:

24 byte 802.11b header + 20 byte IP header + 8 byte UDP header + 12 byte RTP header + 160 byte payload = 224 bytes

With the 20ms sampling rate, the total full-duplex bandwidth required is:

224 byte x 100 packet/second x 8 bits/byte = 179,200 bps

For a single 802.11b RF channel, the maximum theoretical throughput for a 256-byte packet is 2,596,588 bps. Dividing this number by 179,200 bps yields 14 as the maximum number of G.711 clients. To take other application traffic and 802.11b control traffic into consideration, the total reserveable bandwidth for VoIP is typically set to 40 percent to 60 percent of the total available bandwidth. This effectively reduces the maximum number of VoIP clients to 5 to 8.

In practical deployment, the actual number of wireless clients supported is typically the theoretical maximum number of clients multiplied by an over-subscription ratio appropriate to that particular application. For example, using the Erlang ratio, the total number of wireless VoIP clients supported per 802.11b RF channel is typically 3 to 5 times more than the theoretical limit.

Besides the maximum throughput limitation per 802.11b RF channel as shown, another limiting factor for the total number of wireless clients supported is the overall radio contention for a particular RF channel. The general rule is that no more than 15 to 25 802.11b endpoints should be deployed per access point to minimize the transmission delay.

**Designing & Planning…**

## Handling WLAN AP Congestion

Congestion on an access point is one of the most common reasons for packet loss and increased latency for wireless networks. To minimize it, the same wired QoS design principle can be applied. As shown in the previous section, appropriate capacity planning can achieve predictable traffic thus minimizing congestion as much as possible. This is especially true for radio upstream traffic since there is neither queuing nor policing among clients on the 802.11 wireless side of the access point.

However, the bandwidth available on the wireless link is largely dependent on the signal quality and signal strength. The actual throughput will degrade if either or both are reduced. Because of the uncertainty of signal transmission performance over a wireless medium, quantifying latency and jitter for 802.11 wireless networks is much more difficult than that for wired networks. A good approach is to conduct a

Continued

staging test and a trial deployment before large-scale deployment. This normally helps network designers to identify and resolve persistent congestion on access points commonly introduced by:

- RF impairments such as antenna and cable problems
- Incorrect software configuration problems
- Firmware and driver problems associated with a particular hardware model or software release

Rate limiting and queuing discipline during traffic congestion are key QoS enforcement techniques to ensure a bounded end-to-end delay and jitter WLAN network as a whole. For WLAN wireless networks, EDCF-based frame prioritization support can help minimize the delay and jitter for time-sensitive applications such as VoIP. Rate limiting can also play a vital role in ensuring that downstream traffic from aggregation LAN switches or routers to access points stays within certain thresholds thus decreasing the possibility of overloading the access point with excessive downstream traffic.

# Handling Roaming Introduced Delay

One common source of delay and jitter in wireless networks is roaming. The hand-off due to roaming among access points can introduce up to 400 to 500 ms of delay, which will have a very audible impact on the call quality. However, if roaming is between same Layer 2 segments (i.e., same VLAN), then the changes in the MAC forwarding tables can be taken care of by Inter Access Point Protocol (IAPP) between the two access points. Thus, from an IP perspective there is no change and packet flow will continue as normal. Although the actual time for Layer 2 roaming will vary depending on whether static wired equivalent privacy (WEP) or extensible authentication protocol (EAP) is being used, in general, the roaming delay can be reduced down to sub 100 ms range.

In the case of Layer 3 roaming among different IP subnets or roaming via mobile IP, the hand-off process is more complicated and no good solutions are available today to reduce the roaming delay. As a general rule, to support delay-sensitive applications, it is advantageous to design your WLAN network in such a way that the likelihood of Layer 3 roaming is minimized.

# Configuring for Wireless QoS in IOS

Cisco Aironet wireless product support QoS is based on the IEEE 802.11e Draft standard specification as of November 2002. Previously, QoS feature support came in two forms. For Aironet 1200 and 350 series access points and bridges, QoS is supported in VxWorks firmware release 12.00T or later, the real-time operating system used in these products; for the Aironet 1100 series, QoS features are available in native IOS mode since release 12.2(4)JA. However, as of this writing, Cisco IOS Software Release 12.2(11)JA or later supports both Cisco Aironet 1200 and 1100 Series access points. A new conversion kit including a Cisco Aironet conversion tool and conversion upgrade image file is also available to convert a Cisco Aironet 1200 Series access point from VxWorks operating system to Cisco IOS software.

The following section uses MetroWiFi's QoS design as an example, and covers configuration steps for its implementation on the access point. The Aironet 1200 series access point running 12.01T1 and IOS version 12.2(11)JA1 is used as an example.

## MetroWiFi integrated QoS Design

As shown in Figure 10.3, MetroWiFi implements an integrated QoS architecture that supports premium voice, premium II multimedia, premium I data, and standard best-effort services. Table 10.2 summarizes MetroWiFi's detailed QoS provisioning for both wireless and wired networks.

**Table 10.2** MetroWiFi QoS Design for its Bundled IP Services

| Services | Premium Voice | Premium II Multimedia | Premium I Data | Standard |
|---|---|---|---|---|
| Traffic Characteristics | Small packets, delay and jitter sensitive | Delay sensitive | Sensitive to packet loss | Best effort |
| Edge QoS Function (DiffServ Traffic Control Block) | Configure DSCP value "5"; drop out-of-profile traffic | Configure DSCP value "2"; drop out-of-profile traffic | Configure DSCP value "7"; accept out-of-profile traffic; provide an option to remark at the edge to re-classify traffic as standard traffic | Configure DSCP value "0"; drop out-of-profile traffic |
| Core QoS Function (DiffServ Per-Hop-Behavior) | | | | |
| Scheduling | Priority queuing to provide highest priority, and allow customers to choose priority between Premium Voice and Premium II; DWRR or CBWFQ to provide highest weight/deficit so that the queue is serviced often | Priority queuing to provide highest priority, and allow customers to choose priority between Premium Voice and Premium II; DWRR or CBWFQ to provide highest weight/deficit so that the queue is serviced often | Priority queuing to provide as medium priority; DWRR or CBWFQ to provide medium weight/deficit so that the queue is not starved | With priority queuing, provide as low priority; with DWRR or CBWFQ, provide low weight/deficit so that the queue is serviced last |

Continued

**Table 10.2** MetroWiFi QoS Design for its Bundled IP Services

| Services | Premium Voice | Premium II Multimedia | Premium I Data | Standard |
|---|---|---|---|---|
| Queue Management | Maintain input queue depth to be lower than average <50ms; maintain average output queue depth to reduce jitter | Maintain queue depth to be lower than average < 50ms; maintain low output queue depth | Maintain queue depth to be very high > 200 ms; maintain high output queue depth > 200 ms | Maintain input/output queue depth depending on availability of buffers; preferably high to prevent traffic loss |
| Congestion Control | Tail drop, avoid excessive buffering the traffic and allow higher layer to retransmit | Tail drop, avoid excessive buffering the traffic and allow higher layer to retransmit | RED or WRED, but keep the drops to absolute minimum | RED or WRED |

# Configuring EDCF Frame Prioritization Scheme

To support Premium Voice, Premium II Multimedia, Premium I Data, and Best Effort over WLAN wireless networks, the access point's EDCF configuration options (from the main menu, click **Setup** | **Protocol Filters** | **Quality of Service**) allows customizable $CW_{min}$ and $CW_{max}$ tuning to prioritize these services accordingly. By default, the access point supports eight traffic categories with a predefined set of EDCF $CW_{min}$ and $CW_{max}$ values. Figure 10.5 shows the default $CW_{min}$ and $CW_{max}$ values that are based on those proposed in the 802.11e Draft.

**Figure 10.5** Default EDCF $CW_{min}$ and $CW_{max}$ Values of Different Traffic Categories

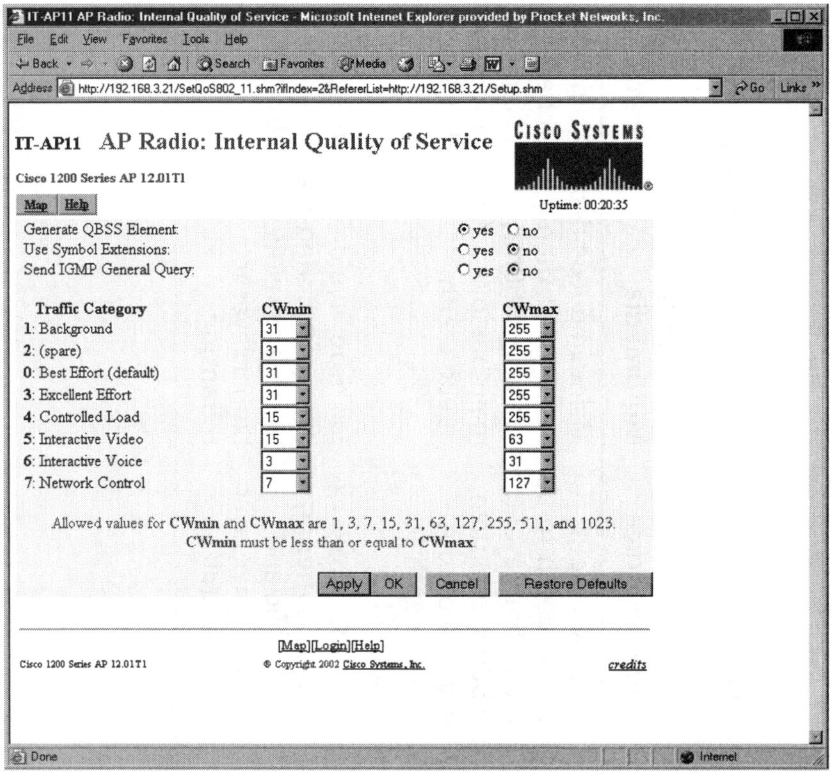


As shown in Figure 10.5, high priority traffic for real-time applications uses a small $CW_{min}$ so that they will be served faster than low priority best-effort traffic with a larger value of $CW_{min}$.

## Configuring & Implementing...

### Tuning $CW_{min}$ and $CW_{max}$ for High Priority Traffic

Changing these settings for production networks could have adverse affects and should be followed by significant tests specific to the applications in question to get the most optimal results. Like any priority-queuing configuration on wired network devices, setting up the appropriate priority parameters ($CW_{min}$ and $CW_{max}$ values in this case) is critical to avoid undesirable queuing behavior. For example, configuring a $CW_{max}$ value less than the $CW_{min}$ of another class may cause "starvation" of the other traffic class, as the worst-case random backoff of the preferred class would be better than the best-case random backoff of the less favored class. Thus, the preferred class gets served first all of the time. It should also be noted that the $CW_{min}$ and $CW_{max}$ settings are applied after the traffic has already been queued based on its traffic classification by the access point.

# Configuring Traffic Classification for EDCF Prioritization Scheme

As shown in Figure 10.6, an access point supports five classification schemes to map a particular type of traffic into one of eight priority categories. In a typical WLAN design, the access point aggregation Ethernet LAN switch sends IP packets to the access point with one of three QoS settings to mark their priority, namely VLAN (802.1Q) tag, 802.1p, and DSCP value. The access point provides configuration options (from the main menu, click **Setup** | **Protocol Filters**) for these three classification schemes.

**Figure 10.6** Traffic Classification Options

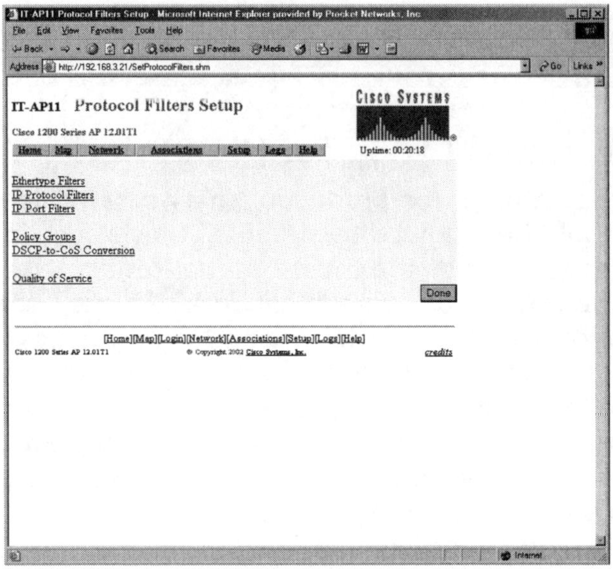


In the MetroWiFi example, as shown in Table 10.2, Premium Voice, Premium Multimedia, Premium Data, and Best Effort services are identified by the DSCP value. To configure them in the access point, from the main menu, click **Setup** | **Protocol Filters** | **DSCP to CoS Conversion**. Figure 10.7 illustrates the configuration menu for DSCP-based traffic classification.

**Figure 10.7** Configuring Traffic Classification based on DSCP

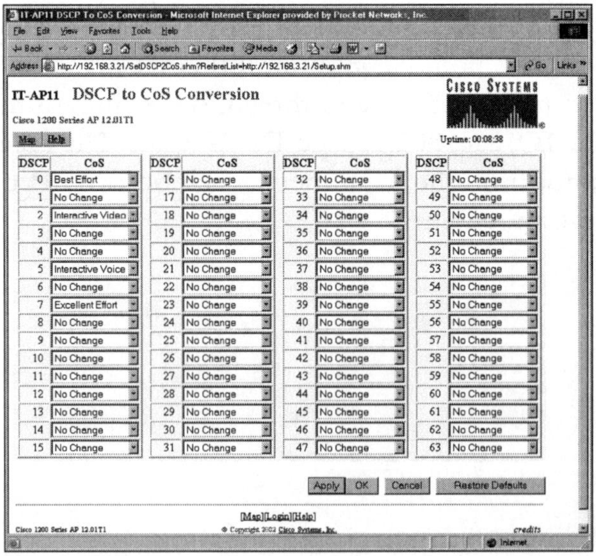

In the configuration shown in Figure 10.7, Premium Voice is mapped to DSCP value 5, Premium Multimedia to 2, Premium Data to7, and Best Effort to 0.

By default, if the switch infrastructure does not mark frames or packets with IEEE 802.1p CoS or DSCP, the VLAN default CoS on the access point is used to apply a specific wireless CoS. To configure traffic classification using VLAN, from the main menu, click **Setup** | **VLAN**. Figures 10.8 and 10.9 show a VLAN-based classification configuration example for MetroWiFi's QoS design.

**Figure 10.8** Configuring Traffic Classification Based on VLAN ID

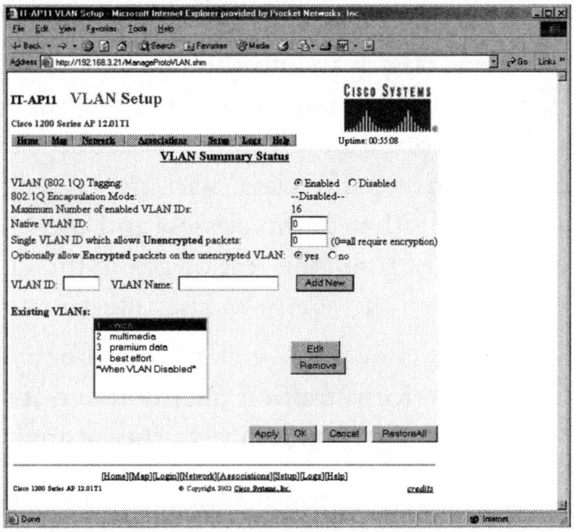


**Figure 10.9** Configuring Each VLAN Traffic Classification via Priority Setting or Pre-defined Policy Group

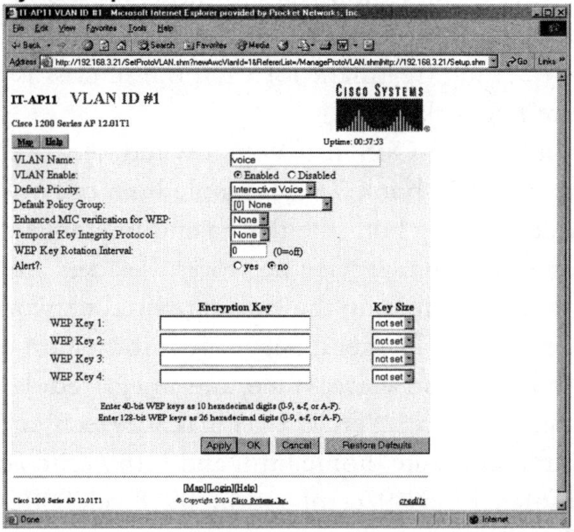

# Using Existing Network QoS Configuration

The previous section exclusively discussed wireless access point QoS configuration. To implement an entire QoS design as illustrated in Table 10.3, network engineers also must configure QoS on WLAN aggregation LAN switches as well as edge and core routers. Under the IOS DiffServ QoS model, configuring QoS on these devices is mainly comprised of two tasks:

- Based on the traffic classification scheme deployed in access points, define an equivalent class map and then construct a QoS policy via policy map for each traffic class to perform the queuing function. A common practice is to use Low Latency Queuing (LLQ) to serve the real-time delay-sensitive traffic class (with guaranteed bandwidth and strict priority over all other traffic classes) and use Class Based Weighted Fair Queuing (CBWFQ) for the rest of the traffic classes. Finally, apply this QoS policy map to the WAN-facing interface.
- Based on the same class map, construct QoS policy via policy map for each traffic class to perform traffic policing functions, and apply it to the access point facing the LAN interface in the desired traffic direction.

If reusing existing QoS configuration on these devices, extra care should be exercised to make sure the traffic classification in the wireless network segment is consistent with what is configured on these devices. This can be accomplished by applying the same traffic classification rules as those already defined by the service class map on the aggregation routers to the access points. Equally important is to ensure that the queuing treatment for each traffic class is consistent throughout the entire network.

The subject of configuring QoS in IOS in a wired network environment, however, is out of the scope of this book. There is a plethora of books and white papers written on this subject covering both the design and implementation of end-to-end QoS for Cisco-powered IP multi-service networks. Readers are encouraged to refer to them when configuring or reusing the existing wired network QoS configuration for a WLAN network deployment. For more information on Cisco IOS QoS technology and its application in voice, video, and integrated data multi-service IP network, refer to: http://cisco.com/en/US/tech/tk543/tech_topology_and_network_serv_and_protocol_suite_home.html and http://cisco.com/application/pdf/en/us/guest/netsol/ns17/c649/ccmigration_09186a00800d67ed.pdf.

# Summary

In recent years, we have seen rapid WLAN adoption in both residential and commercial networks. As more applications are converging onto an all-IP network, the ability to provide adequate QoS support end-to-end in an integrated wireless and wired network is becoming a key requirement for a successful WLAN deployment.

With the DiffServ QoS model, supporting bundled IP services over a WLAN network require appropriate QoS design on both the wireless and wired portions of the WLAN network. There are many tools available in IOS that allow QoS fine tuning. To optimize them, a sound QoS design principle should be followed. In a nutshell, QoS techniques minimize congestion and provide predictable network performance. QoS does not create new bandwidth. The basic design principle is accurate capacity planning followed by appropriate enforcement tools for classification, rate limiting, and queuing.

Configuring for wireless QoS mainly involves setting up traffic classification categories in access points and defining a priority scheme for various applications. 801.lp and DSCP support by access point allows network designers to fully leverage the existing QoS configuration for wired networks.

With the addition of the 802.11 wireless network, a new set of media access protocol, 802.11e, is enhancing QoS support to provide a seamless extension of wired QoS into wireless. However, more fundamentally, the new applications that are being developed for the increasing number of wireless clients will inevitably add new challenges to QoS with their new and unique requirements.

# Solutions Fast Track

## The Requirement for Service Quality

- ☑ Real time applications require strict boundaries for latency, jitter, and packet loss level for acceptable performance.
- ☑ Latency consists of both fixed and variable components.
- ☑ Minimizing congestion is the key to achieve low latency, jitter, and packet loss.

## QoS in General

- ☑ QoS refers to the ability to provide differentiated forwarding treatment for different services based on a predefined set of SLAs.
- ☑ Multiplexing in packetized IP networks inherently can lead to congestion. QoS techniques provide predictable network performance by managing and minimizing congestion.
- ☑ The DiffServ model uses traffic conditioner and PHB to provide a coarse grade of QoS support and scale better than the per-flow based IntServ model.

## How Wireless Changes QoS

- ☑ The uncertainty of signal transmission over wireless medium adds a higher level of difficulty for quantifying a QoS budget than over a wired network.
- ☑ 802.11e EDCF enables frame prioritization based on its traffic category classification over 802.11 radio access network.
- ☑ 802.1p and DSCP support in access points allows seamless integration of QoS configurations for wired and wireless networks.

## WLAN QoS Design Guideline

- ☑ Predictable traffic pattern and QoS enforcement are critical for achieving predictable network performance.
- ☑ Rate limiting and queuing during congestion are key to design a bounded delay and jitter network.
- ☑ Adequate throughput capacity and radio coverage are both critical when designing a WLAN network for achieving optimal network performance.

## Configuring for Wireless QoS in IOS

- ☑ Access point QoS configuration includes two tasks: configure the EDCF frame prioritization scheme and configure the EDCF traffic classification mapping between the aggregation access point and the upstream switch infrastructure.

☑ Changing the $CW_{min}$ and $CW_{max}$ for EDCF for production network may have an undesired effect on network performance and should be supported by significant testing.

## Using Existing Network QoS Configuration

☑ Traffic classification configuration should be consistent between wireless and wired network segments.

# Frequently Asked Questions

The following Frequently Asked Questions, answered by the authors of this book, are designed to both measure your understanding of the concepts presented in this chapter and to assist you with real-life implementation of these concepts. To have your questions about this chapter answered by the author, browse to **www.syngress.com/solutions** and click on the **"Ask the Author"** form. You will also gain access to thousands of other FAQs at ITFAQnet.com.

**Q:** Our WLAN network supports VoIP service. VoIP calls sometime get slow because of the large delay and jitter caused by the reduced available bandwidth and increased number of retransmissions when radio link quality degrades. Is there any solution to handle low link quality?

**A:** You can minimize the performance degradation due to low link quality by increasing the priority of time-sensitive applications so that they receive more radio access than other low priority traffic. You can also utilize the features that are available from the application layer to more gracefully handle the performance hit. For example, for your VoIP application, you can use a high quality, high bit-rate, and low complexity codec such as G.711, and choose a more adaptive jitter buffer algorithm in media gateways.

**Q:** I designed my WLAN network after a thorough site survey and made sure radio coverage was adequate and appropriate QoS hooks were in place. However, I'm still getting people complaining about network performance. How can I revise my design to improve it?

**A:** QoS does not create bandwidth. Because the radio portion of a WLAN network is a shared media rather than a switched media, in most cases sufficient capacity planning is just as important as adequate radio coverage when it comes to designing a WLAN network.

If RF coverage is adequate, a short-term fix you can try to improve the throughput is to force users to associate with a new access point whenever their throughput reduces to a pre-defined threshold, by setting a minimum association rate.

The long-term fix is to derive accurate capacity requirements for your WLAN network based on the total number of users and bandwidth requirement for their applications. As a general rule, when calculating the number of access points necessary to support the aggregated total bandwidth, use 5 Mbps and 30 Mbps as average throughput per radio channel for 802.11b and 802.11a access points, respectively. For 802.11g, use 6 Mbps for mixed-mode operation access points, and 20 Mpbs when there are no 802.11b clients in the same cell.

**Q:** Having fast roaming support for wireless VoIP phones is part of the QoS provisioning in my WLAN network. However, when the phone roams, I can see that authentication works fine but the phone keeps indefinitely trying to register into Call Manager and gets stuck in IP Network configuration, until it finally drops the call. What is the cause?

**A:** This is likely a problem with DHCP and/or IP addressing. This typically happens when you attempt to roam at Layer 3 (i.e., across subnets). Keeping voice calls up with Layer 3 roaming is not possible. Roaming to another Layer 2 network is equivalent to Layer 3 roaming. This can also happen if you are roaming into another Layer 2 network (thus the phone needs to obtain a new IP address).

To minimize roaming delay you should design your WLAN in such a way that Layer 2 roaming is utilized so that you are able to keep a call active when roaming occurs.

**Q:** I followed your recommended 802.11b network dimensioning guidelines and implemented a dedicated VLAN for wireless VoIP phones in my campus WLAN network. I also set up a QoS mechanism to give both VoIP control and data traffic higher priority. However, when I generate background data traffic to test QoS effectiveness, the quality of voice is very poor. Why doesn't the QoS mechanism work?

**A:** Conducting QoS testing on a WLAN causes bi-directional data traffic to saturate the access point on the upstream forwarding path. This is because the QoS support in the current 802.11 access point is for downstream traffic only while all upstream traffic is treated equally, causing congestion which leads to immense latency for all traffic including high priority VoIP traffic.

**Q:** I have implemented a wireless point-to-point link for a customer, using 350 bridges and 2600 routers for the QoS. I am running data along with voice traffic. The VoIP products use DSCP EF to mark the IP packets for QoS. What is the best QoS mechanism to use on the routers?

**A:** In general, a QoS design principle for IP Telephony specified in Cisco's Architecture for Voice, Video, and Integrated Data (AVVID), is the best guideline. In particular, there are two QoS techniques commonly used on the routers that serve as the gateway to the wireless access points and bridges. On the outbound WAN interface, CBWFQ is sufficient to handle upstream traffic congestion, while a LLQ for VoIP traffic class can provide expedited forwarding for delay-sensitive voice packets.

On a router's LAN interface to wireless access points or bridges, CAR is normally needed to rate limit downstream traffic. This is because the router's FastEthernet LAN interface can transmit and receive at 100 Mbps. On the other hand, 802.11b access points or bridges only have a practical throughput of 6 Mbps or less. This throughput mismatch means that with a burst of traffic, the access point needs to do packet drops, which would add an excessive processing burden to the access point and affect performance. By taking advantage of CAR's packet policing capability in the router, the task of dropping excessive packets is removed from the access point. Typically, CAR rate limits the practical throughput of 6 Mbps and guarantees 1 Mbps for high-priority voice and control traffic.

# Index

## Numerals


2.4-GHz antennas, 219–224
5-GHz antennas, 224–225
802 standards initiatives, 75–77
802.1d (Spanning Tree Protocol), 75, 398
802.1p, 425
802.1q trunk encapsulation protocol, 399
802.1x standard
  authentication, configuring, 371–374
  description, 316, 369
  devices, roles of, 370–371
  Extensible Authentication Protocol (EAP), 313–314, 316–317
  port statistics, 373–374
  wired network devices, 385
  *see also* Port-based security, 802.1x
802.11 standards
  choosing among, 132
  compatibility among vendors, 77–78
  data rates, 85
  description, 73–74, 292–293
802.11a channels, 80–81
802.11a OFDM physical layer, 80
802.11b direct sequence channels, 78–79
802.11d wireless local area network (WLAN) standards, 78
802.11f Inter-Access Point Protocol, 78
802.11g, 78
802.11h, 78
802.11i, 136
7920 WVoIP phones, 135, 153–159, 165–166, 225–226


## A


AAP (Authoritative access points), 170
Absolute horizon, 64
Access, logging, 309
Access, unauthorized
  access points, 304
  remote, 284–286
  wireless, 301–303
Access control lists (ACLs), 192, 337–338
Access Control Server (ACS)
  adding access points, 321–323
  description, 216
  mutual (dual) authentication, 368
  RADIUS and 802.1x, support for, 371
  Token cards, compatibility with, 317
Access points (APs)
  Aironet 1100 series, 430
  Authoritative (AAPs), 170
  beacon frames, 137–143, 149
  Cisco products, 214
  configuration, duplicating, 310
  configuring, 207–208, 323–326, 336, 402–407
  congestion, 433–434
  filters, 313, 353
  hot standby, 309–311
  mutual authentication, 305–306, 367–369
  probe frames, 143–149
  scanning for better APs, 151
  in site surveys, 103, 107–108
  unauthorized access, 304
  wireless, 334
  *see also* Aironet 1200 access points; Rogue access points
ACLs (access control lists), 337–338
ACS. *see* Access Control Server (ACS)
Active and passive modes, 119–121
ACU. *see* Aironet Client Utility (ACU)
Ad-hoc mode in WLANs, 27–29
Adapters, client. *see* Client adapters
Addresses. *see* IP addresses
Adhesive tape, 114
Advanced Encryption Standard (AES), 288
Advanced Research Projects Agency (ARPA), 279–280

AES (Advanced Encryption Standard), 288
AH (Authentication Header), 331–332
Aironet 1200 access points
  classification schemes, 431
  integrated 5-GHz antenna, 224
  multiple radios, 229–230, 236, 301, 338, 345, 405
  OFDM rates to *require/basic,* 141
  scalability and performance, 280–281
  upgrading to IOS, 227
Aironet Client Utility (ACU)
  AP name and address, 143
  configuration, 326–328
  core features, 257–258
  exterior site surveys, 124
  interior site surveys, 118–121
  RF Network configuration screen, 150
  roaming configuration, 151
  tracking rogue access points, 362–363
  verifying signal rates, 116
Aironet Wireless Communications, Inc., 212
AirSnort tool, 343
Algorithm, Diffie-Hellman, 329
Amplifiers
  for receivers, 69
  for transmitters, 67
Amplitude *(a)* of waves, 49–50
Amplitude modulation (AM), 52
Analog modulation, 52–53
Analyzers, spectrum, 93, 113, 131
Antennas
  2.4 GHz, 219–224
  5-GHz, 224–225
  choosing for site surveys, 108–109
  Cisco products, 217–225
  coverage, 109
  dipole, 88
  directional, 88–92, 109
  diversity, 83
  external, 297
  half-wavelength (half-λ) dipole, 51, 87–88
  height, 128
  isotropic, 217
  omnidirectional, 87–88, 109
  parabolic, 88, 92
  phased array, 88
  planar arrays, 90–91
  quarter-wavelength (quarter-λ) dipole, 88
  receiver, 69
  sectorized arrays, 91–92
  splitters, 123–124
  transmitter, 68–69
  war driving, 297–298
  Yagi arrays, 89–90
AP. *see* Access points (APs)
AP mode in WLANs, 27–29
Apparent horizon, 64
Architecture for Voice, Video, and Integrated Data (AVVID), 447
ARPA (Advanced Research Projects Agency), 279–280
Assessment, risk, 308
Assisted site surveys, 35–36, 93
Association flapping, 151
Asynchronous Traffic Mode (ATM) traffic management functions (802.1p), 425
ATM (Asynchronous Traffic Mode) traffic management functions (802.1p), 425
Attacks. *see* Threats, security
Attenuation
  antenna cabling, 13
  atmospheric absorption, 14–15
  description, 12, 58–60, 217–218
  Earth bulge, 19
  exterior considerations, 13–18
  Fresnel zone, 18–19
  multipath distortion, 15–17
  rain, snow, and fog, 13–14, 61
  refraction, 17–18
Authentication
  EAP types, 317–318

infrastructure, 164
Media Access Control (MAC), 302–303
multiple hosts, 372–373
multiple methods in EAP, 314
mutual (dual), 305–306, 359, 367–369
open, 301
servers, 316, 319, 370–371
shared, 301–302
virtual private networks (VPNs), 303
wireless networks, 295
*see also* Extensible Authentication Protocol (EAP); Lightweight Extensible Authentication Protocol (LEAP)
Authentication Header (AH), 331–332
Authenticator Port Access Entities (PAEs), 316, 319
Authoritative access points (AAPs), 170
Auto rate negotiation, 116
Autoconnecting, 299
Availability, 308–309, 424
AVVID (Architecture for Voice, Video, and Integrated Data), 447

## B

Backoff mechanisms
CSMA/CA, 186, 190
random (contention window), 428–430, 439
Bandwidth QoS, 423
Battery packs, 109–110
Beacon frames, 137–143, 149
Beam clamps, 113, 115
Bit streams, 54
Bluetooth, 79
Bouncing electromagnetic (EM) waves, 61–63
Bridge Virtual Interface (BVI), 170
Bridges
Cisco products, 214–215
LAN/MAN, 75
point-to-multipoint, 194
point-to-point, 193–194
in site surveys, 104, 107–108, 124–125
workgroup bridges, 195–196
Broadband communication network, cable-TV-based (802.14), 76
Broadband LAN standards (802.7), 75
Broadband wireless access standards (802.16), 77
Broadcast communication process
description, 184
minimum data rate settings in IOS, 188–190
Broadcast domain segmentation
broadcast traffic, 408
multicast traffic, 409
unicast traffic, 408
wireless networks, 409–410
Broadcast WEP keys, 325
Broadcasts, disabling, 122, 300–301
Brute force, 284
Bugtraq security mailing list, 291
Buildings, WLANs in, 24
BVI (Bridge Virtual Interface), 170

## C

C-clamps, 113, 115
Cable-TV-based broadband standards (802.14), 76
Cables
Category 5 (Cat5) or 5e (Cat5e), 101
console, for access point and bridge, 107–108
plenum, 99, 101
receiver, 69
rogue APs, tracking, 375
sheaths, 101
transmitter, 67–68
wire types, 102

CAM (Constant Awake Mode), 122
Cameras, digital, 112
Canada, Royal Society of, 73
CAR (Committed Rate Limiting), 426, 447
Card, wireless, for war driving, 297–298
Care-of-Address (CoA), 167–169
Carrier waves, 50–51, 51
Catalyst switches
configuring 802.1x protocol, 371–374
configuring port security, 378–381
filters, 376–381
limiting MAC addresses, 376–378
Category 5 (Cat5) or 5e (Cat5e) cables, 101
CBWFQ-LLQ (Class Based Weighted Fair Queuing with Low Latency Queue), 426, 442, 447
CCKM (Centralized Key Management), 160, 162–164
CCoA (Co-located Care-of-Address), 167–169
CDMA (Code Division Multiple Accessing), 63
Ceiling-mounted omnidirectional antennas, 219
Cellular architecture, 81–82
Center frequency, 20
Centralized Key Management (CCKM), 160, 162–164
Certificates, digitally signed, 318
Channel filter, receiver, 69
Channel scanning, 160–162
Channels
802.11a, 80–81
802.11b, 78–79
annotating, 117
Cisco
7920 WVoIP phones, 135, 153–159, 165–166, 225–226
Access Control Server (ACS), 216, 317
access points (APs), 214
advisory releases, 291
affordable wireless solutions, 280–281
Aironet 1100 series access points, 430
Aironet Client Utility (ACU), 116, 118–121
Aironet Software Release 12.00T, 430
Aironet Wireless Communications, Inc., 212
bridges, 214–215
client adapters, 215
current bugs and patches, 291
default SSID ("tsunami"), 122, 300, 411
IOS operating system, 215
Linksys, 212
security features, 278
structured wireless-aware network products, 213–226
switches combining layers, 391
vision, 212–213
wireless antennas and accessories, 217–225
Wireless LAN Solution Engine (WLSE), 32, 215, 307, 364–367
wireless security suite, 216
WLAN switches and routers, 216
*see also* Aironet 1200 access points; Catalyst switches; Lightweight Extensible Authentication Protocol (LEAP)
CiscoWorks WLAN Solution Engine (WLSE), 32, 215, 307, 364–367
Clamps, 113, 115
Class-Based Weighted Fair Queuing with Low Latency Queue (CBWFQ-LLQ), 426, 442, 447
Class of Service (CoS), 425, 431
Client adapters
Cisco products, 215
configuring for WEP, 345–346
for site surveys, 106
for VPNs in WLANs, 334
for war driving, 297
Clients
802.1x, 370
channel scanning, 160–162
initial authentication, 164

maximum-clients formula, 432–433
reassociation, 161
reauthentication, 162–164
VPN, 334, 339–341
Co-located Care-of-Address (CCoA), 167–169
CoA (Care-of-Address), 167–169
Coaxial cable, 68
Code Division Multiple Accessing (CDMA), 63
Codec processing time, 423
Collisions, IV, 342
Command line interface (CLI)
IOS, 310–311, 337–338
service set IDs (SSIDs), 300–301
Committed Rate Limiting (CAR), 426, 447
Communication methods, 182–185
Compatibility among vendors, 77–78
Compression, 54
Concentrators, VPN, 303, 334
Confidentiality, 290
Configuration
access points, 323–326, 336
access points (APs), 402–407
Aironet clients, 326–328
filters for access points, 337–339
service set IDs (SSIDs), 122
VLANs and IOS, 401–407
VPN clients, 339–341
VPN gateways, 334–336
WEP with TKIP, 344–346
Congestion, transient, 425
Congestion in access points (APs), 433–434
Connectivity, WLAN, validating, 333
Constant Awake Mode (CAM), 122
Consumption, power, 109, 122
Contention window (CW), 428–430, 438–439
Cordless drills, 110
CoS (Class of Service), 425, 431
Coverage
antennas, 109
measuring, 116–117
wireless local area networks (WLANs), 81–85
CSMA/CD access method (Ethernet) standards (802.3), 75
Current flow, alternating, 48–49
CW (contention window), 428–430, 438–439

# D

Data communication methods, 182–185
Data Link layer
defined in 802.2 standard, 74
protecting, 294–296
Data-manipulation attacks, 288
Data rates, 85
Data retry counter, 150
Data-sniffing attacks, 287–288
dBi value, 217
DBm (decibels above 1 milliwatt), 57
DCF (Distributed Coordination Function), 428–429
Decibels (dB), 217
Decoder, receiver, 69–70
Default SSIDs, 285–286
Delay
queuing, 423–424
roaming-introduced, 434
Delay-sensitive applications, 434
Demand Priority Access Method standards (802.12), 76
Demodulator, receiver, 69
Denial of service (DoS) attacks, 286–287
Detecting
interference, 93
rogue access points, 305–309, 361–367
DHCP (dynamic host control protocol), 104, 337, 399

Differentiated Service Code Point (DSCP) bytes, 426
Differentiated Service (DiffServ) QoS schemes, 425–427, 442
Diffie-Hellman key exchange algorithm, 329
DIFS (Distributed IFS), 429
Digital cameras, 112
Digital modulation, 53–58
Digital signatures, 331
Digitally signed certificates, 318
Dipole antennas
 half-wavelength (half-λ), 51, 87–88
 quarter-wavelength (quarter-λ), 88
Direct sequence spread spectrum (DSSS)
 description, 78
 FSK modulation, 57
 functionality, 74
 IEEE 802.11a channels, 80–81
 IEEE 802.11a OFDM physical layer, 80
 IEEE 802.11b direct sequence channels, 78–79
 physical layer (PHY), 293
Directional antennas, 88–92, 109
Dish antennas, 222, 224–225
Distance, link, 125
Distributed Coordination Function (DCF), 428–429
Distributed denial of service (DDoS) attacks, 286–287
Distributed IFS (DIFS), 429
Diversity antennas
 ceiling-mounted omnidirectional patch, 221
 description, 83
 directional wall-mounted patch, 222
 pillar-mounted omnidirectional, 220
 POS dipole omnidirectional, 220
Diversity ceiling-mounted omnidirectional patch antennas, 221
Diversity wall-mounted directional patch antennas, 222
DMZ, wireless, 358–359
Downstream, radio, 430
Drills, cordless, 110
Driven element, 90
Drivers in site surveys, 103
DSCP (Differentiated Service Code Point) bytes, 426
DSSS. *see* Direct sequence spread spectrum (DSSS)
Dual (mutual) authentication, 305–306, 359, 367–369
Dynamic host control protocol (DHCP), 104, 337, 399
Dynamic IP assignment, 399
Dynamic MAC addresses, 377
Dynamic WEP keys, 289, 316, 318

# E

EAP. *see* Extensible Authentication Protocol (EAP)
EAPOW (Extensible Authentication Protocol Over Wireless), 316
EDCF. *see* Enhanced Distributed Coordination Function (EDCF)
802 standards initiatives, 75–77
802.1d (Spanning Tree Protocol), 75, 398
802.1p, 425
802.1q trunk encapsulation protocol, 399
802.1x standard
 authentication, configuring, 371–374
 description, 316, 369
 devices, roles of, 370–371
 Extensible Authentication Protocol (EAP), 313–314, 316–317
 port statistics, 373–374
 wired network devices, 385
 *see also* Port-based security, 802.1x
802.11 standards
 choosing among, 132
 compatibility among vendors, 77–78

data rates, 85
description, 73–74, 292–293
802.11a channels, 80–81
802.11a OFDM physical layer, 80
802.11b direct sequence channels, 78–79
802.11d wireless local area network (WLAN) standards, 78
802.11f Inter-Access Point Protocol, 78
802.11g, 78
802.11h, 78
802.11i, 136
Electromagnetic (EM) waves
amplitude modulation (AM), 52
attenuation, 58–61
bouncing, 61–63
frequency modulation (FM), 52–53
line of sight, 64
modulating signals, 50–51
modulation, analog, 52–53
modulation, digital, 53–58
penetration, 64–66
refraction, 63–64
transmitting radio signals, 48–49
waveforms, 49–50
Employee-installed access points, 356–358, 360
Encapsulating Security Payload (ESP), 331–332
Encoder, transmitter, 67
Enhanced Distributed Coordination Function (EDCF)
description, 429–430
frame prioritization configuration, 438–439
traffic classification, 439–441
Environmental considerations, 72–73
Equipment cases, foam, 106
ESP (Encapsulating Security Payload), 331–332
ESS (LAN Extended Sub System), 399
Ethernet frame prioritization (802.1p), 425
Ethernet standards (802.3), 75
Ethernet VLAN numbers, 399
Extended VLAN numbers, 399
Extensible Authentication Protocol (EAP)
802.1x standard, 313–314, 316–317
authentication types, 317–318
Cisco. *see* Lightweight Extensible Authentication Protocol (LEAP)
description, 302
multiple authentication methods, 314
mutual (dual) authentication, 369
packet formats, 314–315
requirements, 107
Transport Layer Security (TLS), 317–318, 369
variants, 314
Extensible Authentication Protocol Over Wireless (EAPOW), 316
Exterior site surveys, 124–128
External antennas, 297


# F


Faraday, Michael, 65
Fast authentication, 164
Fast Power Save Mode (FastPSP), 122
Fast Roam, 161
Fast Secure Roaming (FSR), 33, 160, 176
FastPSP (Fast Power Save Mode), 122
FCC (Federal Communications Commission), 70–71
Federal Communications Commission (FCC), 70–71
FHSS (frequency hopping spread spectrum), 57, 74, 78, 293
Fiber Distributed Data Interface (FDDI), 399
Fiber optics standards (802.8), 76
File Transfer Protocol (FTP), 287–288
Filters for access points
Catalyst switches, 376–381
configuring, 337–339
description, 313

firewalls, compared to, 353
guest WLANs, 208
Fine for unlicensed operation, 71
Fire codes, 98–99
Firewalls
IPSec passing through, 330
Private Internet Exchange (PIX), 303
security, 290, 311–312
as VPN termination points, 353
5-GHz antennas, 224–225
Flapping, 151
FM (frequency modulation), 52–53
FMS attacks, 342
Foam equipment cases, 106
Fog attenuation, 14
Form, pre-site survey, 95–97
Freezer, walk-in, 123
Frequency Allocations, Table of, 70
Frequency (*f*) of waves, 50
Frequency hopping spread spectrum (FHSS), 57, 74, 78, 293
Frequency modulation (FM), 52–53
Frequency shift keying (FSK), 55, 57
Fresnel zone radius, 125–127
FSK (frequency shift keying), 55, 57
FSR (Fast Secure Roaming), 33, 160, 176
FTP (File Transfer Protocol), 287–288

## G

Gateways, VPN, 334–336
Generic Routing Encapsulation (GRE), 169
Generic Token Card authentication, 317
Global Positioning System (GPS), 113, 296, 298
Graph paper, 111
GRE (Generic Routing Encapsulation), 169
Green laser pointers, 112
Ground wire, 50
Guest SSIDs, 411
Guest WLANs
configuration, 205–207
deployment, 204
description, 202
design, 202–203
filtering, 208
limiting coverage area, 210
quantity, 210
recommendations, 204
switch and AP configuration, 207–208
topology, 203–204

## H

Half-wavelength (half-λ) διπολε αντεννασ, 51, 87∠88
Harmonics, interference from, 21–22
Harnesses, safety, 112
Health risk, 73, 131
Hertz, Heinrich, 48, 50
High-frequency radar, 61
High-gain omnidirectional antennas, 219–220
High Rate Direct Sequence (HR/DSS), 293
Historical sites, permits for, 99
Hopping, VLAN, 397, 406, 412–413, 416
Horizon
absolute, 64
apparent, 64
bouncing past, 62
signal constraint, 60
Hospital, deploying in, 131
Hospitals, WLANs in, 23–24
Hot standby access points, 309–311
HR/DSS (High Rate Direct Sequence), 293
HTTP access, 304
Hubs in site surveys, 103
Hybrid mode in WLANs, 29–30

## I

IANA (Internet Assigned Numbers Authority), 282–283

IAPP (Inter-Access Point Protocol), 78, 136, 434
ICMP (Internet Control Message Protocol), 282
Identifying intruders, 294
IDS (Intrusion Detection Systems), 290, 353
IEEE (Institute of Electrical and Electronic Engineers). *see* 802.11, etc.
IFS (interframe space), 428–429
IGMP (Internet Group Management Protocol), 204
IKE (Internet Key Exchange), 329–331, 335
In-building surveys, 97
Industrial, Scientific, and Medical (ISM) frequency bands, 71–72
Infrared (IR), 49, 74
Infrastructure authentication, 164
Initial authentication, 164
Initialization vectors (IVs), 341–343
Inline power, 104
Institute of Electrical and Electronic Engineers (IEEE). *see* 802.11, etc.
Integrated Service (IntServ) QoS schemes, 425–428
Integrity, 308–309
Inter-Access Point Protocol (802.11f) (IAPP), 78, 136, 434
Inter-Switch Link (ISL), 83
Interference
  center frequency, 20
  description, 19
  detection, 35, 93
  harmonics, 21–22
  radio transmitters, 20–21
Interframe space (IFS), 428–429
Interior site surveys, 115–124
International Standard Organization (ISO), 391
  *see also* Data Link layer; Physical layer (PHY)
International Telecommunication Union (ITU), 70
Internet Assigned Numbers Authority (IANA), 282–283
Internet Control Message Protocol (ICMP), 282
Internet Group Management Protocol (IGMP), 204
Internet Key Exchange (IKE), 329–331, 335
Internet Router Discovery Protocol (IRDP), 168
Internet SA and Key Management Protocol (ISAKMP), 329
Internet scanning, 296
Interoperable LAN security (SILS) standards (802.10), 76
Intruder-installed rogue access points, 359
Intruders, tracking, 294
Intrusion Detection Systems (IDS), 290, 353
IOS operating system
  command line interface (CLI), 310–311, 337–338
  configuring port security, 378–381
  description, 215, 336
  local RADIUS feature, 323
  quality of service (QoS), 435
  VLAN support, 347, 401–407
  Web browser interface, 338–339
IP addresses
  for multicasting, 200
  for site survey devices, 108
IP Security (IPSec)
  Authentication Header (AH), 331–332
  description, 328
  Encapsulating Security Payload (ESP), 331–332
  firewalls, passing through, 330
  implementing with VPNs over WLANs, 332–333
  Incompatibilities, 332
  Internet Key Exchange (IKE), 329–331
IR (infrared), 49, 74
IRDP (Internet Router Discovery Protocol), 168

ISAKMP (Internet SA and Key Management Protocol), 329
ISL (Inter-Switch Link), 83
ISLAN (isochronous services local area network) standards (802.9), 76
ISM (Industrial, Scientific, and Medical) frequency bands, 71–72
ISO (International Standard Organization), 391
*see also* Data Link layer; Physical layer (PHY)
Isochronous services local area network (ISLAN) standards (802.9), 76
Isotropic antennas, 217
ITU (International Telecommunication Union), 70
IVs (initialization vectors), 341–343

# J

Jamming, 295
Jitter QoS, 424

# K

Keys
broadcast WEP, 325
cracking, 301–302
exchange algorithm, Diffie-Hellman, 329
Internet Key Exchange (IKE), 329–331, 335
preshared, 330
Public Key Infrastructure (PKI), 317
renewal, 318, 321–322
static WEP, 285, 341–346

# L

L2 roaming
Centralized Key Management (CCKM), 160, 162–164
channel scanning, 160–162
description, 134–136
Fast Secure Roaming (FSR), 160, 176
reauthentication, 162
L3 roaming
description, 134
disruptive process, 165
Mobile IP (MIP), 166–170, 176
Proxy Mobile IP (PMIP), 170–171
superset of L2, 135
WLAN design considerations, 171–173
WVoIP phones, 177
Ladders, 112
LAN, isochronous services (ISLAN) (802.9), 76
LAN Extended Sub System (ESS), 399
LAN/MAN bridging and management standard, 75
LAN security, interoperable (SILS) (802.10), 76
LANs (local area networks), 66
Laptop computer for site surveys, 105
Laser pointers, 112
"Last-mile" local-loop issues, 77
Latency QoS, 423–424
Layer 3 switches, 334
LEAP. *see* Lightweight Extensible Authentication Protocol (LEAP)
Lightweight Extensible Authentication Protocol (LEAP)
authentication process, 319–320
description, 318
implementing, 320–328
mutual (dual) authentication, 367–369
requirements, 107
WVoIP phones, 159
Line-of-sight path, 64, 217
Link distance, 125
Link quality, 445
Link setup and testing, 127–128
Linksys, 212

LLC (logical link control) standards (802.2), 75
LLQ (Low Latency Queuing), 442
Local area networks (LANs), 66
Local-loop issues, 77
Logging access, 309
Logical link control (LLC) standards (802.2), 75
Low Latency Queuing (LLQ), 442
Low link quality, 445
Low-power, unlicensed transmitters, 71–72

# M

MAC. *see* Media Access Control (MAC)
Macintosh operating system, 103
Man-in-the-middle (MITM) attackers, 279, 288, 344
Man-lifts, 112, 115
MANs (metropolitan area networks), 75
Marconi, Guglielmo, 48
Markers, 111
Mast-mounted antennas
  dish, 224–225
  omnidirectional, 219, 224–225
  sector, 224–225
Maximum-clients formula, 432–433
Maximum queuing delay, 423–424
MD5 (Message Digest 5) authentication, 317
MDRR (Modified Deficit Round Robin), 426
Measuring devices, 111–112
Media Access Control (MAC)
  address authentication, 302–303
  address types, 376–377
  addresses, finding, 361
  addresses, limiting per port, 376–378
  addresses of access points, 121
  layer, 73–74
  relative MAC addresses, 375
  *see also* Data Link layer
Media types, network, 101–102
Message Digest 5 (MD5) authentication, 317
Message Integrity Check (MIC), 143, 254, 344, 396
Metal construction, WLANs in, 24
Metropolitan area networks (MANs), 75
MetroWiFi example
  availability problems, 424
  classification configuration, 441
  description, 422–423
  DSCP value, 440
  network architecture, 422
  quality of service (QoS) design, 435–437
MIC (Message Integrity Check), 143, 254, 344, 396
MIP (Mobile IP), 166–170, 176
MITM (man-in-the-middle) attackers, 279, 288, 344
Mobile IP (MIP), 166–170, 176
Mobile Wi-Fi VoIP (MoWLAN), 423–424, 427
Modified Deficit Round Robin (MDRR), 426
Modulating signals, 50–51
Modulation
  amplitude (AM), 52
  analog, 52–53
  description, 50–51
  digital, 53–58
  frequency (FM), 52–53
Modulator, transmitter, 67
Mounting equipment, temporary, 113–115
MoWLAN (Mobile Wi-Fi VoIP), 423–424, 427
Multicast communication process
  bandwidth's adverse effect, 200
  considerations for, 199
  deployment recommendations, 185–190
  description, 184–185
  IP addresses, 200
  minimum data rate settings in IOS, 188–190

peer-to-peer WLAN with bridges, 193–196
reasons for, 199
WLAN configuration, 190–191
WLAN filtering, 191–192
workgroup bridges, 195–196
Multicast traffic, segmenting, 409
Multipath scattering, 63
Multiple host authentication, 372–373
Mutual (dual) authentication, 305–306, 359, 367–369

# N

NAT (Network Address Translation), 331
National Electric Code (NEC), 99
National Telecommunications and Information Administration (NTIA), 70–71
Native VLANs, 399
NEC (National Electric Code), 99
Nessus tool, 291
Network Address Translation (NAT), 331
Network availability, 308–309, 424
Network media types, 101–102
Network scanners. *see* Port scanning
Network security. *see* Security, network
Network Stumbler (NetStumbler), 296, 298–299, 363
NMAP tool, 282–283, 291, 374
Nonces, RSA-encrypted, 330
NTIA (National Telecommunications and Information Administration), 70–71

# O

Oakley protocol, 329
Occupational Safety and Health Administration (OSHA), 98
OFDM (Orthogonal Frequency Division Multiplexing), 80, 293
Omnidirectional antennas
ceiling-mounted, 219, 221
description, 87–88, 109
diversity ceiling-mounted patch, 221
high-gain, 219–220
mast-mounted, 219, 224–225
pillar-mounted diversity, 220
POS diversity, 220
On/off keying (OOK), 54
One-Time Password (OTP) implementation, 285, 317
OOK (on/off keying), 54
Open authentication, 301
Open System Interconnection (OSI) network layers, 180–182, 292–293, 391
*see also* Data Link layer; Physical layer (PHY)
Operating systems in site surveys, 102–103
Organizationally Unique Identifiers (OUIs), 361
Orthogonal Frequency Division Multiplexing (OFDM), 80, 293
Oscillating fields, 49
OSHA (Occupational Safety and Health Administration), 98
OSI. *see* Data Link layer; Open System Interconnection (OSI) network layers; Physical layer (PHY)
OTP (One-Time Password) implementation, 285, 317
OUIs (Organizationally Unique Identifiers), 361

# P

Packet formation time, 423
Packet formats, EAP, 314–315
Packet loss, 424
PAEs (Port Access Entities), 316, 319
PAM (pulse amplitude modulation), 55
Paper, graph, 111

Parabolic antennas, 88, 92
Passive and active modes, 119–121
PAT (Port Address Translation), 331
Patch antennas
diversity ceiling-mounted omnidirectional, 221
diversity wall-mounted directional, 222
Patches, news and security, 291
Peer-to-peer mode in WLANs, 27–29
Peer-to-peer WLAN with bridges, 193–196
Penalties for unlicensed operation, 71
Penetration by electromagnetic (EM) waves, 64–66
Per-Hop Behavior (PHB), 426–427
Per-packet-keying, 254
Per-VLAN features
authentication and encryption, 401
filters, 400
quality of service (QoS), 401
settings, 396–397
Period (τ) of waves, 50
Permits for historical sites, 99
Phase shift keying (PSK), 55–56
Phased array antennas, 88
PHB (Per-Hop Behavior), 426–427
Physical detection, 281, 307
Physical layer (PHY)
contents, 293
defined in 802.11 standards, 73–74
DSSS channels, 78–79
protecting, 295–296
Pigtail cable, 298
Pillar-mounted diversity omnidirectional antennas, 220
Ping sweeps, 282
PKI (Public Key Infrastructure), 317
Planar array antennas, 90–91
Plenum cable, 99, 101
PMIP (Proxy Mobile IP), 170–171
Point-to-multipoint bridging, 193–194
Point-to-point bridging, 193–194
Point-to-point surveys, 97
Pointers, laser, 112
Policy, security, 308, 360
Polyvinyl chloride (PVC), 101
Port, Universal Serial Bus (USB), 298
Port 53, 283
Port Access Entities (PAEs), 316, 319
Port Address Translation (PAT), 331
Port-based security, 802.1x
description, 367
preventing rogue AP connection, 369–374
preventing rogue AP use, 367–369
Port numbers, 374
Port scanning
extra traffic, 376
NMAP Scanner, 282–283, 291, 374
wired detection, 307, 374–376
Port security
configuring in IOS Catalyst switches, 378–381
restricting communication input, 295–296
Port statistics, 373
POS diversity dipole omnidirectional antennas, 220
Power consumption, 109, 122
Power Save Mode (PSP), 122
Power supplies in site surveys, 104–105, 109–110
PQ (Priority Queuing), 426
Pre-site survey form, 95–97
Priority Queuing (PQ), 426
Private Internet Exchange (PIX) firewall, 303
Probe frames, 143–149
Propagation, wireless, 69
Propagation time, 423
"Protect" mode of port security, 378
Protocol filters (access lists), 192
Protocols in site surveys, 102–103
Proxy Mobile IP (PMIP), 170–171
PSK (phase shift keying), 55–56
PSP (Power Save Mode), 122

Public Key Infrastructure (PKI), 317
Public Secure Packet Forwarding (PSPF), 312–313, 418
Pulse amplitude modulation (PAM), 55
PVC (polyvinyl chloride), 101

# Q

QoS Basis Service Set (QBSS), 154, 157–158, 430
Quality of service (QoS)
  bandwidth, 423
  description, 425–427
  Differentiated Service (DiffServ), 425–427, 442
  existing QoS configuration, 442
  filters, 313
  Integrated Service (IntServ), 425–428
  integration, 430–432
  IOS, 435
  jitter, 424
  latency, 423–424
  need for plan, 422
  network availability, 424
  packet loss, 424
  virtual local area networks (VLANs), 401
  WLAN, testing on, 427
  WLAN design guidelines, 432–434
  WLAN effect on, 427
  WLAN support, 428–430
  *see also* MetroWiFi example
Quantization error, 53
Quarter-wavelength (quarter-λ) dipole antennas, 88
Queuing delay, 423–424

# R

Radar, 61
Radiation, radio frequency (RF), 72–73
Radio components, 66–70
Radio downstream, 430
Radio frequency (RF) communication
  basics, 48
  interference, 19–22
  radiation, 72–73
  *see also* Electromagnetic (EM) waves; Modulation
Radio frequency (RF) deployment
  description, 81
  seamless roaming, 83–85
  throughput, 85–87
  WLAN coverage, 81–85
  WLAN data rates, 85
  *see also* Antennas
Radio transmitters, interference from, 20–21
RADIUS. *see* Remote Authentication Dial-in User Service (RADIUS)
Rain attenuation, 13, 61
Random backoff (contention window), 428–430, 439
Rate-limiting features, 432
Rate shifting, 116
Rates, data, 85
Ratio, signal-to-noise (S/N), 57–58
RC4 (Rivest Cipher Four) algorithm, 279, 341–342
Reassociation, 161
Reauthentication, 162–164
  L2 roaming, 162
Received Signal Strength Indicator (RSSI), 154, 157–158, 366
Receiver components, 69–70
Reconnaissance attacks, 281–284
Reflector element, 90
Refraction, 63–64
Refraction of electromagnetic (EM) waves, 63–64
Regulations, 71–72
Regulatory agencies, 70–71
Remote access, unauthorized, 284–286

Remote Authentication Dial-in User Service (RADIUS)
Access Control Server (ACS), support by, 371
accounting for logins, 290, 334
control list, configuring, 413–414
survey laptops, 107
VLAN access control, 347, 411–413
Repeaters, 82
Request and response packets, 315
Resource Reservation Protocol (RSVP), 425
"Restrict" mode of port security, 378
Retransmit counter, 150–151
RF-producing devices, 25
Risk assessment, 308
Risk to health, 73, 131
Rivest, Ron, 341
Rivest Cipher Four (RC4) algorithm, 279, 341–342
Roaming
association flapping, 151
beacon frames, 149
Centralized Key Management (CCKM), 160, 162–164
channel scanning, 160–162
Cisco 7920 WVoIP phones, 135, 153–159, 165–166
data retry counter, 150
decisions and criteria, 149–151
delay introduced by, 434
DHCP or IP problem, 446
Fast Roam and Very Fast Roam, 161
Fast Secure Roaming (FSR), 160, 176
periodic scans for better APs, 151
reassociation, 161
reauthentication, 162–164
retransmit counter, 150–151
seamless, 83–85
switches, 103–104
target selection process, 151–153
*see also* L2 roaming; L3 roaming
Rogue access points
Aironet Client Utility (ACU), 362–363
Catalyst switch filters, 376–381
dangers of, 45, 356–359
description, 304–305
detecting, 34, 305–309, 361–367, 374–376
employee-installed, 356–358, 360
intruder-installed, 359
preventing connection, 369–374
preventing introduction, 360
preventing use, 367–369
protection from, 317, 348
Routers in site surveys, 104
Royal Society of Canada, 73
RSA Data Security Inc., 341
RSA-encrypted nonces, 330
RSSI (Received Signal Strength Indicator), 154, 157–158, 366
RSVP (Resource Reservation Protocol), 425
"Rubber ducky" antennas, 108

# S

Safety harnesses, 112
SAs (Security Associations), 329
Scanners, network. *see* Port scanning
Scanning, channel, 160–162
Scanning for better APs, 151
Scanning ports. *see* Port scanning
Scattering, multipath, 63
Scattering signals, 61–62
Seamless roaming, 82, 83–85
Sector antennas, 224–225
Sectorized array antennas, 91–92
Secure Shell (SSH), 289, 304
Security, interoperable LAN (SILS) standards (802.10), 76
Security, network
choosing a design, 353
description, 279–280
designing for, 308–309

firewalls, 290, 311–312
jamming, 295
logging access, 309
need for, 280–281
physical layer, protecting, 295–296
policy, 308, 360
risk assessment, 308
site survey area, 100
threats to, 281–288
wireless technology effects, 292–296
WVoIP phones, 176
*see also* Rogue access points; Security cycle; Threats, security
Security, password, 285
Security Associations (SAs), 329
Security cycle
description, 288–289
improving, 291
monitoring, 290
securing, 289–290
testing, 290–291
Security issues with WLANs, 25–26
Security mailing lists, 291
Security patches, 291
Segmentation. *see* Broadcast domain segmentation
Service Level Agreements (SLA), 424, 427
Service set IDs (SSIDs)
access point, matching, 103
broadcasts, disabling, 122, 300–301
command line interface (CLI), 300–301
configuration, 122
description, 410–411
guest SSIDs, 411
RADIUS SSID control list, 413–414
security risk, 285–286
SSID-based RADIUS verification, 412
Services, unwanted, 304
7920 WVoIP phones, 135, 153–159, 165–166, 225–226
Shared authentication, 301–302
Shared network model, 293
Sheaths, cable, 101
Shielding, 66
Short IFS (SIFS), 429
Shutdown mode of port security, 378
SIFS (Short IFS), 429
Sight, line of, 64
Signal-to-noise (S/N) ratio, 57–58
Signal waves, 51
Signals
bouncing or scattering, 61–62
through walls, 356
*see also* Attenuation; Interference
SILS (standard for interoperable LAN security) (802.10), 76
Simple Network Management Protocol (SNMP)
restricting access, 289
trap for rogue access points, 306–307, 364–365
Sinusoidal waveform, 49
Site surveys
access points, 103, 107–108
antenna height, 128
assisted, 93
bridges, 104, 107–108, 124–125
Cisco ACU, 118–121
Cisco Aironet Client Utility (ACU), 116, 118–121
client adapters, 106
description, 93–94
drivers, 103
equipment cases, foam, 106
exterior surveys, 124–128
Fresnel zone radius, 125–127
hubs, 103
infrastructure awareness, 100–105
interior surveys, 115–124
laptop computer, 105
Macintosh operating system, 103
mounting equipment, temporary, 113–115

network media types, 101–102
operating systems, 102–103
passive and active modes, 119–121
power consumption, 122
power supplies, 104–105, 109–110
pre-site survey form, 95–97
preparation, 94–100
protocols, 102–103
rate shifting, 116
routers, 104
service set IDs (SSIDs), 103, 122
switches, 103–104
tools, 110–114
Site surveys, assisted, 35–36
SLA (Service Level Agreements), 424, 427
S/N (signal-to-noise) ratio, 57–58
Sniffers
attacks with, 287–288
description, 306
detecting rogue access points, 361
SNMP. *see* Simple Network Management Protocol (SNMP)
Snow attenuation, 13–14
Social engineering, 281–282
Spanning Tree Protocol (802.1d), 75, 398
Spectrum, RF, regulation of, 70
Spectrum analyzers, 93, 113, 131
Splitters, antenna, 123–124
Spoofing, 282
SSH (Secure Shell), 289, 304
SSID. *see* Service set IDs (SSIDs)
Static, 58
Static MAC addresses, 377
Static WEP keys, 285, 341–346
Sticky MAC addresses, 377
Structured wireless-aware network products, 213–226
Structures, WLANs in, 24
Stub devices, 398
Stub networks, 425
Stumbler (NetStumbler), 296, 298–299, 363
Success packets, 315
Supplicant PAEs, 316, 319
Surveyor's tape, 111
Surveys. *see* Site surveys
Sweeps, 282
Switches
802.1x, 370
configuration for guest WLANs, 207–208
in site surveys, 103–104
trunking, 83
System logic (SYSLOG) protocol, 290


# T


Table of Frequency Allocations, 70
TACACS (Terminal Access Controller Access Control System), 334
Tape
adhesive, 114
surveyor's, 111
TCB (Traffic Conditioner Block), 426–427
Telnet
disabling, 304
disallowing access by, 313
in port scans, 307
SSH alternative, 289
Temporal Key Integrity Protocol (TKIP)
configuring WEP, 344–346
description, 143, 343–344
IPSec incompatibility, 332
per-packet-keying, 254
RC4 encryption implementation, 288
Terminal Access Controller Access Control System (TACACS), 334
Termination points, firewalls as, 353
Tesla, Nikola, 48
Threats, security
data manipulation, 288
data sniffing, 287–288
denial of service (DoS), 286–287

distributed denial of service (DDoS), 286–287
FMS attacks, 342
man-in-the-middle (MITM) attackers, 279, 288, 344
reconnaissance attacks, 281–284
unauthorized remote access, 284–286
unauthorized wireless access, 301–303
Throughput, RF, 85–87
Tie wraps, 114–115
TKIP. *see* Temporal Key Integrity Protocol (TKIP)
TLS (Transport Layer Security), 317–318, 369
Token-passing bus standards (802.4), 75
Token Ring standards (802.5), 75
Tools
for site surveys, 110–114
for war driving, 297–300
Topology of virtual private networks (VPNs), 333
Tracking intruders, 294
Traffic classification, configuring, 439–441
Traffic Conditioner Block (TCB), 426–427
Traffic management functions in Asynchronous Traffic Mode (ATM), 425
Transient congestion, inherent, 425
Transmission time, 423
Transmitters
components, 67–68
unlicensed, 71–72
Transmitting radio signals, 48–49
Transport Layer Security (TLS), 317–318, 369
Triangulation, 365
Trunk ports, 394–395, 398–399
Trunking, 83
"Tsunami" (Cisco default SSID), 122, 300, 411
Tunnels
Authentication Header (AH), 331
encrypted, 328
lack of Cisco support, 303
Secure Shell (SSH), 304
security, 339
VPN client and gateway, 334
Twisted pair, unshielded (UTP), 101
2.4-GHz antennas, 219–224

# U

U-bolts, 114
Unauthorized access. *see* Access, unauthorized
Unicast communication process, 182–184
Unicast traffic, segmenting, 408
Uninterruptible power supply (UPS), 105
Union facilities, 97, 99
Universal Serial Bus (USB) port, 298
Unlicensed transmitters, 71–72
Unshielded twisted pair (UTP), 101
Unwanted services, 304
UPS (uninterruptible power supply), 105
USB (Universal Serial Bus) port, 298
UTP (unshielded twisted pair), 101

# V

Velcro, 114
Velocity *(v)* of waves, 49–50
Vendor compatibility, 77–78
Vendor Specific Attribute (VSA), 413–414
Vendors' news and security patches, 291
Very Fast Roam, 161
Virtual local area networks (VLANs)
description, 388–392, 389–390
hopping, 397, 406, 412–413, 416
individual authentication and encryption, 401
individual filters, 400
individual quality of service (QoS), 401
individual settings, 396–397

IOS configuration, 401–407
limit on quantity, 418
multiple switches, 390
native, 392, 399
numbering, 392
routing between, 399
switch ports, 103
switch trunking, 84
VLAN-based RADIUS verification, 412
wireless deployment, 399–401
wireless environment, 395–397
WLAN security, 346–347
*see also* Broadcast domain segmentation; Guest WLANs
Virtual private networks (VPNs)
authentication, 303
client, configuring, 339–341
client, description of, 334
concentrator, 303, 334
device list in WLAN, 334
firewalls as termination points, 353
gateway, configuring, 334–336
implementing IPSec over WLANs, 332–333
topology, 333
VLAN Trunking Protocol (VTP)
description, 389
wired networks, 393–394
wireless networks, 398–399
VLANs. *see* Virtual local area networks (VLANs)
Voice over Internet Protocol (VoIP)
maximum-clients formula, 432–433
packet loss, 424
roundtrip delay, 423–424
Voice over IP over Wireless LAN (VoIPoWLAN), 423–424
Voice over Wireless LAN (VoWLAN), 427
VoIP. *see* Voice over Internet Protocol (VoIP)
VoIPoWLAN (Voice over IP over Wireless LAN), 423–424
VoWLAN (Voice over Wireless LAN), 427
VPN. *see* Virtual private networks (VPNs)
VSA (Vendor Specific Attribute), 413
VTP. *see* VLAN Trunking Protocol (VTP)
Vulnerabilities of WEP
with initialization vectors (IVs), 342
mitigating, 343–344


# W


Walk-in freezer, 123
Walls, signals passing through, 356
War dialers, 284
War driving
antenna, 297–298
client adapters, 297
description, 296
protecting against, 300–301
software, 299
SSID broadcasts, disabling, 300–301
tools, 297–300
Waveforms, 49–50
Wavelength ($\lambda$) of waves, 50
WDS (wireless domain services), 32, 163–164, 176
Web browser interface
access point administrator, 405
HTTP access, 304
IOS, 338–339
WEP, 344–345
Web sites
AirSnort tool, 343
Bugtraq security mailing list, 291
Cisco IOS QoS technology, 442
Cisco wireless product listings, 281
Federal Communications Commission (FCC), 70
FMS attacks, 342
International Telecommunication Union (ITU), 70
IV collisions, 342

Nessus tool, 291
NMAP tool, 282, 291, 374
Royal Society of Canada, 73
WECA (Wireless Ethernet Compatibility Alliance), 77
Weighted Red Early Drop (WRED), 427
WEP. *see* Wired Equivalent Privacy (WEP) Protocol
Wire types, cable, 102
Wired Equivalent Privacy (WEP) Protocol
broadcast domain segmentation, 409–410
broadcast WEP keys, 325
confidentiality, 290
configuring with TKIP, 344–346
cracking keys, 301–302
description, 341–342
dynamic WEP keys, 289, 316, 318
initialization vectors (IVs), 341–343
key renewal, 318, 321–322
LEAP authentication, 319–320
open or shared authentication, 301–302
Rivest Cipher Four (RC4) algorithm, 279, 341–342
static WEP keys, 285, 341–346
Temporal Key Integrity Protocol (TKIP), 332, 343–344
vulnerabilities, mitigating, 343–344
vulnerabilities with IVs, 342
Wireless access
access points, 334
elements, 66–70
unauthorized, 301–303
Wireless-aware LANs
benefits, 31–32
description, 30–31
design considerations, 39
security solutions support, 38–39
value, 44
Wireless-aware network products, 213–226
Wireless card for war driving, 297–298
Wireless domain services (WDS), 33, 163–164, 176
Wireless Ethernet Compatibility Alliance (WECA), 77
Wireless LAN Solution Engine (WLSE), CiscoWorks, 32, 215, 307, 364–367
Wireless local area networks (WLANs)
ad-hoc mode, 27–29
AP mode, 27–29
applications, 22
benefits, 9–12
broadcast domain segmentation, 409–410
capacity planning, 432–433
Cisco switches and routers, 216
coverage, 81–85
data rates, 85
deployment, simplified, 35–36
description, 2–3
design considerations, 12–22, 171–173
frame format, 3–9
hospitals, 23–24
hybrid mode, 29–30
implementing IPSec with VPNs, 332–333
interference detection, 35
multicast, considerations for, 199
multicast, reasons for, 199
multicast configuration, 190–191
multicast filtering, 191–192
network management, 26–27
peer-to-peer mode, 27–29
peer-to-peer with bridges, 193–196
QoS, effect on, 427
QoS design guidelines, 432–434
QoS support, 428–430
RF-producing devices, 25
rogue AP detection, 34
security, choosing, 353
security issues, 25–26
security with VLANs, 346–347
site surveys, assisted, 35–36
standards (802.11), 76, 78

static WEP keys, 341–346
streamlined management and operations support, 36–37
structural considerations, 22–25
validating connectivity, 333
VPN device lists, 334
warehouses, 24
*see also* Access points (APs); Guest WLANs; Rogue access points
Wireless personal area network (WPAN) standards (802.15), 76
Wireless propagation, 69
Wireless Protected Access (WPA), 187
Wireless security suite, Cisco, 216
Wireless Voice over Internet Protocol (WVoIP) phones
Cisco 7920, 135, 153–159, 165–166
L3 roaming, 177
most demanding users, 173
security, 176
WLAN. *see* Wireless local area networks (WLANs)
WLSE (CiscoWorks WLAN Solution Engine), 32, 215, 307, 364–367
Workgroup bridges, 195–196
WPA (Wireless Protected Access), 187
WPAN (wireless personal area network) standards (802.15), 76
WRED (Weighted Red Early Drop), 427
WVoIP. *see* Wireless Voice over Internet Protocol (WVoIP) phones

## X

X-rays, 64
XPower300 battery pack/inverter combination, 110

## Y

Yagi, Hidetsugu, 89
Yagi antennas, 89–90, 222

## Z

Zone, Fresnel, 125–127